Felipe Lenz
June 2017

PLASMA PHYSICS

An Introduction

PLASMA PHYSICS
An Introduction

Richard Fitzpatrick
University of Texas at Austin

CRC Press is an imprint of the
Taylor & Francis Group, an **informa** business

CRC Press
Taylor & Francis Group
6000 Broken Sound Parkway NW, Suite 300
Boca Raton, FL 33487-2742

Printed on acid-free paper
Version Date: 20140701

International Standard Book Number-13: 978-1-4665-9426-5 (Hardback)

Visit the Taylor & Francis Web site at
http://www.taylorandfrancis.com

and the CRC Press Web site at
http://www.crcpress.com

Contents

Preface

This textbook is intended to accompany a single-semester, introductory, graduate-level course on plasma physics. Students are assumed to have a thorough grasp of undergraduate classical mechanics and classical electrodynamics, as well as the mathematics of waves and oscillations, integral and differential calculus, vector fields, complex analysis, and Fourier and Laplace transforms. This book is not geared to any particular application of plasma physics, and should be suitable for students whose primary interest is either magnetic fusion, ionospheric physics, space plasma physics, or astrophysics. There are, in fact, many different types of plasma (e.g., strongly coupled plasma, dusty plasma, non-neutral plasma, degenerate plasma, weakly ionized plasma). However, this book only discusses non-relativistic, fully ionized, non-degenerate, quasi-neutral, weakly coupled plasma, which is, by far, the most common occurring plasma type in nature. The aim of the book is to set out the theoretical framework conventionally used to describe such plasma in a clear and concise manner. Chapter 1 introduces the fundamental parameters that characterize plasmas, and, in the process, makes clear exactly what is meant by a non-degenerate, quasi-neutral, weakly coupled plasma. Chapter 2 outlines the theory of charged particle motion in weakly inhomogeneous electric and magnetic fields. This theory is central to understanding how the magnetic confinement of a collisionless plasma works at an individual particle level, and is used to investigate the Van Allen radiation belts. Chapter 3 derives the ensemble-averaged kinetic equation that is the basis of all descriptions of collective plasma motion. This derivation incorporates a detailed treatment of binary Coulomb collisions in a weakly coupled plasma. Chapter 4 describes how fluid equations are obtained by taking low-order moments of the kinetic equation, and also discusses the various asymptotic closure schemes that allow such equations to form complete sets. Particular care is taken to explain the circumstances in which it is legitimate to adopt the well-known cold-plasma and MHD subsets of the complete fluid equations. Finally, the fluid approach to plasma physics is illustrated via an investigation of Langmuir sheaths. Chapter 5 investigates the propagation of low-amplitude electromagnetic waves through uniform cold plasmas. Chapter 6 extends this investigation to deal with waves propagating through weakly inhomogeneous plasmas. Chapter 7 is devoted to the theory of magnetohydrodynamical fluids. Applications of this theory that are discussed in detail include the solar wind, dynamo theory, magnetic reconnection, and MHD shocks. Finally, Chapter 8 analyses low-amplitude wave propagation through warm collisionless plasmas. Major sources for the material appearing in this book include the *The Framework of Plasma Physics* by my colleagues Richard Hazeltine and Françis Waelbroeck (particle drift theory, fundamental fluid theory, collision theory, ray tracing), *Plasma Physics* by Alan Cairns (waves in cold plasmas, waves in warm plasmas), *The Physics of Plasmas* by Thomas Boyd and Jeffrey Sanderson (MHD shock theory, solar wind), *Solar Magnetohydrodynamics* by Eric Priest (solar wind), and *The Theory of Plasma Waves* by Tom Stix (waves in cold plasmas).

Working through exercises is a vital stage in mastering any branch of physics. Hence, every chapter in this book ends with a selection of exercises that range from simply filling in the inevitable gaps in the material presented in the chapter (such exercises are tedious, but mandatory for the serious student) to interesting further applications of this material. A complete set of solutions to all of the exercises appearing in the book is available to adopting professors, upon request, from the publisher.

Author

Richard Fitzpatrick is a professor of physics at the University of Texas at Austin, where he has been a faculty member since 1994. He is a member of the Royal Astronomical Society, a fellow of the American Physical Society, and the author of *Maxwell's Equations and the Principles of Electromagnetism* (2008), *An Introduction to Celestial Mechanics* (2012), and *Oscillations and Waves: An Introduction* (2013). He earned a Master's degree in physics from the University of Cambridge and a DPhil in astronomy from the University of Sussex.

1

Introduction

1.1 What is Plasma?

In essence a plasma is an ionized gas. However, as this book is intended to demonstrate, the behavior of ionized gases is sufficiently different from their non-ionized cousins that it is meaningful to talk of plasma as a fourth state of matter. (The other three states being solid, liquid, and gas.) As is well known, a liquid is produced when a crystalline solid is heated sufficiently that the thermal motions of its constituent atoms disrupt its interatomic bonds. Likewise, a neutral gas is produced when a liquid is heated sufficiently that atoms evaporate from its surface at a faster rate than they recondense. Finally, a plasma is produced when a neutral gas is heated until interatomic collisions become sufficiently violent that they detach electrons from colliding atoms. Heating a plasma does not, however, produce a fifth state of matter. Plasmas resulting from ionization of neutral gases consist of myriads of positive and negative charge carriers whose relative numbers are in the inverse proportion to the magnitude of their individual charges. In this situation, the oppositely charged fluids, which are strongly coupled electrostatically, tend to electrically neutralize one another on macroscopic lengthscales. Such plasmas are termed *quasi-neutral* ("quasi" because the small deviations from exact neutrality can have important consequences—see Section 4.16). Strongly non-neutral plasmas, which may even contain charge carriers of one sign only, occur primarily in laboratory experiments, and are not discussed in this book. (Interested readers are referred to Davidson 2001.) In earlier epochs of the universe, all (baryonic) matter was in the plasma state (Longair 2008). In the present epoch, most (baryonic) matter remains in this state. For instance, stars, nebulae, and even interstellar space, are filled with plasma. The solar system is also permeated with plasma in the form of the solar wind, and the Earth is completely surrounded by plasma trapped within its magnetic field (Kallenrode 2010). Terrestrial plasmas occur in lightning, fluorescent lamps, a variety of laboratory experiments, and a growing array of industrial processes. Indeed, the glow discharge has recently become the mainstay of the micro-circuit fabrication industry (Lieberman and Lichtenberg 2005).

1.2 Brief History of Plasma Physics

When blood is cleared of its various corpuscles there remains a transparent liquid, which was termed *plasma* (after the ancient Greek word πλάσμα, which means "that which is formed or molded") by the great Czech medical scientist Johannes Purkinje (1787-1869). The American Nobel laureate chemist Irving Langmuir first used this term to describe an ionized gas in 1927—Langmuir was reminded of the way that blood plasma carries red and white corpuscles by the way that an electrified fluid carries electrons and ions. Langmuir, along with his colleague Lewi Tonks, was investigating the physics and chemistry of tungsten-filament light-bulbs, with a view to finding a way to greatly extend the lifetime of the filament (a goal that he eventually achieved). In the process, he developed the theory of *plasma sheaths*—the boundary layers that form between plasmas and solid surfaces. (See Section 4.16.) He also discovered that certain regions of a plasma discharge tube exhibit periodic variations of the electron density, which we nowadays term *Langmuir waves*. (See Section 8.2.) This was the genesis of plasma physics. Interestingly enough, Langmuir's research nowadays forms the theoretical basis of most plasma processing techniques for fabricating integrated circuits (Lieberman and Lichtenberg 2005). After Langmuir, plasma research gradually spread in other directions, of which five are particularly significant. Firstly, the development of radio broadcasting in the early twentieth century led to the discovery of the Earth's *ionosphere*—a layer of partially ionized gas in the upper atmosphere that reflects radio waves, and is responsible for the fact that radio signals can be received on the surface of the Earth when the transmitter lies over the horizon. Unfortunately, the ionosphere also occasionally absorbs and distorts radio waves. For instance, the Earth's magnetic field causes waves with different polarizations (relative to the orientation of the magnetic field) to propagate at different velocities, an effect that can give rise to "ghost signals" (in other words, signals that arrive a little before, or a little after, the main signal). In order to understand, and possibly correct, some of the deficiencies in radio communication, various scientists, such as E.V. Appleton and K.G. Budden, systematically developed the theory of electromagnetic wave propagation through nonuniform magnetized plasmas (Budden 1985). Secondly, in the first half of the twentieth century, astrophysicists recognized that much of the universe consists of plasma, and, thus, that a better understanding of astrophysical phenomena requires a better grasp of plasma physics. The pioneer in this field was the Swedish Nobel laureate Hannes Alfvén, who around 1940 developed the theory of *magnetohydrodynamics*, or MHD, in which plasma is treated essentially as a conducting fluid (Cowling 1957a). This theory has been successfully employed to investigate sunspots, solar flares, the solar wind, star formation, and a host of other topics in astrophysics (Kallenrode 2010). Two topics of particular interest in MHD theory are *magnetic reconnection* and *dynamo theory*. (See Sections 7.9–7.17.) Magnetic reconnection is a process by which magnetic field-lines suddenly change their topology. It can give rise to the rapid conversion of a great deal of magnetic energy into thermal energy, as well as the acceleration of some charged particles to extremely high energies, and is thought to be the basic mechanism behind solar flares (Priest 1984; Priest and Forbes 2007). Dynamo theory studies how the motion of an MHD fluid can give rise to the generation of a macroscopic magnetic field. This process is important because the terrestrial and solar magnetic fields would both decay away comparatively rapidly (in astrophysical terms) were they not maintained by dynamo action (Kulsrud 2004). The Earth's magnetic field is maintained by the motion of its molten core, which can be treated as an MHD fluid to a reasonable approximation. Thirdly, the creation of the hydrogen bomb in 1952 generated a great deal of interest in *controlled thermonuclear fusion* as a possible power source for the future (Fowler 1997). At first, this research was carried out secretly, and independently, by the United States, the Soviet Union, Great Britain, and France. However, thermonuclear fusion research was declassified in 1958, leading to the publication of a number of immensely important and influential papers in the late 1950s and the early 1960s. Broadly speaking, theoretical plasma physics first emerged as a mathematically rigorous discipline in these years. Not surprisingly, fusion physicists are mostly concerned with understanding how a thermonuclear plasma can be trapped—in most cases by a magnetic field—and investigating the many plasma instabilities that may allow it to escape (Freidberg 2008). Fourthly, in 1958 James A. Van Allen discovered the so-called *Van Allen radiation belts* surrounding the Earth, using data transmitted by the U.S. Explorer satellite. This discovery marked the start of the systematic exploration of the Earth's magnetosphere via satellite observations, and opened up the field of *space plasma physics* (Baumjohan and Treumann 1996). Finally, the development of high powered lasers in the 1960s opened up the field of *laser plasma physics* (Kruer 2003). When a high powered laser beam strikes a solid target, material is immediately ablated, and a plasma forms at the boundary between the beam and the target. Laser plasmas tend to have fairly extreme properties (for instance, densities characteristic of solids) not found in more conventional plasmas. A major application of laser plasma physics is the approach to fusion energy known as *inertial confinement fusion*. In this approach, tightly focused laser beams are used to implode a small solid target until the densities and temperatures characteristic of nuclear fusion (which are similar to those at the center of a hydrogen bomb) are achieved (Atzeni and Meyer-ter-Vehn 2009). Another interesting application of laser plasma physics is the use of the extremely strong electric fields generated when a high intensity laser pulse passes through a plasma to accelerate charged particles (Joshi 2006). High-energy physicists hope to use plasma acceleration techniques to dramatically reduce the size and cost of particle accelerators.

1.3 Fundamental Parameters

Consider an idealized plasma consisting of an equal number of electrons, with mass m_e and charge $-e$ (here, e denotes the magnitude of the electron charge), and ions, with mass m_i and charge $+e$. Without necessarily assuming that the system has attained thermal equilibrium, we shall employ the symbol

$$T_s \equiv \frac{1}{3}\, m_s \langle v_s^2 \rangle \tag{1.1}$$

to denote a *kinetic temperature* measured in units of energy. Here, v is a particle speed, and the angular brackets denote an ensemble average (Reif 1965). The kinetic temperature of species s is a measure of the mean kinetic energy of particles of that species. (Here, s represents either e for electrons, or i for ions.) In plasma physics, kinetic temperature is invariably measured in *electron-volts* (1 joule is equivalent to 6.24×10^{18} eV).

Quasi-neutrality demands that

$$n_i \simeq n_e \equiv n, \tag{1.2}$$

where n_s is the *particle number density* (that is, the number of particles per cubic meter) of species s.

Assuming that both ions and electrons are characterized by the same temperature, T (which is, by no means, always the case in plasmas), we can estimate typical particle speeds in terms of the so-called *thermal speed*,

$$v_{t\,s} \equiv \left(\frac{2\,T}{m_s}\right)^{1/2}. \tag{1.3}$$

Incidentally, the ion thermal speed is usually far smaller than the electron thermal speed. In fact,

$$v_{t\,i} \sim \left(\frac{m_e}{m_i}\right)^{1/2} v_{t\,e}. \tag{1.4}$$

Of course, n and T are generally functions of position in a plasma.

1.4 Plasma Frequency

The *plasma frequency*,

$$\Pi = \left(\frac{n\,e^2}{\epsilon_0\, m}\right)^{1/2}, \tag{1.5}$$

is the most fundamental timescale in plasma physics. There is a different plasma frequency for each species. However, the relatively large electron frequency is, by far, the most important of these, and references to "the plasma frequency" in textbooks invariably mean the electron plasma frequency.

It is easily seen that Π corresponds to the typical electrostatic oscillation frequency of a given species in response to a small charge separation. For instance, consider a one-dimensional situation in which a slab (whose bounding planes are normal to the x-axis) consisting entirely of particles of one species (with charge e and mass m) is displaced from its quasi-neutral position by an infinitesimal distance δx (parallel to the x-axis). The resulting charge density that develops on the leading face of the slab is $\sigma = e\,n\,\delta x$. An equal and opposite charge density develops on the opposite face. The x-directed electric field generated inside the slab is $E_x = -\sigma/\epsilon_0 = -e\,n\,\delta x/\epsilon_0$ (Fitzpatrick 2008). Thus, Newton's second law of motion applied to an individual particle inside the slab yields

$$m\,\frac{d^2\delta x}{dt^2} = e\,E_x = -m\,\Pi^2\,\delta x, \tag{1.6}$$

giving $\delta x = (\delta x)_0\,\cos(\Pi\,t)$.

Plasma oscillations are observed only when the plasma system is studied over time periods, τ, longer than the plasma period, $\tau_p \equiv 2\pi/\Pi$, and when external influences modify the system at a rate no faster than Π. In the opposite case, one is obviously studying something other than plasma physics (for instance, nuclear reactions), and the system cannot usefully be considered to be a plasma. Similarly, observations over lengthscales L shorter than the distance $v_t\,\tau_p$ traveled by a typical plasma particle during a plasma period will also not detect plasma behavior. In this case, particles will exit the system before completing a plasma oscillation. This distance, which is the spatial equivalent to τ_p, is called the *Debye length*, and is defined

$$\lambda_D \equiv \frac{1}{\Pi}\left(\frac{T}{m}\right)^{1/2}. \tag{1.7}$$

It follows that

$$\lambda_D = \left(\frac{\epsilon_0\,T}{n\,e^2}\right)^{1/2} \tag{1.8}$$

is independent of mass, and therefore generally comparable for different species.

According to the preceding discussion, our idealized system can usefully be considered to be a plasma only if

$$\frac{\lambda_D}{L} \ll 1, \tag{1.9}$$

and

$$\frac{\tau_p}{\tau} \ll 1. \tag{1.10}$$

Here, τ and L represent the typical timescale and lengthscale of the process under investigation.

It should be noted that, despite the conventional requirement given in Equation (1.9), plasma physics is actually capable of describing structures on the Debye scale (Hazeltine and Waelbroeck 2004). The most important example of this ability is the theory of the Langmuir sheath, which is the boundary layer that surrounds a plasma confined by a material surface. (See Section 4.16.)

1.5 Debye Shielding

Plasmas generally do not contain strong electric fields in their rest frames. The shielding of an external electric field from the interior of a plasma can be viewed as a result of high plasma conductivity. According to this explanation, electrical current can generally flow freely enough through a plasma to short out any interior electric fields. However, it is more useful to consider the shielding as a dielectric phenomenon. According to this explanation, it is the polarization of the plasma medium, and the associated redistribution of space charge, that prevents penetration by an external electric field. Not surprisingly, the lengthscale associated with such shielding is the Debye length.

Let us consider the simplest possible example. Suppose that a quasi-neutral plasma is sufficiently close to thermal equilibrium that the number densities of its two species are distributed according to the Maxwell-Boltzmann law (Reif 1965),

$$n_s = n_0 \exp(-e_s\,\Phi/T), \tag{1.11}$$

where $\Phi(\mathbf{r})$ is the electrostatic potential, and n_0 and T are constant. From $e_i = -e_e = e$, it is clear that quasi-neutrality requires the equilibrium potential to be zero. Suppose that the equilibrium potential is perturbed, by an amount $\delta\Phi(\mathbf{r})$, as a consequence of a small, localized, perturbing charge density, $\delta\rho_{\rm ext}$. The total perturbed charge density is written

$$\delta\rho = \delta\rho_{\rm ext} + e\,(\delta n_i - \delta n_e) \simeq \delta\rho_{\rm ext} - 2\,e^2\,n_0\,\delta\Phi/T. \tag{1.12}$$

Thus, Poisson's equation yields

$$\nabla^2\delta\Phi = -\frac{\delta\rho}{\epsilon_0} = -\left(\frac{\delta\rho_{\rm ext} - 2\,e^2\,n_0\,\delta\Phi/T}{\epsilon_0}\right), \tag{1.13}$$

which reduces to

$$\left(\nabla^2 - \frac{2}{\lambda_D^2}\right)\delta\Phi = -\frac{\delta\rho_{\rm ext}}{\epsilon_0}. \tag{1.14}$$

If the perturbing charge density actually consists of a point charge q, located at the origin, so that $\delta\rho_{\rm ext} = q\,\delta(\mathbf{r})$, then the solution to the previous equation is written

$$\delta\Phi(r) = \frac{q}{4\pi\,\epsilon_0\,r}\,\exp\left(-\frac{\sqrt{2}\,r}{\lambda_D}\right). \tag{1.15}$$

This expression implies that the Coulomb potential of the perturbing point charge q is shielded over distances longer than the Debye length by a shielding cloud of approximate radius λ_D that consists of charge of the opposite sign.

By treating n as a continuous function, the previous analysis implicitly assumes that there are many particles in the shielding cloud. Actually, Debye shielding remains statistically significant, and physical, in the opposite limit in which the cloud is barely populated. In the latter case, it is the probability of observing charged particles within a Debye length of the perturbing charge that is modified (Hazeltine and Waelbroeck 2004).

1.6 Plasma Parameter

Let us define the average distance between particles,

$$r_d \equiv n^{-1/3}. \tag{1.16}$$

We can also define a *mean distance of closest approach*,

$$r_c \equiv \frac{e^2}{4\pi\,\epsilon_0\,T}, \tag{1.17}$$

by balancing the one-dimensional thermal energy of a particle against the repulsive electrostatic potential of a binary pair. In other words,

$$\frac{1}{2}\,m\,v_t^2 = \frac{e^2}{4\pi\,\epsilon_0\,r_c}. \tag{1.18}$$

The significance of the ratio r_d/r_c is readily understood. If this ratio is small then charged particles are dominated by one another's electrostatic influence more or less continuously, and their kinetic energies are small compared to the interaction potential energies. Such plasmas are termed *strongly coupled*. On the other hand, if the ratio is large then strong electrostatic interactions between individual particles are occasional, and relatively rare, events. A typical particle is electrostatically influenced by all of the other particles within its Debye sphere, but this interaction very rarely causes any sudden change in its motion. Such plasmas are termed *weakly coupled*. It is possible to describe a weakly coupled plasma using a modified Boltzmann equation (in other words, the same type of equation that is conventionally used to describe a neutral gas). (See Chapter 3.) Understanding the strongly coupled limit is far more difficult, and will not be attempted in this book. (Interested readers are directed to Fortov, Iakubov, and Khrapak 2007.) Actually, a strongly coupled plasma has more in common with a liquid than a conventional weakly coupled plasma.

Let us define the *plasma parameter*,

$$\Lambda = \frac{4\pi}{3}\,n\,\lambda_D^3. \tag{1.19}$$

This dimensionless parameter is obviously equal to the typical number of particles contained in a Debye sphere. However, Equations (1.8), (1.16), (1.17), and (1.19) can be combined to give

$$\Lambda = \frac{\lambda_D}{3\,r_c} = \frac{1}{3\,\sqrt{4\pi}}\left(\frac{r_d}{r_c}\right)^{3/2} = \frac{4\pi\,\epsilon_0^{3/2}}{3\,e^3}\,\frac{T^{3/2}}{n^{1/2}}. \tag{1.20}$$

It can be seen that the case $\Lambda \ll 1$, in which the Debye sphere is sparsely populated, corresponds to a strongly coupled plasma. Likewise, the case $\Lambda \gg 1$, in which the Debye sphere is densely populated, corresponds to a weakly coupled plasma. It can

Plasma	$n(\text{m}^{-3})$	$T(\text{eV})$	$\Pi(\text{sec}^{-1})$	$\lambda_D(\text{m})$	Λ
Solar wind (1AU)	10^7	10	2×10^5	7×10^0	5×10^{10}
Tokamak	10^{20}	10^4	6×10^{11}	7×10^{-5}	4×10^8
Interstellar medium	10^6	10^{-2}	6×10^4	7×10^{-1}	4×10^6
Ionosphere	10^{12}	10^{-1}	6×10^7	2×10^{-3}	1×10^5
Inertial confinement	10^{28}	10^4	6×10^{15}	7×10^{-9}	5×10^4
Solar chromosphere	10^{18}	2	6×10^{10}	5×10^{-6}	2×10^3
Arc discharge	10^{20}	1	6×10^{11}	7×10^{-7}	5×10^2

Table 1.1
Key parameters for some typical weakly coupled plasmas.

also be appreciated, from Equation (1.20), that strongly coupled plasmas tend to be cold and dense, whereas weakly coupled plasmas tend to be diffuse and hot. Examples of strongly coupled plasmas include solid density laser ablation plasmas, the very "cold" (i.e., with kinetic temperatures similar to the ionization energy) plasmas found in "high pressure" arc discharges, and the plasmas that constitute the atmospheres of collapsed objects such as white dwarfs and neutron stars. On the other hand, the hot diffuse plasmas typically encountered in ionospheric physics, astrophysics, nuclear fusion, and space plasma physics are invariably weakly coupled. Table 1.1 lists the key parameters for some typical weakly coupled plasmas. In conclusion, characteristic plasma behavior is only observed on timescales longer than the plasma period, and on lengthscales larger than the Debye length. The statistical character of this behavior is controlled by the plasma parameter. Although Π, λ_D, and Λ are the three most fundamental plasma parameters, there are a number of other parameters that are worth mentioning.

1.7 Collisions

Collisions between charged particles in a plasma differ fundamentally from those between molecules in a neutral gas because of the long range of the Coulomb force. In fact, the discussion in Section 1.6 implies that binary collision processes can only be defined for weakly coupled plasmas. However, binary collisions in weakly coupled plasmas are still modified by collective effects, because the many-particle process of Debye shielding enters in a crucial manner. (See Chapter 3.) Nevertheless, for large Λ we can speak of binary collisions, and therefore of a *collision frequency*, denoted by $\nu_{ss'}$. Here, $\nu_{ss'}$ measures the rate at which particles of species s are scattered by those of species s'. When specifying only a single subscript, one is generally referring to the total collision rate for that species, including impacts with all other species. Very roughly,

$$\nu_s \simeq \sum_{s'} \nu_{ss'}. \tag{1.21}$$

The species designations are generally important. For instance, the relatively small electron mass implies that, for unit ionic charge and comparable species temperatures [see Equation (1.27)],

$$\nu_e \sim \left(\frac{m_i}{m_e}\right)^{1/2} \nu_i. \tag{1.22}$$

The collision frequency, ν, measures the frequency with which a particle trajectory undergoes a major angular change due to Coulomb interactions with other particles. Coulomb collisions are, in fact, predominately small angle scattering events, so the collision frequency is not the inverse of the typical time between collisions. (See Chapter 3.) Instead, it is the inverse of the typical time needed for enough collisions to occur that the particle trajectory is deviated through 90°. For this reason, the collision frequency is sometimes termed the 90° *scattering rate*.

It is conventional to define the *mean-free-path*,

$$\lambda_{\rm mfp} \equiv \frac{v_t}{\nu}. \tag{1.23}$$

Clearly, the mean-free-path measures the typical distance a particle travels between "collisions" (i.e., 90° scattering events). A collision-dominated, or *collisional*, plasma is simply one in which

$$\lambda_{\rm mfp} \ll L, \tag{1.24}$$

where L is the observation lengthscale. The opposite limit of long mean-free-path is said to correspond to a *collisionless* plasma. Collisions greatly simplify plasma behavior by driving the system toward statistical equilibrium, characterized by Maxwellian distribution functions. (See Section 3.6.) Furthermore, short mean-free-paths generally ensure that plasma transport is local (i.e., diffusive) in nature, which is a considerable simplification.

The typical magnitude of the collision frequency is (see Section 3.12)

$$\nu \sim \frac{\ln \Lambda}{\Lambda}\, \Pi. \tag{1.25}$$

Note that $\nu \ll \Pi$ in a weakly coupled plasma. It follows that collisions do not seriously interfere with plasma oscillations in such systems. On the other hand, Equation (1.25) implies that $\nu \gg \Pi$ in a strongly coupled plasma, suggesting that collisions effectively prevent plasma oscillations in such systems. This accords well with our basic picture of a strongly coupled plasma as a system, dominated by Coulomb interactions, that does not exhibit conventional plasma dynamics.

Equations (1.7), (1.23), and (1.25) imply that the ratio of the mean-free-path to the Debye length can be written

$$\frac{\lambda_{\rm mfp}}{\lambda_D} \sim \frac{\Lambda}{\ln \Lambda}. \tag{1.26}$$

It follows that the mean-free-path is much larger than the Debye length in a weakly coupled plasma. This is a significant result because the effective range of the inter-particle force (i.e., the Coulomb force) in a plasma is of approximately the same magnitude as the Debye length. We conclude that the mean-free-path is much larger than the effective range of the inter-particle force in a weakly coupled plasma.

Equations (1.5) and (1.20) yield

$$\nu \sim \frac{3\,e^4\,\ln\Lambda}{4\pi\,\epsilon_0^2\,m^{1/2}}\,\frac{n}{T^{3/2}}. \tag{1.27}$$

Thus, diffuse, high temperature plasmas tend to be collisionless, whereas dense, low temperature plasmas are more likely to be collisional.

While collisions are crucial to the confinement and dynamics of neutral gases, they play a far less important role in plasmas. In fact, in many plasmas the magnetic field effectively plays the role that collisions play in a neutral gas. In such plasmas, charged particles are constrained from moving perpendicular to the field by their small Larmor orbits, rather than by collisions. Confinement along the field-lines is more difficult to achieve, unless the field-lines form closed loops (or closed surfaces). Thus, it makes sense to talk about a "collisionless plasma," whereas it makes little sense to talk about a "collisionless neutral gas." Many plasmas are collisionless to a very good approximation, especially those encountered in astrophysics and space plasma physics contexts.

1.8 Magnetized Plasmas

A *magnetized* plasma is one in which the ambient magnetic field, $\mathbf{B}$, is strong enough to significantly alter particle trajectories. In particular, magnetized plasmas are highly anisotropic, responding differently to forces that are parallel and perpendicular to the direction of $\mathbf{B}$. Incidentally, a magnetized plasma moving with mean velocity $\mathbf{V}$ contains an electric field $\mathbf{E} = -\mathbf{V}\times\mathbf{B}$ that is not affected by Debye shielding. Of course, the electric field is essentially zero in the rest frame of the plasma.

As is well known, charged particles respond to the Lorentz force,

$$\mathbf{F} = q\,\mathbf{v}\times\mathbf{B}, \tag{1.28}$$

by freely streaming in the direction of $\mathbf{B}$, while executing circular *Larmor orbits*, or *gyro-orbits*, in the plane perpendicular to $\mathbf{B}$ (Fitzpatrick 2008). As the field-strength increases, the resulting helical orbits become more tightly wound, effectively tying particles to magnetic field-lines.

The typical *Larmor radius*, or *gyroradius*, of a charged particle gyrating in a magnetic field is given by

$$\rho \equiv \frac{v_t}{\Omega}, \tag{1.29}$$

where

$$\Omega = \frac{eB}{m} \tag{1.30}$$

is the *cyclotron frequency*, or *gyrofrequency*, associated with the gyration. As usual, there is a distinct gyroradius for each species. When species temperatures are comparable, the electron gyroradius is distinctly smaller than the ion gyroradius:

$$\rho_e \sim \left(\frac{m_e}{m_i}\right)^{1/2} \rho_i. \tag{1.31}$$

A plasma system, or process, is said to be magnetized if its characteristic length-scale, L, is large compared to the gyroradius. In the opposite limit, $\rho \gg L$, charged particles have essentially straight-line trajectories. Thus, the ability of the magnetic field to significantly affect particle trajectories is measured by the *magnetization parameter*,

$$\delta \equiv \frac{\rho}{L}. \tag{1.32}$$

There are some cases of interest in which the electrons are magnetized, but the ions are not. However, a "magnetized" plasma conventionally refers to one in which both species are magnetized. This state is generally achieved when

$$\delta_i \equiv \frac{\rho_i}{L} \ll 1. \tag{1.33}$$

1.9 Plasma Beta

The fundamental measure of a magnetic field's effect on a plasma is the magnetization parameter, δ. The fundamental measure of the inverse effect is called β, and is defined as the ratio of the thermal energy density, $n\,T$, to the magnetic energy density, $B^2/(2\,\mu_0)$. It is conventional to identify the plasma energy density with the pressure,

$$p \equiv n\,T, \tag{1.34}$$

as in an ideal gas, and to define a separate β_s for each plasma species. Thus,

$$\beta_s = \frac{2\,\mu_0\,p_s}{B^2}. \tag{1.35}$$

The total β is written

$$\beta = \sum_s \beta_s. \tag{1.36}$$

1.10 De Broglie Wavelength

Quantum effects become important if the mean inter-particle distance, r_d, becomes comparable, or less than, the *de Broglie wavelength*,

$$\lambda\!\!\!^{-} \equiv \frac{h}{m\,v_t}, \tag{1.37}$$

where h is Planck's constant. According to Equations (1.3) and (1.16), the condition $r_d \ll \lambda\!\!\!^{-}$ is equivalent to

$$\frac{T^{3/2}}{n} \ll \frac{h^3}{(2\,m)^{3/2}}. \tag{1.38}$$

A plasma that satisfies this condition is said to be *degenerate*, whereas a plasma that does not is said to be *non-degenerate*. The behavior of degenerate plasmas is fundamentally different to that of the non-degenerate plasmas discussed in this book (because the former plasmas are governed by quantum mechanics, whereas the latter are governed by classical mechanics). It can be seen that if both species have comparable temperatures then the condition, given in Equation (1.38), for degeneracy is more easily satisfied by the electrons than by the ions. Moreover, it is evident that degenerate plasmas tend to be cold and dense, whereas non-degenerate plasmas are generally hot and diffuse. (See Haas 2011 for a comprehensive discussion of degenerate plasmas.)

It is actually possible for quantum effects to modify collisions in non-degenerate plasmas that satisfy the inequality $r_d \gg \lambda\!\!\!^{-}$. In fact, the criterion for quantum effects not to modify collisions is $r_c \gg \lambda\!\!\!^{-}$, where r_c is the mean distance of closest approach during collisions. However, it follows from Equation (1.20) that $r_c \sim r_d/\Lambda^{2/3}$. Hence, the criterion for classical collisions becomes $r_d \gg \Lambda^{2/3}\,\lambda\!\!\!^{-}$. In a weakly coupled plasma, for which $\Lambda \gg 1$, this criterion is harder to satisfy that the criterion, $r_d \gg \lambda\!\!\!^{-}$, for non-degeneracy.

1.11 Exercises

1.1 Consider a quasi-neutral plasma consisting of electrons of mass m_e, charge $-e$, temperature T_e, and mean number density, n_e, as well as ions of mass m_i, charge $Z\,e$, temperature T_i, and mean number density $n_i = n_e/Z$.

(a) Generalize the analysis of Section 1.4 to show that the effective plasma frequency of the plasma can be written

$$\Pi = \left(\Pi_e^2 + \Pi_i^2\right)^{1/2},$$

where $\Pi_e = (e^2\,n_e/\epsilon_0\,m_e)^{1/2}$ and $\Pi_i = (Z^2\,e^2\,n_i/\epsilon_0\,m_i)^{1/2}$. Furthermore, demonstrate that the characteristic ratio of ion to electron displacement in a plasma oscillation is $\delta x_i/\delta x_e = -Z\, m_e/m_i$.

(b) Generalize the analysis of Section 1.5 to show that the effective Debye length, λ_D, of the plasma can be written

$$\left(\frac{1}{\lambda_D}\right)^2 = \frac{1}{2}\left[\left(\frac{1}{\lambda_{D\,e}}\right)^2 + \left(\frac{1}{\lambda_{D\,i}}\right)^2\right],$$

where $\lambda_{D\,e} = (\epsilon_0\, T_e/n_e\, e^2)^{1/2}$ and $\lambda_{D\,i} = (\epsilon_0\, T_i/n_i\, Z^2\, e^2)^{1/2}$.

1.2 The perturbed electrostatic potential $\delta\Phi$ due to a charge q placed at the origin in a plasma of Debye length λ_D is governed by

$$\left(\nabla^2 - \frac{2}{\lambda_D^2}\right)\delta\Phi = -\frac{q\,\delta(\mathbf{r})}{\epsilon_0}.$$

Show that the non-homogeneous solution to this equation is

$$\delta\Phi(r) = \frac{q}{4\pi\,\epsilon_0\, r}\,\exp\left(-\frac{\sqrt{2}\,r}{\lambda_D}\right).$$

Demonstrate that the charge density of the shielding cloud is

$$\delta\rho(r) = -\frac{2\,q}{4\pi\, r\,\lambda_D^2}\,\exp\left(-\frac{\sqrt{2}\,r}{\lambda_D}\right),$$

and that the net shielding charge contained within a sphere of radius r, centered on the origin, is

$$Q(r) = -q\left[1 - \left(1 + \frac{\sqrt{2}\,r}{\lambda_D}\right)\exp\left(-\frac{\sqrt{2}\,r}{\lambda_D}\right)\right].$$

1.3 A quasi-neutral slab of cold (i.e., $\lambda_D \to 0$) plasma whose bounding surfaces are normal to the x-axis consists of electrons of mass m_e, charge $-e$, and mean number density n_e, as well as ions of charge e, and mean number density n_e. The ions can effectively be treated as stationary. The slab is placed in an externally generated, x-directed electric field that oscillates sinusoidally at the angular frequency ω. By generalizing the analysis of Section 1.4, show that the relative dielectric constant of the plasma is

$$\epsilon = 1 - \frac{\Pi^2}{\omega^2},$$

where $\Pi = (e^2\, n_e/\epsilon_0\, m_e)^{1/2}$.

1.4 A capacitor consists of two parallel plates of cross-sectional area A and spacing $d \ll \sqrt{A}$. The region between the capacitors is filled with a uniform hot plasma of Deybe length λ_D. By generalizing the analysis of Section 1.5, show that the d.c. capacitance of the device is

$$C = \frac{\epsilon_0 A}{d} \frac{(d/\sqrt{2}\,\lambda_D)}{\tanh(d/\sqrt{2}\,\lambda_D)}.$$

1.5 A uniform isothermal quasi-neutral plasma with singly-charged ions is placed in a relatively weak gravitational field of acceleration $\mathbf{g} = -g\,\mathbf{e}_z$. Assuming, first, that both species are distributed according to the Maxwell-Boltzmann statistics; second, that the perturbed electrostatic potential is a function of z only; and, third, that the electric field is zero at $z = 0$ (and well behaved as $z \to \infty$), demonstrate that the electric field in the region $z > 0$ takes the form $\mathbf{E} = E_z\,\mathbf{e}_z$, where

$$E_z(z) = E_0\left[1 - \exp\left(\frac{\sqrt{2}\,z}{\lambda_D}\right)\right],$$

and

$$E_0 = \frac{m_i\,g}{2\,e}.$$

Here, λ_D is the Debye length, e the magnitude of the electron charge, and m_i the ion mass.

1.6 Consider a charge sheet of charge density σ immersed in a plasma of unperturbed particle number density n_0, ion temperature T_i, and electron temperature T_e. Suppose that the charge sheet coincides with the y-z plane. Assuming that the (singly-charged) ions and electrons obey Maxwell-Boltzmann statistics, demonstrate that in the limit $|e\,\Phi/T_{i,e}| \ll 1$ the electrostatic potential takes the form

$$\Phi(x) = \frac{\sigma\,\lambda_D}{2\,\epsilon_0}\,\mathrm{e}^{-|x|/\lambda_D},$$

where $\lambda_D = [(\epsilon_0/e^2\,n_0)\,T_i\,T_e/(T_i + T_e)]^{1/2}$.

1.7 Consider the previous exercise again. Let $T_i = T_e = T$. Suppose, however, that $|e\,\Phi/T|$ is not necessarily much less than unity. Demonstrate that the potential, V, of the charge sheet (relative to infinity) satisfies

$$\frac{e\,V}{T} = \cosh^{-1}\left(1 + \frac{\sigma^2}{16\,\epsilon_0\,n_0\,T}\right).$$

Furthermore, show that

$$\tanh(e\,\Phi/4\,T) = \tanh(e\,V/4\,T)\,\mathrm{e}^{-|x|/\lambda_D},$$

where $\lambda_D = \sqrt{\epsilon_0\,T/2\,e^2\,n_0}$. Let x_s be the distance from the sheet at which the potential has fallen to V/e, where $\ln \mathrm{e} = 1$. Sketch x_s/λ_D versus $e\,V/T$.

1.8 A long cylinder of plasma of radius a consists of cold (i.e., $T_i = T_e = 0$) singly-charged ions and electrons with uniform number density n_0. The cylinder of electrons is perturbed a distance δ (where $\delta \ll a$) in a direction perpendicular to its axis.

(a) Assuming that the ions are immobile, show that the oscillation frequency of the electron cylinder is

$$\Pi = \left(\frac{e^2\, n_0}{2\, \epsilon_0\, m_e}\right)^{1/2},$$

where m_e is the electron mass.

(b) Assuming that the ions have the finite mass m_i, show that the oscillation frequency is

$$\Pi = \left[\frac{e^2\, n_0}{2\, \epsilon_0}\left(\frac{1}{m_e} + \frac{1}{m_i}\right)\right]^{1/2}.$$

1.9 A sphere of plasma of radius a consists of cold (i.e., $T_i = T_e = 0$) singly-charged ions and electrons with uniform number density n_0. The sphere of electrons is perturbed a distance δ (where $\delta \ll a$). Assuming that the ions are immobile, show that the oscillation frequency of the electron sphere is

$$\Pi = \left(\frac{e^2\, n_0}{3\, \epsilon_0\, m_e}\right)^{1/2},$$

where m_e is the electron mass.

2

Charged Particle Motion

2.1 Introduction

All descriptions of plasma behavior are based, ultimately, on the motions of the constituent particles. For the case of an unmagnetized plasma, the motions are fairly trivial because the constituent particles move essentially in straight-lines between collisions. The motions are also trivial in a magnetized plasma in which the collision frequency, ν, greatly exceeds the gyrofrequency, Ω. In this case, the particles are scattered after executing only a small fraction of a gyro-orbit, and, therefore, still move essentially in straight-lines between collisions. The situation of primary interest in this chapter is that of a magnetized, but collisionless (i.e., $\nu \ll \Omega$), plasma, in which the gyroradius, ρ, is much smaller than the typical variation lengthscale, L, of the $\mathbf{E}$ and $\mathbf{B}$ fields, and the gyroperiod, Ω^{-1}, is much less than the typical timescale, τ, on which these fields change. In such a plasma, we expect the motion of the constituent particles to consist of a rapid gyration perpendicular to magnetic field-lines, combined with free streaming parallel to the field-lines. We are particularly interested in calculating how this motion is affected by the spatial and temporal gradients in the $\mathbf{E}$ and $\mathbf{B}$ fields. In general, the motion of charged particles in spatially and temporally nonuniform electromagnetic fields is extremely complicated. However, we hope to considerably simplify this motion by exploiting the assumed smallness of the parameters ρ/L and $(\Omega\,\tau)^{-1}$. What we are essentially trying to understand, in this chapter, is how the magnetic confinement of a collisionless plasma works at an individual particle level. The type of collisionless, magnetized plasma investigated here occurs primarily in magnetic fusion and space plasma physics contexts.

2.2 Motion in Uniform Fields

Let us, first of all, consider the motion of a particle of mass m and charge e in spatially and temporally uniform electromagnetic fields. The particle's equation of motion takes the form

$$m\,\frac{d\mathbf{v}}{dt} = e\,(\mathbf{E} + \mathbf{v}\times\mathbf{B}). \tag{2.1}$$

The component of this equation parallel to the magnetic field,

$$\frac{dv_{\parallel}}{dt} = \frac{e}{m}\,E_{\parallel}, \tag{2.2}$$

predicts uniform acceleration along magnetic field-lines. Consequently, plasmas close to equilibrium generally have either small or vanishing $E_{\parallel}$.

As can easily be verified by substitution, the perpendicular (to the magnetic field) component of Equation (2.1) yields

$$\mathbf{v}_{\perp} = \frac{\mathbf{E}\times\mathbf{B}}{B^2} + \rho\,\Omega\left[\sin(\Omega\,t+\gamma_0)\,\mathbf{e}_1 + \cos(\Omega\,t+\gamma_0)\,\mathbf{e}_2\right], \tag{2.3}$$

where $\Omega = eB/m$ is the gyrofrequency, ρ is the gyroradius, $\mathbf{e}_1$ and $\mathbf{e}_2$ are unit vectors such that $\mathbf{e}_1$, $\mathbf{e}_2$, $\mathbf{B}$ form a right-handed, mutually orthogonal set, and γ_0 is the particle's initial *gyrophase*. The motion consists of gyration around the magnetic field at the frequency Ω, superimposed on a steady drift with velocity

$$\mathbf{v}_E = \frac{\mathbf{E}\times\mathbf{B}}{B^2}. \tag{2.4}$$

This drift, which is termed the *E-cross-B drift*, is identical for all plasma species, and can be eliminated entirely by transforming to a new inertial frame in which $\mathbf{E}_{\perp} = \mathbf{0}$. This frame, which moves with velocity $\mathbf{v}_E$ with respect to the old frame, can properly be regarded as the rest frame of the plasma.

We can complete the previous solution by integrating the velocity to find the particle position. Thus,

$$\mathbf{r}(t) = \mathbf{R}(t) + \boldsymbol{\rho}(t), \tag{2.5}$$

where

$$\boldsymbol{\rho}(t) = \rho\left[-\cos(\Omega\,t+\gamma_0)\,\mathbf{e}_1 + \sin(\Omega\,t+\gamma_0)\,\mathbf{e}_2\right], \tag{2.6}$$

and

$$\mathbf{R}(t) = \left(v_{0\,\parallel}\,t + \frac{e}{m}\,E_{\parallel}\,\frac{t^2}{2}\right)\mathbf{b} + \mathbf{v}_E\,t. \tag{2.7}$$

Here, $\mathbf{b} \equiv \mathbf{B}/B$. Of course, the trajectory of the particle describes a spiral. The *gyrocenter*, $\mathbf{R}$, of this spiral, which is termed the *guiding center*, drifts across the magnetic field with the velocity $\mathbf{v}_E$, and also accelerates along field-lines at a rate determined by the parallel electric field.

The concept of a guiding center gives us a clue as to how to proceed. Perhaps, when analyzing charged particle motion in nonuniform electromagnetic fields, we can somehow neglect the rapid, and relatively uninteresting, gyromotion, and focus, instead, on the far slower motion of the guiding center? In order to achieve this goal, we need to somehow average the equation of motion over gyrophase, so as to obtain a reduced equation of motion for the guiding center.

2.3 Method of Averaging

In many dynamical problems, the motion consists of a rapid oscillation superimposed on a slow secular drift. For such problems, the most efficient approach is to describe the evolution in terms of the average values of the dynamical variables. The method outlined below is adapted from a classic paper by Morozov and Solov'ev (Morozov and Solev'ev 1966; Hazeltine and Waelbroeck 2004).

Consider the equation of motion

$$\frac{d\mathbf{z}}{dt} = \mathbf{f}(\mathbf{z}, t, \tau), \tag{2.8}$$

where $\mathbf{f}$ is a periodic function of its last argument, with period 2π, and

$$\tau = t/\epsilon. \tag{2.9}$$

Here, the small parameter ϵ characterizes the separation between the short oscillation period and the timescale for the slow secular evolution of the "position" $\mathbf{z}$.

The basic idea of the averaging method is to treat t and τ as distinct independent variables, and to look for solutions of the form $\mathbf{z}(t, \tau)$ that are periodic in τ. Thus, we replace Equation (2.8) by

$$\frac{\partial \mathbf{z}}{\partial t} + \frac{1}{\epsilon}\frac{\partial \mathbf{z}}{\partial \tau} = \mathbf{f}(\mathbf{z}, t, \tau), \tag{2.10}$$

and reserve Equation (2.9) for substitution into the final result. The indeterminacy introduced by increasing the number of variables is lifted by the requirement of periodicity in τ. All of the secular drifts are thereby attributed to the variable t, while the oscillations are described entirely by the variable τ.

Let us denote the τ-average of $\mathbf{z}$ by $\mathbf{Z}$, and seek a change of variables of the form

$$\mathbf{z}(t, \tau) = \mathbf{Z}(t) + \epsilon\zeta(\mathbf{Z}, t, \tau). \tag{2.11}$$

Here, ζ is a periodic function of τ with vanishing mean. Thus,

$$\langle\zeta(\mathbf{Z}, t, \tau)\rangle \equiv \frac{1}{2\pi}\oint \zeta(\mathbf{Z}, t, \tau)\, d\tau = 0, \tag{2.12}$$

where $\oint$ denotes the integral over a full period in τ.

The evolution of $\mathbf{Z}$ is determined by substituting the expansions

$$\zeta = \zeta_0(\mathbf{Z}, t, \tau) + \epsilon\zeta_1(\mathbf{Z}, t, \tau) + \epsilon^2\zeta_2(\mathbf{Z}, t, \tau) + \cdots, \tag{2.13}$$

$$\frac{d\mathbf{Z}}{dt} = \mathbf{F}_0(\mathbf{Z}, t) + \epsilon\mathbf{F}_1(\mathbf{Z}, t) + \epsilon^2\mathbf{F}_2(\mathbf{Z}, t) + \cdots, \tag{2.14}$$

into the equation of motion, Equation (2.10), and solving order by order in ϵ.

To lowest order, we obtain

$$\mathbf{F}_0(\mathbf{Z}, t) + \frac{\partial \zeta_0}{\partial \tau} = \mathbf{f}(\mathbf{Z}, t, \tau). \tag{2.15}$$

The solubility condition for this equation is

$$\mathbf{F}_0(\mathbf{Z}, t) = \langle \mathbf{f}(\mathbf{Z}, t, \tau) \rangle \equiv \langle \mathbf{f} \rangle (\mathbf{Z}, t). \tag{2.16}$$

Integrating the oscillating component of Equation (2.15) yields

$$\zeta_0(\mathbf{Z}, t, \tau) = \int_0^\tau [\mathbf{f}(\mathbf{Z}, t, \tau') - \langle \mathbf{f} \rangle (\mathbf{Z}, t)]\, d\tau'. \tag{2.17}$$

To first order, Equation (2.10) gives,

$$\mathbf{F}_1 + \frac{\partial \zeta_0}{\partial t} + \mathbf{F}_0 \cdot \nabla \zeta_0 + \frac{\partial \zeta_1}{\partial \tau} = \zeta_0 \cdot \nabla \mathbf{f}(\mathbf{Z}, t, \tau). \tag{2.18}$$

The solubility condition for this equation yields

$$\mathbf{F}_1(\mathbf{Z}, t) = \langle \zeta_0(\mathbf{Z}, t, \tau) \cdot \nabla \mathbf{f}(\mathbf{Z}, t, \tau) \rangle \equiv \langle \zeta_0 \cdot \nabla \mathbf{f} \rangle (\mathbf{Z}, t). \tag{2.19}$$

The final result is obtained by combining Equations (2.14), (2.16), and (2.19):

$$\frac{d\mathbf{Z}}{dt} = \langle \mathbf{f} \rangle (\mathbf{Z}, t) + \epsilon \langle \zeta_0 \cdot \nabla \mathbf{f} \rangle (\mathbf{Z}, t) + O(\epsilon^2). \tag{2.20}$$

Evidently, the secular motion of the "guiding center" position $\mathbf{Z}$ is determined to lowest order by the average of the "force" $\mathbf{f}$, and to next order by the correlation between the oscillation in the "position" $\mathbf{z}$ and the oscillation in the spatial gradient of the "force."

2.4 Guiding Center Motion

Consider the motion of a charged particle of mass m and charge e in the limit in which the electromagnetic fields experienced by the particle do not vary much in a gyroperiod, so that

$$\rho\,|\nabla \mathbf{B}| \ll B, \tag{2.21}$$

$$\frac{1}{\Omega}\frac{\partial B}{\partial t} \ll B. \tag{2.22}$$

The electric force is assumed to be comparable to the magnetic force. To keep track of the order of the various quantities, we introduce the parameter ϵ as a book-keeping device, and make the substitution $\rho \to \epsilon \rho$, as well as $(\mathbf{E}, \mathbf{B}, \Omega) \to \epsilon^{-1}(\mathbf{E}, \mathbf{B}, \Omega)$. The parameter ϵ is set to unity in the final answer.

In order to make use of the technique described in the previous section, we write the dynamical equations in the first-order differential form,

$$\frac{d\mathbf{r}}{dt} = \mathbf{v}, \tag{2.23}$$

$$\frac{d\mathbf{v}}{dt} = \frac{e}{\epsilon m}(\mathbf{E} + \mathbf{v} \times \mathbf{B}), \tag{2.24}$$

and seek a change of variables (Hazeltine and Waelbroeck 2004),

$$\mathbf{r} = \mathbf{R}(t) + \epsilon \boldsymbol{\rho}(\mathbf{R}, \mathbf{U}, t, \gamma), \tag{2.25}$$

$$\mathbf{v} = \mathbf{U}(t) + \mathbf{u}(\mathbf{R}, \mathbf{U}, t, \gamma), \tag{2.26}$$

such that the new guiding center variables $\mathbf{R}$ and $\mathbf{U}$ are free of oscillations along the particle trajectory. Here, γ is a new independent variable describing the phase of the gyrating particle. The functions $\boldsymbol{\rho}$ and $\mathbf{u}$ represent the gyration radius and velocity, respectively. We require periodicity of these functions with respect to their last argument, with period 2π, and with vanishing mean, so that

$$\langle \boldsymbol{\rho} \rangle = \langle \mathbf{u} \rangle = \mathbf{0}. \tag{2.27}$$

Here, the angular brackets refer to the average over a period in γ.

The equation of motion is used to determine the coefficients in the following expansion of $\boldsymbol{\rho}$ and $\mathbf{u}$ (Hazeltine and Waelbroeck 2004):

$$\boldsymbol{\rho} = \boldsymbol{\rho}_0(\mathbf{R}, \mathbf{U}, t, \gamma) + \epsilon \boldsymbol{\rho}_1(\mathbf{R}, \mathbf{U}, t, \gamma) + \cdots , \tag{2.28}$$

$$\mathbf{u} = \mathbf{u}_0(\mathbf{R}, \mathbf{U}, t, \gamma) + \epsilon \mathbf{u}_1(\mathbf{R}, \mathbf{U}, t, \gamma) + \cdots . \tag{2.29}$$

The dynamical equation for the gyrophase is likewise expanded, assuming that $d\gamma/dt \simeq \Omega = O(\epsilon^{-1})$,

$$\frac{d\gamma}{dt} = \epsilon^{-1}\, \omega_{-1}(\mathbf{R}, \mathbf{U}, t) + \omega_0(\mathbf{R}, \mathbf{U}, t) + \cdots . \tag{2.30}$$

In the following, we suppress the subscripts on all quantities except the guiding center velocity $\mathbf{U}$, because this is the only quantity for which the first-order corrections are calculated.

To each order in ϵ, the evolution of the guiding center position, $\mathbf{R}$, and velocity, $\mathbf{U}$, are determined by the solubility conditions for the equations of motion, Equations (2.23) and (2.24), when expanded to that order. The oscillating components of the equations of motion determine the evolution of the gyrophase. The velocity equation, Equation (2.23), is linear. It follows that, to all orders in ϵ, its solubility condition is simply

$$\frac{d\mathbf{R}}{dt} = \mathbf{U}. \tag{2.31}$$

To lowest order [that is, $O(\epsilon^{-1})$], the momentum equation, Equation (2.24), yields

$$\omega \frac{\partial \mathbf{u}}{\partial \gamma} - \Omega\, \mathbf{u} \times \mathbf{b} = \frac{e}{m} \left(\mathbf{E} + \mathbf{U}_0 \times \mathbf{B} \right). \tag{2.32}$$

The solubility condition (that is, the gyrophase average) is

$$\mathbf{E} + \mathbf{U}_0 \times \mathbf{B} = \mathbf{0}. \tag{2.33}$$

This immediately implies that

$$E_{\parallel} \equiv \mathbf{E} \cdot \mathbf{b} \sim \epsilon E. \tag{2.34}$$

In other words, the rapid acceleration caused by a large parallel electric field would invalidate the ordering assumptions used in this calculation. Solving for $\mathbf{U}_0$, we obtain

$$\mathbf{U}_0 = U_{0\,\parallel}\,\mathbf{b} + \mathbf{v}_E, \tag{2.35}$$

where all quantities are evaluated at the guiding center position, $\mathbf{R}$. The perpendicular component of the velocity, $\mathbf{v}_E$, has the same form—namely, Equation (2.4)—as that obtained for uniform fields. The parallel velocity, $U_{0\,\parallel}$, is undetermined at this order.

The integral of the oscillating component of Equation (2.32) yields

$$\mathbf{u} = \mathbf{c} + u_\perp \left[\sin\left(\Omega\,\gamma/\omega\right)\mathbf{e}_1 + \cos\left(\Omega\,\gamma/\omega\right)\mathbf{e}_2\right], \tag{2.36}$$

where $\mathbf{c}$ is a constant vector, and $\mathbf{e}_1$ and $\mathbf{e}_2$ are again mutually orthogonal unit vectors perpendicular to $\mathbf{b}$. All quantities in the previous equation are functions of $\mathbf{R}$, $\mathbf{U}$, and t. The periodicity constraint, combined with Equation (2.27), requires that $\omega = \Omega(\mathbf{R}, t)$ and $\mathbf{c} = \mathbf{0}$. The gyration velocity is thus

$$\mathbf{u} = u_\perp \left(\sin\gamma\,\mathbf{e}_1 + \cos\gamma\,\mathbf{e}_2\right), \tag{2.37}$$

and, from Equation (2.30), the gyrophase is given by

$$\gamma = \gamma_0 + \Omega\,t, \tag{2.38}$$

where γ_0 is the initial gyrophase. The amplitude, $u_\perp$, of the gyration velocity is undetermined at this order.

The lowest order oscillating component of the velocity equation, Equation (2.23), yields

$$\Omega\,\frac{\partial\boldsymbol{\rho}}{\partial\gamma} = \mathbf{u}. \tag{2.39}$$

This is readily integrated to give

$$\boldsymbol{\rho} = \rho\,(-\cos\gamma\,\mathbf{e}_1 + \sin\gamma\,\mathbf{e}_2), \tag{2.40}$$

where $\rho = u_\perp/\Omega$. It follows that

$$\mathbf{u} = \Omega\,\boldsymbol{\rho}\times\mathbf{b}. \tag{2.41}$$

The gyrophase average of the first-order [that is, $O(\epsilon^0)$] momentum equation, Equation (2.24), reduces to

$$\frac{d\mathbf{U}_0}{dt} = \frac{e}{m}\left[E_\parallel\,\mathbf{b} + \mathbf{U}_1\times\mathbf{B} + \langle\mathbf{u}\times(\boldsymbol{\rho}\cdot\nabla)\,\mathbf{B}\rangle\right]. \tag{2.42}$$

All quantities in the previous expression are functions of the guiding center position, $\mathbf{R}$, rather than the instantaneous particle position, $\mathbf{r}$. In order to evaluate the last term, we make the substitution $\mathbf{u} = \Omega\,\boldsymbol{\rho}\times\mathbf{b}$, and calculate

$$\begin{aligned}\langle(\boldsymbol{\rho}\times\mathbf{b})\times(\boldsymbol{\rho}\cdot\nabla)\,\mathbf{B}\rangle &= \mathbf{b}\,\langle\boldsymbol{\rho}\cdot(\boldsymbol{\rho}\cdot\nabla)\,\mathbf{B}\rangle - \langle\boldsymbol{\rho}\,\mathbf{b}\cdot(\boldsymbol{\rho}\cdot\nabla)\,\mathbf{B}\rangle\\ &= \mathbf{b}\,\langle\boldsymbol{\rho}\cdot(\boldsymbol{\rho}\cdot\nabla)\,\mathbf{B}\rangle - \langle\boldsymbol{\rho}\,(\boldsymbol{\rho}\cdot\nabla B)\rangle.\end{aligned} \tag{2.43}$$

The averages are specified by

$$\langle \rho\rho \rangle = \frac{u_\perp^2}{2\,\Omega^2}\,(\mathbf{I} - \mathbf{bb}), \tag{2.44}$$

where $\mathbf{I}$ is the identity tensor. Thus, making use of $\mathbf{I} : \nabla\mathbf{B} = \nabla\cdot\mathbf{B} = 0$, it follows that

$$-e\,\langle \mathbf{u}\times(\boldsymbol{\rho}\cdot\nabla)\,\mathbf{B}\rangle = \frac{m\,u_\perp^2}{2\,B}\,\nabla B. \tag{2.45}$$

This quantity is the secular component of the gyration induced fluctuations in the magnetic force acting on the particle.

The coefficient of ∇B in the previous equation,

$$\mu = \frac{m\,u_\perp^2}{2\,B}, \tag{2.46}$$

plays a central role in the theory of magnetized particle motion. We can interpret this coefficient as a *magnetic moment* by drawing an analogy between a gyrating particle and a current loop. The (vector) magnetic moment of a plane current loop is simply

$$\boldsymbol{\mu} = I\,A\,\mathbf{n}, \tag{2.47}$$

where I is the current, A the area of the loop, and $\mathbf{n}$ the unit normal to the surface of the loop. For a circular loop of radius $\rho = u_\perp/\Omega$, lying in the plane perpendicular to $\mathbf{b}$, and carrying the current $e\,\Omega/2\pi$, we find

$$\boldsymbol{\mu} = I\,\pi\,\rho^2\,\mathbf{b} = \frac{m\,u_\perp^2}{2\,B}\,\mathbf{b}. \tag{2.48}$$

We shall demonstrate, in Section 2.6, that the (scalar) magnetic moment, μ, is a constant of the particle motion. Thus, the guiding center behaves exactly like a particle with a conserved magnetic moment $\boldsymbol{\mu}$ that is always aligned with the magnetic field.

The first-order guiding center equation of motion, Equation (2.42), reduces to

$$m\,\frac{d\mathbf{U}_0}{dt} = e\,E_\parallel\,\mathbf{b} + e\,\mathbf{U}_1\times\mathbf{B} - \mu\,\nabla B. \tag{2.49}$$

The component of this equation along the magnetic field determines the evolution of the parallel guiding center velocity:

$$m\,\frac{dU_{0\,\parallel}}{dt} = e\,E_\parallel - \boldsymbol{\mu}\cdot\nabla B - m\,\mathbf{b}\cdot\frac{d\mathbf{v}_E}{dt}. \tag{2.50}$$

Here, use has been made of Equation (2.35), and $\mathbf{b}\cdot d\mathbf{b}/dt = 0$. The component of Equation (2.49) perpendicular to the magnetic field determines the first-order perpendicular drift velocity:

$$\mathbf{U}_{1\,\perp} = \frac{\mathbf{b}}{\Omega}\times\left(\frac{d\mathbf{U}_0}{dt} + \frac{\mu}{m}\,\nabla B\right). \tag{2.51}$$

The first-order correction to the parallel velocity, the so-called first-order parallel drift velocity, is undetermined to this order. This is not a problem, because the first-order parallel drift is a small correction to a type of motion that already exists at zeroth order, whereas the first-order perpendicular drift is a completely new type of motion. In particular, the first-order perpendicular drift differs fundamentally from the $\mathbf{E} \times \mathbf{B}$ drift, because it is not the same for all species, and, therefore, cannot be eliminated by transforming to a new inertial frame. Thus, without loss of generality, we can absorb the first-order parallel drift into $U_{0\,\parallel}$, and write $\mathbf{U}_1 = \mathbf{U}_{1\,\perp}$.

We can now understand the motion of a charged particle as it moves through slowly varying electric and magnetic fields. The particle always gyrates around the magnetic field at the local gyrofrequency, $\Omega = e B/m$. The local perpendicular gyration velocity, $u_\perp$, is determined by the requirement that the magnetic moment, $\mu = m\,u_\perp^2/(2\,B)$, be a constant of the motion. This, in turn, fixes the local gyroradius, $\rho = u_\perp/\Omega$. The parallel velocity of the particle is determined by Equation (2.50). Finally, the perpendicular drift velocity is the sum of the $\mathbf{E}\times\mathbf{B}$ drift velocity, $\mathbf{v}_E$, and the first-order drift velocity, $\mathbf{U}_{1\,\perp}$.

2.5 Magnetic Drifts

Equations (2.35) and (2.51) can be combined to give

$$\mathbf{U}_{1\,\perp} = \frac{\mu}{m\,\Omega}\,\mathbf{b}\times\nabla B + \frac{U_{0\,\parallel}}{\Omega}\,\mathbf{b}\times\frac{d\mathbf{b}}{dt} + \frac{\mathbf{b}}{\Omega}\times\frac{d\mathbf{v}_E}{dt}. \tag{2.52}$$

The three terms on the right-hand side of the previous expression are conventionally called the *grad-B drift*, the *inertial drift*, and the *polarization drift*, respectively.

The grad-B drift,

$$\mathbf{U}_{\rm grad} = \frac{\mu}{m\,\Omega}\,\mathbf{b}\times\nabla B, \tag{2.53}$$

is caused by the slight variation of the gyroradius with gyrophase as a charged particle rotates in a nonuniform magnetic field. The gyroradius is reduced on the high-field side of the Larmor orbit, whereas it is increased on the low-field side. The net result is that the orbit does not quite close on itself. In fact, the motion consists of the conventional gyration around the magnetic field combined with a slow drift that is perpendicular to both the local direction of the magnetic field and the local gradient of the field-strength.

Given that, to lowest order,

$$\frac{d\mathbf{b}}{dt} = \frac{\partial\mathbf{b}}{\partial t} + \mathbf{U}_0\cdot\nabla\mathbf{b} = \frac{\partial\mathbf{b}}{\partial t} + (\mathbf{v}_E\cdot\nabla)\,\mathbf{b} + U_{0\,\parallel}\,(\mathbf{b}\cdot\nabla)\,\mathbf{b}, \tag{2.54}$$

the inertial drift can be written

$$\mathbf{U}_{\rm int} = \frac{U_{0\,\parallel}}{\Omega}\,\mathbf{b}\times\left[\frac{\partial\mathbf{b}}{\partial t} + (\mathbf{v}_E\cdot\nabla)\,\mathbf{b}\right] + \frac{U_{0\,\parallel}^2}{\Omega}\,\mathbf{b}\times(\mathbf{b}\cdot\nabla)\,\mathbf{b}. \tag{2.55}$$

In the important limit of stationary magnetic fields, and weak electric fields, the previous expression is dominated by the final term,

$$\mathbf{U}_{\rm curv} = \frac{U_{0\parallel}^2}{\Omega}\,\mathbf{b}\times(\mathbf{b}\cdot\nabla)\,\mathbf{b}, \tag{2.56}$$

which is called the *curvature drift*. As is easily demonstrated, the quantity $(\mathbf{b}\cdot\nabla)\,\mathbf{b}$ is a vector that is directed toward the center of the circle that most closely approximates the magnetic field-line at a given point, and whose magnitude is the inverse of the radius of this circle. Thus, the centripetal acceleration imposed by the curvature of the magnetic field on a charged particle following a field-line gives rise to a slow drift that is perpendicular to both the local direction of the magnetic field and the direction to the local center of curvature of the field.

The polarization drift,

$$\mathbf{U}_{\rm polz} = \frac{\mathbf{b}}{\Omega}\times\frac{d\mathbf{v}_E}{dt}, \tag{2.57}$$

reduces to

$$\mathbf{U}_{\rm polz} = \frac{1}{\Omega}\frac{d}{dt}\left(\frac{\mathbf{E}_\perp}{B}\right) \tag{2.58}$$

in the limit in which the magnetic field is stationary, but the electric field varies in time. This expression can be understood as a polarization drift by considering what happens when we suddenly impose an electric field on a particle at rest. The particle initially accelerates in the direction of the electric field, but is then deflected by the magnetic force. Thereafter, the particle undergoes conventional gyromotion combined with $\mathbf{E}\times\mathbf{B}$ drift. The time between the switch-on of the field and the magnetic deflection is approximately $\Delta t \sim \Omega^{-1}$. There is no deflection if the electric field is directed parallel to the magnetic field, so this argument only applies to perpendicular electric fields. The initial displacement of the particle in the direction of the field is of order

$$\delta \sim \frac{e\,\mathbf{E}_\perp}{m}\,(\Delta t)^2 \sim \frac{\mathbf{E}_\perp}{\Omega\,B}. \tag{2.59}$$

Because $\Omega \propto m^{-1}$, the displacement of the ions greatly exceeds that of the electrons. Thus, when an electric field is suddenly switched on in a plasma, there is an initial polarization of the plasma medium caused, predominately, by a displacement of the ions in the direction of the field. If the electric field, in fact, varies continuously in time then there is a slow drift due to the constantly changing polarization of the plasma medium. This drift is essentially the time derivative of Equation (2.59) [in other words, Equation (2.58)].

2.6 Invariance of Magnetic Moment

Let us now demonstrate that the magnetic moment, $\mu = m\,u_\perp^2/(2\,B)$, is indeed a constant of the motion, at least to lowest order. The scalar product of the equation of motion, Equation (2.24), with the velocity $\mathbf{v}$ yields

$$\frac{m}{2}\frac{dv^2}{dt} = e\,\mathbf{v}\cdot\mathbf{E}. \tag{2.60}$$

This equation governs the evolution of the particle energy during its motion. Let us make the substitution $\mathbf{v} = \mathbf{U} + \mathbf{u}$, as before, and then average the preceding equation over gyrophase. To lowest order, we obtain

$$\frac{m}{2}\frac{d}{dt}(u_\perp^2 + U_0^2) = e\,U_{0\parallel}\,E_\parallel + e\,\mathbf{U}_1\cdot\mathbf{E} + e\,\langle\mathbf{u}\cdot(\boldsymbol{\rho}\cdot\nabla)\,\mathbf{E}\rangle. \tag{2.61}$$

Here, use has been made of the result

$$\frac{d}{dt}\langle f\rangle = \left\langle\frac{df}{dt}\right\rangle, \tag{2.62}$$

which is valid for any f. The final term on the right-hand side of Equation (2.61) can be written

$$e\,\Omega\,\langle(\boldsymbol{\rho}\times\mathbf{b})\cdot(\boldsymbol{\rho}\cdot\nabla)\,\mathbf{E}\rangle = -\mu\,\mathbf{b}\cdot\nabla\times\mathbf{E} = \boldsymbol{\mu}\cdot\frac{\partial\mathbf{B}}{\partial t} = \mu\,\frac{\partial B}{\partial t}, \tag{2.63}$$

where use has been made of Equation (2.44). Thus, Equation (2.61) reduces to

$$\frac{dK}{dt} = e\,\mathbf{U}\cdot\mathbf{E} + \boldsymbol{\mu}\cdot\frac{\partial\mathbf{B}}{\partial t} = e\,\mathbf{U}\cdot\mathbf{E} + \mu\,\frac{\partial B}{\partial t}. \tag{2.64}$$

Here, $\mathbf{U}$ is the guiding center velocity, evaluated to first order, and

$$K = \frac{m}{2}(U_{0\parallel}^2 + \mathbf{v}_E^2 + u_\perp^2) \tag{2.65}$$

is the lowest order kinetic energy of the particle. Evidently, the kinetic energy can change in one of two different ways. First, by motion of the guiding center along the direction of the electric field, and, second, by acceleration of the gyration due to the electromotive force generated around the Larmor orbit by a changing magnetic field.

Equation (2.64) yields

$$m\,U_{0\parallel}\,\frac{dU_{0\parallel}}{dt} + m\,\mathbf{v}_E\cdot\frac{d\mathbf{v}_E}{dt} + \frac{d(B\,\mu)}{dt} = e\,U_{0\parallel}\,E_\parallel + e\,\mathbf{U}_1\cdot\mathbf{E} + \mu\,\frac{\partial B}{dt}. \tag{2.66}$$

It follows from Equation (2.50) that

$$-m\,U_{0\parallel}\,\mathbf{b}\cdot\frac{d\mathbf{v}_E}{dt} + m\,\mathbf{v}_E\cdot\frac{d\mathbf{v}_E}{dt} + B\,\frac{d\mu}{dt} + \mu\,\mathbf{v}_E\cdot\nabla B = e\,\mathbf{U}_1\cdot\mathbf{E}, \tag{2.67}$$

where use has been made of $dB/dt = \partial B/\partial t + U_{0\parallel}\,\mathbf{b}\cdot\nabla B + \mathbf{v}_E\cdot\nabla B$. However, $\mathbf{U}_1 = \mathbf{U}_{1\perp}$. Moreover, according to Equations (2.4), (2.35), and (2.51),

$$e\,\mathbf{U}_1\cdot\mathbf{E} = m\,\mathbf{v}_E\cdot\left[\frac{d}{dt}\,(U_{0\parallel}\,\mathbf{b} + \mathbf{v}_E) + \frac{\mu}{m}\,\nabla B\right]. \tag{2.68}$$

Hence, Equation (2.67) reduces to

$$B\,\frac{d\mu}{dt} = m\,\frac{d}{dt}\,(U_{0\,\|}\,\mathbf{v}_E\cdot\mathbf{b}) = 0, \tag{2.69}$$

which implies that

$$\frac{d\mu}{dt} = \frac{d}{dt}\left(\frac{m\,u_\perp^2}{2\,B}\right) = 0. \tag{2.70}$$

Thus, to lowest order, the magnetic moment, μ, is a constant of the motion. Kruskal has shown that $m\,u_\perp^2/(2\,B)$ is, in fact, the lowest order approximation to a quantity that is a constant of the motion to all orders in the perturbation expansion (Kruskal 1962). Such a quantity is termed an *adiabatic invariant*.

2.7 Poincaré Invariants

An adiabatic invariant is an approximation to a more fundamental type of invariant known as a *Poincaré invariant* (Hazeltine and Waelbroeck 2004). A Poincaré invariant takes the form

$$I = \oint_{C(t)} \mathbf{p}\cdot d\mathbf{q}, \tag{2.71}$$

where all points on the closed curve $C(t)$ in phase-space move according to the equations of motion.

In order to demonstrate that I is a constant of the motion, we introduce a periodic variable s parameterizing the points on the curve C. The coordinates of a general point on C are thus written $q_i = q_i(s,t)$ and $p_i = p_i(s,t)$. The rate of change of I is then

$$\frac{dI}{dt} = \oint\left(p_i\,\frac{\partial^2 q_i}{\partial t\,\partial s} + \frac{\partial p_i}{\partial t}\,\frac{\partial q_i}{\partial s}\right)ds. \tag{2.72}$$

Let us integrate the first term by parts, and then use Hamilton's equations of motion to simplify the result (Goldstein, Poole, and Safko 2002). We obtain

$$\frac{dI}{dt} = \oint\left(-\frac{\partial q_i}{\partial t}\,\frac{\partial p_i}{\partial s} + \frac{\partial p_i}{\partial t}\,\frac{\partial q_i}{\partial s}\right)ds = -\oint\left(\frac{\partial H}{\partial p_i}\,\frac{\partial p_i}{\partial s} + \frac{\partial H}{\partial q_i}\,\frac{\partial q_i}{\partial s}\right)ds, \tag{2.73}$$

where $H(\mathbf{p},\mathbf{q},t)$ is the Hamiltonian for the motion. The integrand is now seen to be the total derivative of H along C. Because the Hamiltonian is a single-valued function, it follows that

$$\frac{dI}{dt} = -\oint\frac{dH}{ds}\,ds = 0. \tag{2.74}$$

Thus, I is indeed a constant of the motion.

2.8 Adiabatic Invariants

Poincaré invariants are generally of little practical interest unless the curve C closely corresponds to the trajectories of actual particles. For the motion of magnetized particles, it is evident from Equations (2.25), (2.38), and (2.40) that points having the same guiding center at a certain time will continue to have approximately the same guiding center at later times. An approximate Poincaré invariant may thus be obtained by choosing the curve C to be a circle of points corresponding to a gyrophase period. In other words,

$$I \simeq I = \oint \mathbf{p} \cdot \frac{\partial \mathbf{q}}{\partial \gamma}\, d\gamma. \tag{2.75}$$

Here, I is an adiabatic invariant.

To evaluate I for a magnetized plasma recall that the canonical momentum for charged particles is (Jackson 1998)

$$\mathbf{p} = m\,\mathbf{v} + e\,\mathbf{A}, \tag{2.76}$$

where $\mathbf{A}$ is the vector potential. Let us express $\mathbf{A}$ in terms of its Taylor series about the guiding center position:

$$\mathbf{A}(\mathbf{r}) = \mathbf{A}(\mathbf{R}) + (\boldsymbol{\rho}\cdot\nabla)\,\mathbf{A}(\mathbf{R}) + O(\rho^2). \tag{2.77}$$

The element of length along the curve $C(t)$ is [see Equation (2.39)]

$$d\mathbf{r} = \frac{\partial \boldsymbol{\rho}}{\partial \gamma}\, d\gamma = \frac{\mathbf{u}}{\Omega}\, d\gamma. \tag{2.78}$$

The adiabatic invariant is thus

$$I = \oint \frac{\mathbf{u}}{\Omega}\cdot (m\,[\mathbf{U} + \mathbf{u}] + e\,[\mathbf{A} + (\boldsymbol{\rho}\cdot\nabla)\,\mathbf{A}])\,d\gamma + O(\epsilon), \tag{2.79}$$

which reduces to

$$I = 2\pi\, m\,\frac{u_\perp^2}{\Omega} + 2\pi\,\frac{e}{\Omega}\,\langle \mathbf{u}\cdot(\boldsymbol{\rho}\cdot\nabla)\,\mathbf{A}\rangle + O(\epsilon). \tag{2.80}$$

The final term on the right-hand side is written [see Equations (2.41) and (2.44)]

$$2\pi\, e\,\langle(\boldsymbol{\rho}\times\mathbf{b})\cdot(\boldsymbol{\rho}\cdot\nabla)\,\mathbf{A}\rangle = -2\pi\, e\,\frac{u_\perp^2}{2\,\Omega^2}\,\mathbf{b}\cdot\nabla\times\mathbf{A} = -\pi\, m\,\frac{u_\perp^2}{\Omega}. \tag{2.81}$$

It follows that

$$I = 2\pi\,\frac{m}{e}\,\mu + O(\epsilon). \tag{2.82}$$

Thus, to lowest order, the adiabatic invariant is proportional to the magnetic moment, μ.

2.9 Magnetic Mirrors

Consider the important case in which the electromagnetic fields do not vary in time. It follows that $\mathbf{E} = -\nabla\phi$, where ϕ is the electrostatic potential. Equation (2.64) yields

$$\frac{dK}{dt} = -e\,\mathbf{U}\cdot\nabla\phi = -\frac{d(e\,\phi)}{dt}, \tag{2.83}$$

because $d/dt = \partial/\partial t + \mathbf{U}\cdot\nabla$. Thus, we obtain

$$\frac{d\mathcal{E}}{dt} = 0, \tag{2.84}$$

where

$$\mathcal{E} = K + e\,\phi = \frac{m}{2}\,(U_{\parallel}^2 + \mathbf{v}_E^2) + \mu\,B + e\,\phi \tag{2.85}$$

is the lowest order total particle energy. Not surprisingly, a charged particle neither gains nor loses energy as it moves around in non-time-varying electromagnetic fields. Because $\mathcal{E}$ and μ are constants of the motion, we can rearrange Equation (2.85) to give

$$U_{\parallel} = \pm\left[(2/m)\,(\mathcal{E} - \mu\,B - e\,\phi) - \mathbf{v}_E^2\right]^{1/2}. \tag{2.86}$$

Thus, charged particles can drift in either direction along magnetic field-lines in regions where $\mathcal{E} > \mu\,B + e\,\phi + m\,\mathbf{v}_E^2/2$. However, particles are excluded from regions where $\mathcal{E} < \mu\,B + e\,\phi + m\,\mathbf{v}_E^2/2$ (because they cannot have imaginary parallel velocities). Evidently, charged particles must reverse direction at those points on magnetic field-lines where $\mathcal{E} = \mu\,B + e\,\phi + m\,\mathbf{v}_E^2/2$. Such points are termed *bounce points* or *mirror points*.

Let us now consider how we might construct a device to confine a collisionless (in other words, very high temperature) plasma. Obviously, we cannot use conventional solid walls, because they would melt. However, it is possible to confine a hot plasma using a magnetic field (fortunately, magnetic field-lines cannot melt). This technique is called *magnetic confinement*. The electric field in confined plasmas is usually weak (that is, $E \ll B\,v$), so that the $\mathbf{E}\times\mathbf{B}$ drift is similar in magnitude to the magnetic and curvature drifts. In this case, the bounce point condition, $U_{\parallel} = 0$, reduces to

$$\mathcal{E} = \mu\,B. \tag{2.87}$$

Consider the magnetic field configuration illustrated in Figure 2.1. As indicated, this configuration is most easily produced by two Helmholtz coils. Incidentally, this type of magnetic confinement device is called a *magnetic mirror machine*. The magnetic field configuration obviously possesses axial symmetry. Let z be a coordinate that measures distance along the axis of symmetry. Suppose that $z = 0$ corresponds to the midplane of the device (that is, halfway between the two field-coils).

It is clear, from the figure, that the magnetic field-strength $B(z)$ on a magnetic field-line situated close to the axis of the device attains a local minimum $B_{\rm min}$ at $z =$

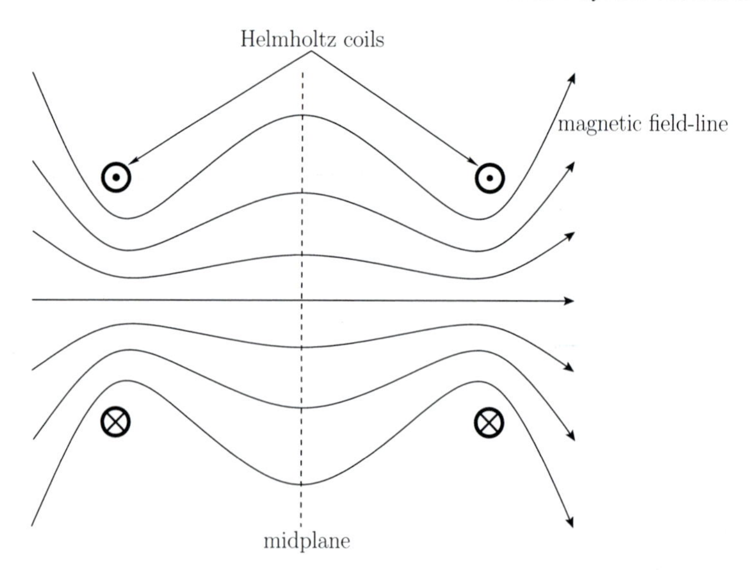

Figure 2.1
Schematic cross-section of a magnetic mirror machine employing two Helmholtz coils.

0, increases symmetrically as $|z|$ increases until reaching a maximum value $B_{\rm max}$ at about the locations of the two field-coils, and then decreases as $|z|$ is further increased. According to Equation (2.87), any particle that satisfies the inequality

$$\mu > \mu_{\rm trap} = \frac{\mathcal{E}}{B_{\rm max}} \tag{2.88}$$

is trapped on such a field-line. In fact, the particle undergoes periodic motion along the field-line between two symmetrically placed (in z) mirror points. The magnetic field-strength at the mirror points is

$$B_{\rm mirror} = \frac{\mu_{\rm trap}}{\mu}\,B_{\rm max} < B_{\rm max}. \tag{2.89}$$

On the midplane, $\mu = m\,v_\perp^2/(2\,B_{\rm min})$ and $\mathcal{E} = m\,(v_\parallel^2 + v_\perp^2)/2$. (From now on, for ease of notation, we shall write $\mathbf{v} = v_\parallel\,\mathbf{b} + \mathbf{v}_\perp$.) Thus, the trapping condition, Equation (2.88), reduces to

$$\frac{|v_\parallel|}{|v_\perp|} < (B_{\rm max}/B_{\rm min} - 1)^{1/2}. \tag{2.90}$$

Particles on the midplane that satisfy this inequality are trapped. On the other hand, particles that do not satisfy the inequality escape along magnetic field-lines. A magnetic mirror machine is incapable of trapping charged particles that are moving parallel, or nearly parallel, to the direction of the magnetic field. In fact, the previous inequality defines a *loss cone* in velocity space. (See Figure 2.2.)

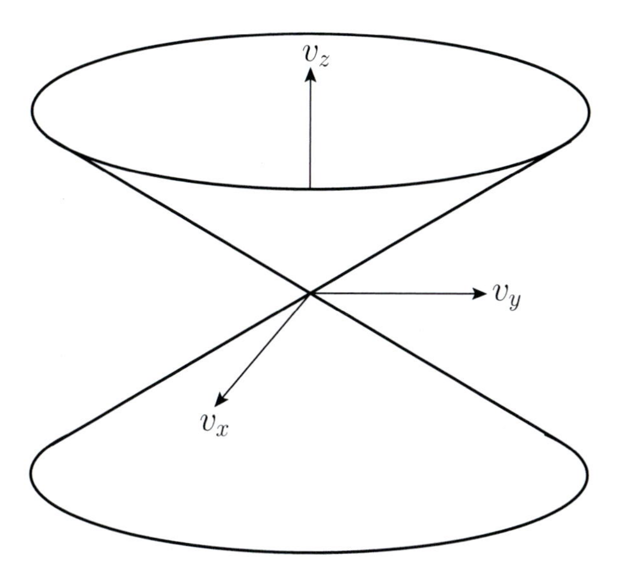

Figure 2.2
A loss cone in velocity space. Particles whose velocity vectors lie inside the cone are not reflected by the magnetic field.

If plasma is placed inside a magnetic mirror machine then all of the particles whose velocities lie in the loss cone promptly escape, but the remaining particles are confined. Unfortunately, that is not the end of the story. There is no such thing as an absolutely collisionless plasma. Collisions take place at a low rate, even in very hot plasmas. One important effect of collisions is to cause diffusion of particles in velocity space (Hazeltine and Waelbroeck 2004). Thus, collisions in a mirror machine continuously scatter trapped particles into the loss cone, giving rise to a slow leakage of plasma out of the device. Even worse, plasmas whose distribution functions deviate strongly from an isotropic Maxwellian (for instance, a plasma confined in a mirror machine) are prone to *velocity-space instabilities* (see Chapter 8) that tend to relax the distribution function back to a Maxwellian. Such instabilities can have a disastrous effect on plasma confinement in a mirror machine.

2.10 Van Allen Radiation Belts

Plasma confinement via magnetic mirroring occurs in nature. For instance, the *Van Allen radiation belts*, which surround the Earth, consist of energetic particles trapped in the Earth's dipole-like magnetic field. These belts were discovered by James A. Van Allen and co-workers using data taken from Geiger counters that flew on the early U.S. satellites, Explorer 1 (which was, in fact, the first U.S. satellite), Explorer 4, and Pioneer 3. Van Allen was actually trying to measure the flux of cosmic rays (high energy particles whose origin is outside the solar system) in outer space, to see if it was similar to that measured on Earth. However, the flux of energetic particles detected by his instruments so greatly exceeded the expected value that it prompted one of his co-workers, Ernie Ray, to exclaim, "My God, space is radioactive!" (Hess 1968). It was quickly realized that this flux was due to energetic particles trapped in the Earth's magnetic field, rather than to cosmic rays.

There are, in fact, two radiation belts surrounding the Earth (Baumjohan and Treumann 1996). The inner belt, which extends from about 1–3 Earth radii in the equatorial plane, is mostly populated by protons with energies exceeding 10 MeV. The origin of these protons is thought to be the decay of neutrons that are emitted from the Earth's atmosphere as it is bombarded by cosmic rays. The inner belt is fairly quiescent. Particles eventually escape due to collisions with neutral atoms in the upper atmosphere above the Earth's poles. However, such collisions are sufficiently uncommon that the lifetime of particles in the inner belt range from a few hours to 10 years. Obviously, with such long trapping times, only a small input rate of energetic particles is required to produce a region of intense radiation.

The outer belt, which extends from about 3–9 Earth radii in the equatorial plane, consists mostly of electrons with energies below 10 MeV. These electrons originate via injection from the outer magnetosphere. Unlike the inner belt, the outer belt is very dynamic, changing on timescales of a few hours in response to perturbations emanating from the outer magnetosphere.

In regions not too far distant (that is, less than 10 Earth radii) from the Earth, the geomagnetic field can be approximated as a dipole field,

$$\mathbf{B} = \frac{\mu_0}{4\pi}\frac{M_E}{r^3}(-2\cos\theta, -\sin\theta, 0), \tag{2.91}$$

where we have adopted conventional spherical coordinates (r, θ, φ) aligned with the Earth's dipole moment, whose magnitude is $M_E = 8.05\times10^{22}$ A m^2 (Baumjohan and Treumann 1996). It is convenient to work in terms of the latitude, $\vartheta = \pi/2 - \theta$, rather than the polar angle, θ. An individual magnetic field-line satisfies the equation

$$r = r_{\rm eq}\cos^2\vartheta, \tag{2.92}$$

where $r_{\rm eq}$ is the radial distance to the field-line in the equatorial plane ($\vartheta = 0°$). It is conventional to label field-lines using the *L-shell parameter*, $L = r_{\rm eq}/R_E$. Here, $R_E = 6.37\times10^6$ m is the Earth's radius (Yoder 1995). Thus, the variation of the magnetic field-strength along a field-line characterized by a given L-value is

$$B = \frac{B_E}{L^3}\frac{(1+3\sin^2\vartheta)^{1/2}}{\cos^6\vartheta}, \tag{2.93}$$

where $B_E = \mu_0 M_E/(4\pi R_E^3) = 3.11\times10^{-5}$ T is the equatorial magnetic field-strength on the Earth's surface (Baumjohan and Treumann 1996).

Consider, for the sake of simplicity, charged particles located on the equatorial plane ($\vartheta = 0°$) whose velocities are predominately directed perpendicular to the magnetic field. The proton and electron gyrofrequencies are written[1]

$$\Omega_p = \frac{e\,B}{m_p} = 2.98\,L^{-3}\ \text{kHz}, \tag{2.94}$$

and

$$|\Omega_e| = \frac{e\,B}{m_e} = 5.46\,L^{-3}\ \text{MHz}, \tag{2.95}$$

respectively. The proton and electron gyroradii, expressed as fractions of the Earth's radius, take the form

$$\frac{\rho_p}{R_E} = \frac{\sqrt{2\,\mathcal{E}\,m_p}}{e\,B\,R_E} = \sqrt{\mathcal{E}(\text{MeV})}\left(\frac{L}{11.1}\right)^3, \tag{2.96}$$

and

$$\frac{\rho_e}{R_E} = \frac{\sqrt{2\,\mathcal{E}\,m_e}}{e\,B\,R_E} = \sqrt{\mathcal{E}(\text{MeV})}\left(\frac{L}{38.9}\right)^3, \tag{2.97}$$

respectively. Thus, MeV energy charged particles in the inner magnetosphere (that is, $L \ll 10$) gyrate at frequencies that are much greater than the typical rate of change of the magnetic field (which varies on timescales that are, at most, a few minutes). Likewise, the gyroradii of such particles are much smaller than the typical variation lengthscale of the magnetospheric magnetic field. Under these circumstances, we expect the magnetic moment to be a conserved quantity. In other words, we expect the magnetic moment to be a good adiabatic invariant. It immediately follows that any MeV energy protons and electrons in the inner magnetosphere that have a sufficiently large magnetic moment are trapped on the dipolar field-lines of the Earth's magnetic field, bouncing back and forth between mirror points located just above the Earth's poles.

It is helpful to define the *pitch-angle*,

$$\alpha = \tan^{-1}(v_\perp/v_\parallel), \tag{2.98}$$

of a charged particle in the magnetosphere. If the magnetic moment is a conserved quantity then a particle of fixed energy drifting along a field-line satisfies

$$\frac{\sin^2\alpha}{\sin^2\alpha_{\text{eq}}} = \frac{B}{B_{\text{eq}}}, \tag{2.99}$$

where α_{eq} is the *equatorial pitch-angle* (that is, the pitch-angle on the equatorial plane), and $B_{\text{eq}} = B_E/L^3$ is the magnetic field-strength on the equatorial plane. According to Equation (2.93), the pitch-angle increases (i.e., the parallel component of the particle velocity decreases) as the particle drifts off the equatorial plane toward the Earth's poles.

[1]It is conventional to take account of the negative charge of electrons by making the electron gyrofrequency Ω_e negative. This approach is implicit in formulae such as Equation (2.52).

The mirror points correspond to $\alpha = 90^\circ$ (i.e., $v_\parallel = 0$). It follows from Equations (2.93) and (2.99) that

$$\sin^2\alpha_{\rm eq} = \frac{B_{\rm eq}}{B_m} = \frac{\cos^6\vartheta_m}{(1+3\,\sin^2\vartheta_m)^{1/2}}, \tag{2.100}$$

where B_m is the magnetic field-strength at the mirror points, and ϑ_m the latitude of the mirror points. It can be seen that the latitude of a particle's mirror point depends only on its equatorial pitch-angle, and is independent of the L-value of the field-line on which it is trapped.

Charged particles with large equatorial pitch-angles have small parallel velocities, and mirror points located at relatively low latitudes. Conversely, charged particles with small equatorial pitch-angles have large parallel velocities, and mirror points located at high latitudes. Of course, if the pitch-angle becomes too small then the mirror points enter the Earth's atmosphere, and the particles are lost via collisions with neutral particles. Neglecting the thickness of the atmosphere with respect to the radius of the Earth, we can say that all particles whose mirror points lie inside the Earth are lost via collisions. It follows from Equation (2.100) that the *equatorial loss cone* is of approximate width

$$\sin^2\alpha_l = \frac{\cos^6\vartheta_E}{(1+3\sin^2\vartheta_E)^{1/2}}, \tag{2.101}$$

where ϑ_E is the latitude of the point at which the magnetic field-line under investigation intersects the Earth. All particles with $|\alpha_{\rm eq}| < \alpha_l$ and $|\pi - \alpha_{\rm eq}| < \alpha_l$ lie in the loss cone. According to Equation (2.92),

$$\cos^2\vartheta_E = L^{-1}. \tag{2.102}$$

It follows that

$$\sin^2\alpha_l = (4\,L^6 - 3\,L^5)^{-1/2}. \tag{2.103}$$

Thus, the width of the loss cone is independent of the charge, the mass, or the energy of the particles drifting along a given field-line, and is a function only of the field-line radius on the equatorial plane. The loss cone is surprisingly small. For instance, at the radius of a geostationary satellite orbit ($6.6\,R_E$), the loss cone is less than 3° wide. The smallness of the loss cone is a consequence of the very strong variation of the magnetic field-strength along field-lines in a dipole field. [See Equations (2.90) and (2.93).]

The *bounce period*, τ_b, is the time it takes a charged particle to move from the equatorial plane to one mirror point, through the equatorial plane to the other mirror point, and then back to the equatorial plane. It follows that

$$\tau_b = 4\int_0^{\vartheta_m}\frac{d\vartheta}{|v_\parallel|}\frac{ds}{d\vartheta}, \tag{2.104}$$

where ds is an element of arc-length along the field-line under investigation, and

$|v_\parallel| = v\,(1 - B/B_m)^{1/2}$. The previous integral cannot be performed analytically. However, it can be solved numerically, and is conveniently approximated as (Baumjohan and Treumann 1996)

$$\tau_b \simeq \frac{L\,R_E}{(\mathcal{E}/m)^{1/2}}\,(3.7 - 1.6\sin\alpha_{\rm eq}). \tag{2.105}$$

Thus, for protons

$$(\tau_b)_p \simeq 2.41\,\frac{L}{\sqrt{\mathcal{E}(\text{MeV})}}\,(1 - 0.43\sin\alpha_{\rm eq})\ \text{ seconds}, \tag{2.106}$$

while for electrons

$$(\tau_b)_e \simeq 5.62\times 10^{-2}\,\frac{L}{\sqrt{\mathcal{E}(\text{MeV})}}\,(1 - 0.43\sin\alpha_{\rm eq})\ \text{ seconds}. \tag{2.107}$$

It follows that MeV electrons typically have bounce periods that are less than a second, whereas the bounce periods for MeV protons usually lie in the range 1 to 10 seconds. The bounce period only depends weakly on equatorial pitch-angle, because particles with small pitch angles have relatively large parallel velocities but a comparatively long way to travel to their mirror points, and vice versa. Naturally, the bounce period is longer for longer field-lines (that is, for larger L).

2.11 Equatorial Ring Current

Up to now, we have only considered the lowest-order motion (in other words, gyration combined with parallel drift) of charged particles in the magnetosphere. Let us now examine the higher-order corrections to this motion. For the case of non-time-varying fields, and a weak electric field, these corrections consist of the following combination of $\mathbf{E}\times\mathbf{B}$ drift, grad-B drift, and curvature drift:

$$\mathbf{v}_{1\perp} = \frac{\mathbf{E}\times\mathbf{B}}{B^2} + \frac{\mu}{m\,\Omega}\,\mathbf{b}\times\nabla B + \frac{v_\parallel^2}{\Omega}\,\mathbf{b}\times(\mathbf{b}\cdot\nabla)\,\mathbf{b}. \tag{2.108}$$

Let us neglect $\mathbf{E}\times\mathbf{B}$ drift, because this motion merely gives rise to the convection of plasma within the magnetosphere, without generating a current. By contrast, there is a net current associated with grad-B drift and curvature drift. In the limit in which this current does not strongly modify the ambient magnetic field (that is, $\nabla\times\mathbf{B}\simeq\mathbf{0}$), which is certainly the situation in the Earth's inner magnetosphere, we can write

$$(\mathbf{b}\cdot\nabla)\,\mathbf{b} = -\mathbf{b}\times(\nabla\times\mathbf{b}) \simeq \frac{\nabla_\perp B}{B}. \tag{2.109}$$

It follows that the higher-order drifts can be combined to give

$$\mathbf{v}_{1\perp} = \frac{\left(v_\perp^2/2 + v_\parallel^2\right)}{\Omega\,B}\,\mathbf{b}\times\nabla B. \tag{2.110}$$

For the dipole magnetic field specified in Equation (2.91), the previous expression yields

$$\mathbf{v}_{1\perp} \simeq -\mathrm{sgn}(\Omega)\,\frac{6\,\mathcal{E}\,L^2}{e\,B_E\,R_E}\,(1 - B/2\,B_m)\,\frac{\cos^5\vartheta\,(1+\sin^2\vartheta)}{(1+3\sin^2\vartheta)^2}\,\mathbf{e}_\varphi. \tag{2.111}$$

It can be seen that the drift is in the azimuthal direction. A positive drift velocity corresponds to eastward motion, whereas a negative velocity corresponds to westward motion. It follows that, in addition to their gyromotion, and their periodic bouncing motion along field-lines, charged particles trapped in the magnetosphere also slowly precess around the Earth. The ions drift westwards and the electrons drift eastwards, giving rise to a net westward current circulating around the Earth. This current is known as the *ring current*.

Although the perturbations to the Earth's magnetic field induced by the ring current are small, they are still detectable. In fact, the ring current causes a slight reduction in the Earth's magnetic field in equatorial regions. The size of this reduction is a good measure of the number of charged particles contained in the Van Allen belts. During the development of so-called *geomagnetic storms*, charged particles are injected into the Van Allen belts from the outer magnetosphere, giving rise to a sharp increase in the ring current, and a corresponding decrease in the Earth's equatorial magnetic field. These particles eventually precipitate out of the magnetosphere into the upper atmosphere at high terrestrial latitudes, giving rise to intense auroral activity, serious interference in electromagnetic communications, and, in extreme cases, disruption of electric power grids. The reduction in the Earth's magnetic field induced by the ring current is measured by the so-called *Dst index*, which is determined from hourly averages of the northward horizontal component of the terrestrial magnetic field recorded at four low-latitude observatories: Honolulu (Hawaii), San Juan (Puerto Rico), Hermanus (South Africa), and Kakioka (Japan). Figure 2.3 shows the Dst index for the month of March 1989. The very marked reduction in the index, centered on March 13, corresponds to one of the most severe geomagnetic storms experienced in recent decades. In fact, this particular storm was so severe that it tripped out the whole Hydro Québec electric distribution system, plunging more than 6 million customers into darkness. Most of Hydro Québec's neighboring systems in the United States came uncomfortably close to experiencing the same cascading power outage scenario. Incidentally, a reduction in the Dst index by 600 nT corresponds to a 2 percent reduction in the terrestrial magnetic field at the equator.

According to Equation (2.111), the precessional drift velocity of charged particles in the magnetosphere is a rapidly decreasing function of increasing latitude (in other words, the ring current is concentrated in the equatorial plane). Because charged particles typically complete many bounce orbits during a full circuit around the Earth, it is convenient to average Equation (2.111) over a bounce period to obtain the average drift velocity. This averaging can only be performed numerically. The final answer is well approximated by (Baumjohan and Treumann 1996)

$$\langle v_d\rangle \simeq \frac{6\,\mathcal{E}\,L^2}{e\,B_E\,R_E}\,(0.35 + 0.15\,\sin\alpha_{\rm eq}). \tag{2.112}$$

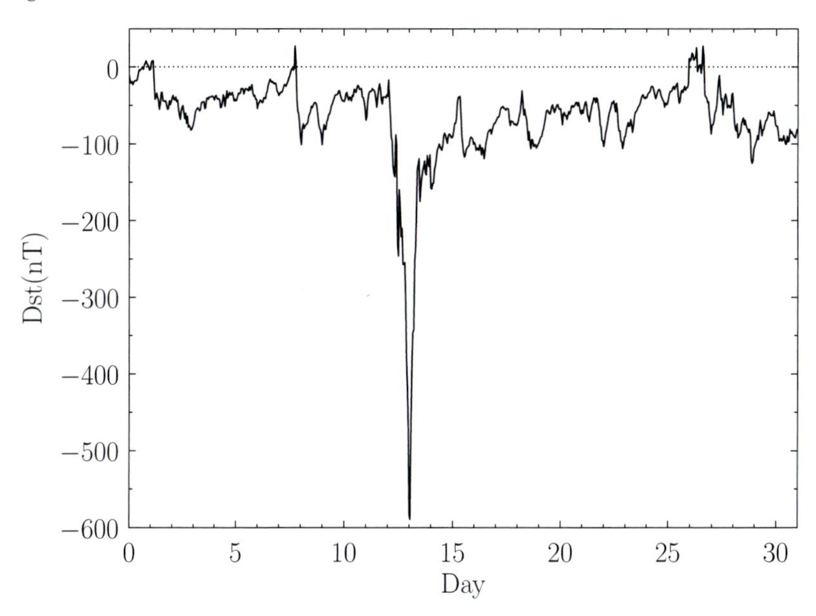

Figure 2.3
Dst data for March 1989 showing an exceptionally severe geomagnetic storm on March 13. Raw data from World Data Center for Geomagnetism, Kyoto.

The average drift period (that is, the time required to perform a complete circuit around the Earth) is simply

$$\langle\tau_d\rangle = \frac{2\pi\,L\,R_E}{\langle v_d\rangle} \simeq \frac{\pi\,e\,B_E\,R_E^2}{3\,\mathcal{E}\,L}\,(0.35 + 0.15\,\sin\alpha_{\rm eq})^{-1}. \tag{2.113}$$

Thus, the drift period for protons and electrons is

$$\langle\tau_d\rangle_p = \langle\tau_d\rangle_e \simeq \frac{1.05}{\mathcal{E}(\text{MeV})\,L}\,(1 + 0.43\,\sin\alpha_{\rm eq})^{-1}\ \text{hours}. \tag{2.114}$$

Note that MeV energy electrons and ions precess around the Earth with about the same velocity, only in opposite directions, because there is no explicit mass dependence in Equation (2.112). It typically takes an hour to perform a full circuit. The drift period only depends weakly on the equatorial pitch angle, as is the case for the bounce period. Somewhat paradoxically, the drift period is shorter on more distant L-shells. Of course, charged particles only get a chance to complete a full circuit around the Earth if the inner magnetosphere remains quiescent on timescales of order an hour. This is, by no means, always the case.

Finally, because the rest mass of an electron is 0.51 MeV, many of the previous formulae require relativistic correction when applied to MeV energy electrons. Fortunately, however, there is no such problem for protons, whose rest mass energy is 0.94 GeV.

2.12 Second Adiabatic Invariant

We have seen that there is an adiabatic invariant associated with the periodic gyration of a charged particle around magnetic field-lines. Thus, it is reasonable to suppose that there is a second adiabatic invariant associated with the periodic bouncing motion of a particle trapped between two mirror points on a magnetic field-line. This is indeed the case.

Recall that an adiabatic invariant is the lowest order approximation to a Poincaré invariant:

$$\mathcal{J} = \oint_C \mathbf{p}\cdot d\mathbf{q}. \tag{2.115}$$

In this case, let the curve C correspond to the trajectory of a guiding center as a charged particle trapped in the Earth's magnetic field executes a bounce orbit. Of course, this trajectory does not quite close, because of the slow azimuthal drift of particles around the Earth. However, it is easily demonstrated that the azimuthal displacement of the end point of the trajectory, with respect to the beginning point, is similar in magnitude to the gyroradius. Thus, in the limit in which the ratio of the gyroradius, ρ, to the variation lengthscale of the magnetic field, L, tends to zero, the trajectory of the guiding center can be regarded as being approximately closed, and the actual particle trajectory conforms very closely to that of the guiding center. Thus, the adiabatic invariant associated with the bounce motion can be written

$$\mathcal{J} \simeq J = \oint p_\parallel\, ds, \tag{2.116}$$

where the path of integration is along a field-line, from the equatorial plane to the upper mirror point, back along the field-line to the lower mirror point, and then back to the equatorial plane. Furthermore, ds is an element of arc-length along the field-line, and $p_\parallel \equiv \mathbf{p}\cdot\mathbf{b}$. Using $\mathbf{p} = m\,\mathbf{v} + e\,\mathbf{A}$, the previous expression yields

$$J = m\oint v_\parallel\, ds + e\oint A_\parallel\, ds = m\oint v_\parallel\, ds + e\,\Phi. \tag{2.117}$$

Here, Φ is the total magnetic flux enclosed by the curve—which, in this case, is obviously zero. Thus, the so-called *second adiabatic invariant*, or *longitudinal adiabatic invariant*, takes the form

$$J = m\oint v_\parallel\, ds. \tag{2.118}$$

In other words, the second invariant is proportional to the loop integral of the parallel (to the magnetic field) velocity taken over a bounce orbit. The preceding "proof" of the invariance of J is not particularly rigorous. In fact, the rigorous proof that J is an adiabatic invariant was first given by Northrop and Teller (Northrop and Teller 1960). Of course, J is only a constant of the motion for particles trapped in the inner magnetosphere provided the magnetospheric magnetic field varies on timescales that are much longer than the bounce time, τ_b. Because the bounce time for MeV energy

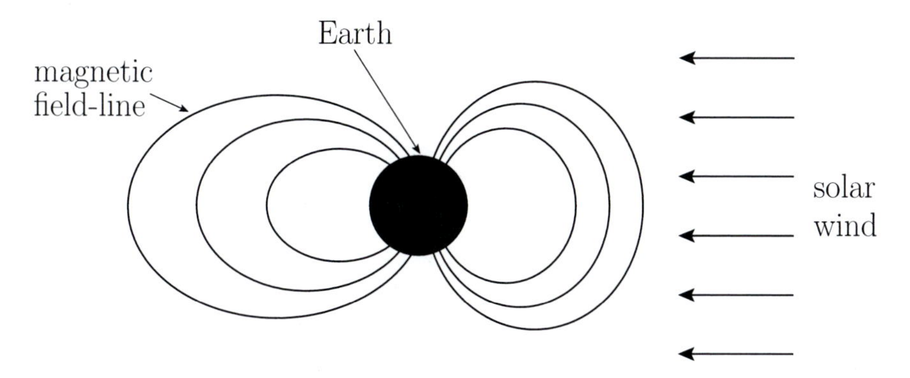

Figure 2.4
Schematic diagram showing the distortion of the Earth's magnetic field by the solar wind.

protons and electrons is, at most, a few seconds, this is not a particularly onerous constraint.

The invariance of J is of great importance for charged particle dynamics in the Earth's inner magnetosphere. It turns out that the Earth's magnetic field is distorted from pure axisymmetry by the action of the solar wind, as illustrated in Figure 2.4. Because of this asymmetry, there is no particular reason to believe that a particle will return to its earlier trajectory as it makes a full circuit around the Earth. In other words, the particle may well end up on a different field-line when it returns to the same azimuthal angle. However, at a given azimuthal angle, each field-line has a different length between mirror points, and a different variation of the field-strength, B, between the mirror points (for a particle with given energy, $\mathcal{E}$, and magnetic moment, μ). Thus, each field-line represents a different value of J for that particle. So, if J is conserved, as well as $\mathcal{E}$ and μ, then the particle must return to the same field-line after precessing around the Earth. In other words, the conservation of J prevents charged particles from spiraling radially in or out of the Van Allen belts as they rotate around the Earth. This helps to explain the persistence of these belts.

2.13 Third Adiabatic Invariant

It is clear, by now, that there is an adiabatic invariant associated with every periodic motion of a charged particle in an electromagnetic field. We have just demonstrated that, as a consequence of J-conservation, the drift orbit of a charged particle precessing around the Earth is approximately closed, despite the fact that the Earth's magnetic field is non-axisymmetric. Thus, there must be a third adiabatic invariant associated with the precession of particles around the Earth. Just as we can define a guiding center associated with a particle's gyromotion around field-lines, we can also define a *bounce center* associated with a particle's bouncing motion between mirror points. The bounce center lies on the equatorial plane, and orbits the Earth once every drift period, τ_d. We can write the third adiabatic invariant as

$$K \simeq \oint p_\phi\, ds, \tag{2.119}$$

where the path of integration is the trajectory of the bounce center around the Earth. Incidentally, the drift trajectory effectively collapses onto the trajectory of the bounce center in the limit that $\rho/L \to 0$, because all of the particle's gyromotion and bounce motion averages to zero. Now, $p_\phi = m\, v_\phi + e\, A_\phi$ is dominated by its second term, as the drift velocity v_ϕ is very small. Thus,

$$K \simeq e \oint A_\phi\, ds = e\, \Phi, \tag{2.120}$$

where Φ is the total magnetic flux enclosed by the drift trajectory (that is, the flux enclosed by the orbit of the bounce center around the Earth). The previous "proof" of the invariance of Φ is, again, not particularly rigorous. In fact, the invariance of Φ was first demonstrated rigorously by Northrop (Northrop 1963). Of course, Φ is only a constant of the motion for particles trapped in the inner magnetosphere provided the magnetospheric magnetic field varies on timescales that are much longer than the drift period, τ_d. Because the drift period for MeV energy protons and electrons is of order an hour, this is only likely to be the case when the magnetosphere is relatively quiescent (in other words, when there are no geomagnetic storms in progress).

The invariance of Φ has interesting consequences for charged particle dynamics in the Earth's inner magnetosphere. Suppose, for instance, that the strength of the solar wind were to increase slowly (that is, on timescales significantly longer than the drift period), thereby, compressing the Earth's magnetic field. The invariance of Φ would cause the charged particles that constitute the Van Allen belts to move radially inwards, toward the Earth, in order to conserve the magnetic flux enclosed by their drift orbits. Likewise, a slow decrease in the strength of the solar wind would cause an outward radial motion of the Van Allen belts.

2.14 Motion in Oscillating Fields

We have seen that charged particles can be confined by a static magnetic field. A somewhat more surprising fact is that charged particles can also be confined by a rapidly oscillating, inhomogeneous electromagnetic wave-field. In order to demonstrate this, we again employ our averaging technique (Hazeltine and Waelbroeck 2004). To lowest order, a particle executes simple harmonic motion in response to an oscillating wave-field. However, to higher order, any weak inhomogeneity in the field causes the restoring force at one turning point to exceed that at the other. On average, this yields a net force that acts on the *center of oscillation* of the particle.

Consider a spatially inhomogeneous electromagnetic wave-field oscillating at frequency ω:

$$\mathbf{E}(\mathbf{r}, t) = \mathbf{E}_0(\mathbf{r}) \cos(\omega t). \tag{2.121}$$

The equation of motion of a charged particle placed in this field is written

$$m \frac{d\mathbf{v}}{dt} = e \left[\mathbf{E}_0(\mathbf{r}) \cos(\omega t) + \mathbf{v} \times \mathbf{B}_0(\mathbf{r}) \sin(\omega t)\right], \tag{2.122}$$

where

$$\mathbf{B}_0 = -\omega^{-1} \nabla \times \mathbf{E}_0, \tag{2.123}$$

in accordance with Faraday's law.

In order for our averaging technique to be applicable, the electric field $\mathbf{E}_0$ experienced by the particle must remain approximately constant during an oscillation. Thus,

$$(\mathbf{v} \cdot \nabla) \mathbf{E}_0 \ll \omega \mathbf{E}_0. \tag{2.124}$$

When this inequality is satisfied, Equation (2.123) implies that the magnetic force experienced by the particle is smaller than the electric force by one order in the expansion parameter.

Let us now apply the averaging technique. We make the substitution $t \to \tau$ in the oscillatory terms, and seek a change of variables,

$$\mathbf{r} = \mathbf{R}(t) + \boldsymbol{\xi}(\mathbf{R}, \mathbf{U}, t, \tau), \tag{2.125}$$

$$\mathbf{v} = \mathbf{U}(t) + \mathbf{u}(\mathbf{R}, \mathbf{U}, t, \tau), \tag{2.126}$$

such that $\boldsymbol{\xi}$ and $\mathbf{u}$ are periodic functions of τ with vanishing mean. Averaging $d\mathbf{r}/dt = \mathbf{v}$ again yields $d\mathbf{R}/dt = \mathbf{U}$ to all orders. To lowest order, the momentum evolution equation, Equation (2.122), reduces to

$$\frac{\partial \mathbf{u}}{\partial \tau} = \frac{e}{m} \mathbf{E}_0(\mathbf{R}) \cos(\omega \tau). \tag{2.127}$$

Moreover, the lowest order oscillating component of $d\mathbf{r}/dt = \mathbf{v}$ gives

$$\frac{\partial \boldsymbol{\xi}}{\partial \tau} = \mathbf{u}. \tag{2.128}$$

The solutions to the previous two equations, taking into account the constraints $\langle \mathbf{u} \rangle = \langle \boldsymbol{\xi} \rangle = \mathbf{0}$, are

$$\mathbf{u} = \frac{e}{m\,\omega} \mathbf{E}_0 \sin(\omega \tau), \tag{2.129}$$

$$\boldsymbol{\xi} = -\frac{e}{m\,\omega^2} \mathbf{E}_0 \cos(\omega \tau), \tag{2.130}$$

respectively. Here, $\langle \cdots \rangle \equiv (2\pi)^{-1} \oint (\cdots)\, d(\omega \tau)$ represents an oscillation average.

It follows that, to lowest order, there is no motion of the center of oscillation. To first order, the oscillation average of Equation (2.122) yields

$$\frac{d\mathbf{U}}{dt} = \frac{e}{m}\,\langle(\boldsymbol{\xi}\cdot\nabla)\,\mathbf{E} + \mathbf{u}\times\mathbf{B}\rangle, \tag{2.131}$$

which reduces to

$$\frac{d\mathbf{U}}{dt} = -\frac{e^2}{m^2\,\omega^2}\left[(\mathbf{E}_0\cdot\nabla)\,\mathbf{E}_0\,\langle\cos^2(\omega\,\tau)\rangle + \mathbf{E}_0\times(\nabla\times\mathbf{E}_0)\,\langle\sin^2(\omega\,\tau)\rangle\right]. \tag{2.132}$$

The oscillation averages of the trigonometric functions are both equal to $1/2$. Furthermore, we have $\nabla(|\mathbf{E}_0|^2/2) \equiv (\mathbf{E}_0\cdot\nabla)\,\mathbf{E}_0 + \mathbf{E}_0\times(\nabla\times\mathbf{E}_0)$. Thus, the equation of motion for the center of oscillation reduces to

$$m\,\frac{d\mathbf{U}}{dt} = -e\,\nabla\Phi_{\rm pond}, \tag{2.133}$$

where

$$\Phi_{\rm pond} = \frac{1}{4}\frac{e}{m\,\omega^2}\,|\mathbf{E}_0|^2. \tag{2.134}$$

It follows that the oscillation center experiences a force, known as the *ponderomotive force*, that is proportional to the gradient in the amplitude of the wave-field. The ponderomotive force is independent of the sign of the charge, so both electrons and ions can be confined in the same potential well.

The total energy of the oscillation center,

$$\mathcal{E}_{\rm osc} = \frac{m}{2}\,U^2 + e\,\Phi_{\rm pond}, \tag{2.135}$$

is conserved by its equation of motion, Equation (2.133). However, it follows from Equation (2.129) that the ponderomotive potential energy is equal to the average kinetic energy of the oscillatory motion: that is,

$$e\,\Phi_{\rm pond} = \frac{m}{2}\,\langle u^2\rangle. \tag{2.136}$$

Thus, the force on the center of oscillation originates in a transfer of energy from the oscillatory motion to the average motion.

Most of the important applications of the ponderomotive force occur in laser plasma physics. For instance, a laser beam can propagate in a plasma provided that its frequency exceeds the plasma frequency. If the beam is sufficiently intense then plasma particles are repulsed from the center of the beam by the ponderomotive force. The resulting variation in the plasma density gives rise to a cylindrical well in the index of refraction that acts as a wave-guide for the laser beam (Kruer 2003).

2.15 Exercises

2.1 Given that $\boldsymbol{\rho} = \rho\,(-\cos\gamma\,\mathbf{e}_1 + \sin\gamma\,\mathbf{e}_2)$, and $\mathbf{u} = \Omega\,\boldsymbol{\rho}\times\mathbf{b}$, where $\rho = u_\perp/\Omega$, and $\mathbf{e}_1$, $\mathbf{e}_2$, $\mathbf{b} \equiv \mathbf{B}/B$ are a right-handed set of mutually perpendicular unit basis vectors, demonstrate that:

(a)

$$\langle \rho\,\rho \rangle = \frac{u_\perp^2}{2\,\Omega^2}\,(\mathbf{I} - \mathbf{b}\,\mathbf{b}).$$

(b)

$$e\,\langle \mathbf{u} \times (\boldsymbol{\rho}\cdot\nabla)\,\mathbf{B}\rangle = -\mu\,\nabla B.$$

(c)

$$e\,\langle \mathbf{u} \cdot (\boldsymbol{\rho}\cdot\nabla)\,\mathbf{E}\rangle = \mu\,\frac{\partial B}{\partial t}.$$

(d)

$$e\,\langle \mathbf{u} \cdot (\boldsymbol{\rho}\cdot\nabla)\,\mathbf{A}\rangle = -\mu\,B.$$

Here, $\mu = m\,u_\perp^2/(2\,B)$, and $\langle\cdots\rangle \equiv \oint(\cdots)\,d\gamma/2\pi$.

2.2 A quasi-neutral slab of cold (i.e., $\lambda_D \to 0$) plasma whose bounding surfaces are normal to the x-axis consists of electrons of mass m_e, charge $-e$, and mean number density n_e, as well as ions of mass m_i, charge e, and mean number density n_e. The slab is fully magnetized by a uniform y-directed magnetic field of magnitude B. The slab is then subject to an externally generated, uniform, x-directed electric field that is gradually ramped up to a final magnitude E_0. Show that, as a consequence of ion polarization drift, the final magnitude of the electric field inside the plasma is

$$E_1 \simeq \frac{E_0}{\epsilon},$$

where

$$\epsilon = 1 + \frac{c^2}{V_A^2},$$

and $V_A = B/\sqrt{\mu_0\,n_e\,m_i}$ is the so-called Alfvén velocity.

2.3 A linear magnetic dipole consists of two infinite straight wires running parallel to the z-axis. The first wire lies at $x = 0$, $y = d/2$ and carries a steady current I. The second lies at $x = 0$, $y = -d/2$ and carries a steady current $-I$. Let $r = (x^2 + y^2)^{1/2}$ and $\theta = \tan^{-1}(y/x)$. Demonstrate that the magnetic field generated by the dipole in the region $r \gg d$ can be written

$$\mathbf{B} = \nabla\psi \times \mathbf{e}_z,$$

where

$$\psi = \frac{\mu_0\,I\,d}{2\pi}\,\frac{\sin\theta}{r}.$$

2.4 Consider a particle of charge e, mass m, and energy $\mathcal{E}$, trapped on a field-line of the linear magnetic dipole discussed in the previous exercise. Let $\vartheta = \pi/2 - \theta$. Suppose that the field-line crosses the "equatorial" plane $\vartheta = 0$ at $r = r_{\rm eq} \gg d/2$, and that the magnetic field-strength at this point is $B_{\rm eq}$. Suppose that the particle's mirror points lie at $\vartheta = \pm\vartheta_m$. Assume that the particle's gyroradius is much smaller than $r_{\rm eq}$, and that the electric field-strength is negligible.

(a) Demonstrate that the variation of the particle's perpendicular and parallel velocity components with the "latitude" ϑ is

$$v_\perp = \left(\frac{2\,\mathcal{E}}{m}\right)^{1/2} \frac{\cos\vartheta_m}{\cos\vartheta},$$

$$v_\parallel = \pm\left(\frac{2\,\mathcal{E}}{m}\right)^{1/2}\left(1 - \frac{\cos^2\vartheta_m}{\cos^2\vartheta}\right)^{1/2},$$

respectively.

(b) Demonstrate that the particle's bounce period is

$$\tau_b = \frac{\sqrt{2}\,\pi\, r_{\rm eq}}{(\mathcal{E}/m)^{1/2}}.$$

(c) Demonstrate that the particle drifts in the z-direction with the mean velocity

$$\langle v_d\rangle = \frac{2\,\mathcal{E}}{e\,B_{\rm eq}\,r_{\rm eq}}.$$

2.5 A charged particle of mass m is trapped in a static magnetic mirror field given by

$$B_z = B_0\left(1 + \frac{z^2}{L^2}\right),$$

and has total kinetic energy $\mathcal{E}$, and pitch angle α at $z = 0$. Assuming that the electric field is negligible, and that the particle's gyroradius is much less than L, use guiding center theory to show that the bounce time is

$$\tau_b = \frac{2\pi\, L}{\sin\alpha\,\sqrt{2\,\mathcal{E}/m}}.$$

2.6 A particle of charge e, mass m, and energy $\mathcal{E}$, is trapped in a one-dimensional magnetic well of the form

$$B(x, t) = B_0\,(1 + k^2\,x^2),$$

where B_0 is constant, and $k(t)$ is a very slowly increasing function of time. Suppose that the particle's mirror points lie at $x = \pm x_m(t)$, and that its bounce time is $\tau_b(t)$. Demonstrate that, as a consequence of the conservation of the first and second adiabatic invariants,

$$x_m(t) = x_m(0)\left[\frac{k(0)}{k(t)}\right]^{1/2},$$

$$\tau_b(t) = \tau_b(0)\left[\frac{k(0)}{k(t)}\right],$$

$$\mathcal{E}(t) = \mathcal{E}_{0\,\perp} + \left[\frac{k(t)}{k(0)}\right]\mathcal{E}_{0\,\parallel}.$$

Here, $\mathcal{E}_{0\perp}$ is the perpendicular energy [i.e., $(1/2)\,m\,v_\perp^2$], and $\mathcal{E}_{0\parallel}$ is the parallel energy [i.e., $(1/2)\,m\,v_\parallel^2$], both evaluated at $x = 0$ and $t = 0$. Assume that the particle's gyroradius is relatively small, and that the electric field-strength is negligible.

2.7 Consider the static magnetic field

$$B_z(y) = \begin{cases} B_0 & y > a \\ B_0\,(y/a) & |y| < a \\ -B_0 & y < -a \end{cases}$$

which corresponds to a current sheet such as that found in the Earth's magnetotail. Let the electric field be negligible. Consider the orbits of charged particles of mass m and charge e whose gyroradii, ρ, are not necessarily much smaller than the shear-length, a, of the magnetic field. In this situation, guiding center theory is inapplicable. The particles' orbits can only be analyzed by directly solving their equations of perpendicular motion. It is easily demonstrated that some orbits do not cross the neutral plane ($y = 0$) and resemble conventional magnetized particle orbits, whereas others meander across the neutral plane and are quite different from conventional orbits.

(a) Consider a particle orbit that does not cross the neutral plane, but is instead confined to the region $y_+ \geq y \geq y_-$, where $a > y_+ > y_- > 0$. Demonstrate that the mean drift velocity of the particle in the x-direction can be written

$$\langle v_x \rangle = -\left(\frac{\Omega_0}{4\,a}\right)(y_+^2 + y_-^2)\,(1 - \alpha),$$

where $\Omega_0 = e\,B_0/m$, and

$$\alpha = \frac{\int_{-1}^{1}(1 + \kappa\,\zeta)^{1/2}\,(1 - \zeta^2)^{-1/2}\,d\zeta}{\int_{-1}^{1}(1 + \kappa\,\zeta)^{-1/2}\,(1 - \zeta^2)^{-1/2}\,d\zeta},$$

with $\kappa = (y_+^2 - y_-^2)/(y_+^2 + y_-^2)$. Show that in the limit $|y_+ - y_-|/a \ll 1$ the previous result is consistent with that obtained from conventional guiding center theory.

(b) Consider a particle orbit that is confined to the region $y_0 \geq y \geq -y_0$, where $a > y_0$, and is such that $v_x = 0$ when $y = 0$. Demonstrate that the mean drift velocity in the x-direction is

$$\langle v_x \rangle = +0.223\left(\frac{\Omega_0}{a}\right)y_0^2.$$

3

Collisions

3.1 Introduction

As was discussed in Chapter 1, collisions do not play as central a role in plasmas as they do in conventional neutral gases. Indeed, relatively hot, diffuse plasmas are essentially collisionless. Probably the most significant effect of collisions is that they act to relax particle distribution functions toward Maxwellian distributions. (See Section 3.6.) The aim of this chapter is to develop a theory of collisions that is applicable to a weakly coupled plasma. The fact that the plasma in question is weakly coupled means that it is a good approximation to treat the collisions as occasional binary events. (See Section 1.6.) As we shall see, the long-range nature of the Coulomb force renders the theory of collisions in a plasma somewhat more complicated than the corresponding theory for a neutral gas (where the inter-particle forces are invariably short-range in nature).

3.2 Collision Operator

Plasma physics can be regarded formally as a closure of Maxwell's equations by means of *constitutive relations*: that is, expressions specifying the charge density, ρ, and the current density, $\mathbf{j}$, in terms of the electric and magnetic fields, $\mathbf{E}$ and $\mathbf{B}$ (Hazeltine and Waelbroeck 2004). Such relations can be expressed in terms of the microscopic distribution functions, $\mathcal{F}_s$, for each plasma species:

$$\rho = \sum_s e_s \int \mathcal{F}_s(\mathbf{r}, \mathbf{v}, t)\, d^3\mathbf{v}, \tag{3.1}$$

$$\mathbf{j} = \sum_s e_s \int \mathbf{v}\, \mathcal{F}_s(\mathbf{r}, \mathbf{v}, t)\, d^3\mathbf{v}. \tag{3.2}$$

Here, $\mathcal{F}_s(\mathbf{r}, \mathbf{v}, t)$ is the exact microscopic phase-space density of plasma species s (with charge e_s and mass m_s) near point $(\mathbf{r},\ \mathbf{v})$ at time t (Reif 1965). The distribution function $\mathcal{F}_s$ is normalized such that its velocity integral is equal to the particle number density in coordinate space. In other words,

$$\int \mathcal{F}_s(\mathbf{r}, \mathbf{v}, t)\, d^3\mathbf{v} = n_s(\mathbf{r}, t), \tag{3.3}$$

where $n_s(\mathbf{r}, t)$ is the number (per unit volume) of species-s particles near point $\mathbf{r}$ at time t.

If we could determine each $\mathcal{F}_s(\mathbf{r}, \mathbf{v}, t)$ in terms of the electromagnetic fields then Equations (3.1) and (3.2) would give us the desired constitutive relations. In fact, the time evolution of the various distribution functions is determined by particle conservation in phase-space, which requires that (Reif 1965)

$$\frac{\partial \mathcal{F}_s}{\partial t} + \mathbf{v} \cdot \frac{\partial \mathcal{F}_s}{\partial \mathbf{r}} + \mathbf{a}_s \cdot \frac{\partial \mathcal{F}_s}{\partial \mathbf{v}} = 0, \tag{3.4}$$

where

$$\mathbf{a}_s = \frac{e_s}{m_s}\,(\mathbf{E} + \mathbf{v} \times \mathbf{B}) \tag{3.5}$$

is the species-s particle acceleration under the influence of the $\mathbf{E}$ and $\mathbf{B}$ fields.

Equation (3.4) is easy to derive because it is exact, taking into account all lengthscales from the microscopic to the macroscopic. Note, in particular, that there is no statistical averaging involved in Equation (3.4). It follows that the microscopic distribution function, $\mathcal{F}_s$, is essentially a sum of Dirac delta-functions, each following the detailed trajectory of a single particle. Consequently, the electromagnetic fields appearing in Equation (3.4) are extremely spiky on microscopic scales. In fact, solving Equation (3.4) is equivalent to solving the classical electromagnetic many-body problem, which is a completely hopeless task.

A much more useful equation can be extracted from Equation (3.4) by ensemble averaging (Reif 1965). The average distribution function,

$$\langle \mathcal{F}_s \rangle \equiv f_s, \tag{3.6}$$

is smooth on microscopic lengthscales, and is closely related to actual experimental measurements. Here, angle brackets denote an ensemble average. Similarly, the ensemble-averaged electromagnetic fields are also smooth. Unfortunately, the extraction of an ensemble-averaged equation from Equation (3.4) is mathematically challenging, and invariably involves some level of approximation. The problem is that, because the exact electromagnetic fields depend on particle trajectories, $\mathbf{E}$ and $\mathbf{B}$ are not statistically independent of $\mathcal{F}_s$. In other words, as a consequence of correlations between the distribution function and the electromagnetic fields on microscopic lengthscales, the ensemble average of the nonlinear acceleration term in Equation (3.4) is such that

$$\left\langle \mathbf{a}_s \cdot \frac{\partial \mathcal{F}_s}{\partial \mathbf{v}} \right\rangle \neq \langle \mathbf{a}_s \rangle \cdot \frac{\partial f_s}{\partial \mathbf{v}}. \tag{3.7}$$

It is convenient to write

$$\left\langle \mathbf{a}_s \cdot \frac{\partial \mathcal{F}_s}{\partial \mathbf{v}} \right\rangle = \langle \mathbf{a}_s \rangle \cdot \frac{\partial f_s}{\partial \mathbf{v}} - C_s(f), \tag{3.8}$$

where C_s is an operator that accounts for the correlations. Because the most important correlations result from close encounters between particles, C_s is known as the *collision operator* (for species s). It is not generally a linear operator, and usually involves the distribution functions of both colliding species (the subscript in the argument of C_s is omitted for this reason). Hence, the ensemble-averaged version of Equation (3.4) is written

$$\frac{\partial f_s}{\partial t} + \mathbf{v}\cdot\frac{\partial f_s}{\partial \mathbf{r}} + \frac{e_s}{m_s}\,(\mathbf{E} + \mathbf{v}\times\mathbf{B})\cdot\frac{\partial f_s}{\partial \mathbf{v}} = C_s(f), \tag{3.9}$$

where $\mathbf{E}$ and $\mathbf{B}$ are now understood to be the smooth, ensemble-averaged electromagnetic fields. Of course, in a weakly coupled plasma, the dominant collisions are two-particle Coulomb collisions. Equation (3.9) is generally known as the *kinetic equation*.

3.3 Two-Body Elastic Collisions

Before specializing to two-body Coulomb collisions, it is convenient to develop a general theory of two-body elastic collisions. Consider an elastic collision between a particle of type 1 and a particle of type 2. Let the mass and instantaneous velocity of the former particle be m_1 and $\mathbf{v}_1$, respectively. Likewise, let the mass and instantaneous velocity of the latter particle be m_2 and $\mathbf{v}_2$, respectively. The velocity of the center of mass is given by

$$\mathbf{U} = \frac{m_1\,\mathbf{v}_1 + m_2\,\mathbf{v}_2}{m_1 + m_2}. \tag{3.10}$$

Moreover, conservation of momentum implies that $\mathbf{U}$ is a constant of the motion. The relative velocity is defined

$$\mathbf{u} = \mathbf{v}_1 - \mathbf{v}_2. \tag{3.11}$$

We can express $\mathbf{v}_1$ and $\mathbf{v}_2$ in terms of $\mathbf{U}$ and $\mathbf{u}$ as follows:

$$\mathbf{v}_1 = \mathbf{U} + \frac{\mu_{12}}{m_1}\,\mathbf{u}, \tag{3.12}$$

$$\mathbf{v}_2 = \mathbf{U} - \frac{\mu_{12}}{m_2}\,\mathbf{u}. \tag{3.13}$$

Here,

$$\mu_{12} = \frac{m_1\,m_2}{m_1 + m_2} \tag{3.14}$$

is the *reduced mass*. The total kinetic energy of the system is written

$$K = \frac{1}{2}\,m_1\,v_1^2 + \frac{1}{2}\,m_2\,v_2^2 = \frac{1}{2}\,(m_1 + m_2)\,U^2 + \frac{1}{2}\,\mu_{12}\,u^2. \tag{3.15}$$

Now, the kinetic energy is the same before and after an elastic collision. Hence, given that U is constant, we deduce that the magnitude of the relative velocity, u, is also the same before and after such a collision. Thus, it is only the direction of the relative velocity vector, rather than its length, that changes during an elastic collision.

3.4 Boltzmann Collision Operator

Let $\sigma(\mathbf{v}_1, \mathbf{v}_2; \mathbf{v}_1', \mathbf{v}_2')$ be the cross-section for a scattering process by which particles of types 1 and 2 (located at position vector $\mathbf{r}$ at time t) are incident with velocities $\mathbf{v}_1$ and $\mathbf{v}_2$, respectively, and are scattered to velocities $\mathbf{v}_1'$ and $\mathbf{v}_2'$, respectively (Reif 1965). Assuming that the scattering process is reversible in time and space (which is certainly the case for two-body Coulomb collisions), the cross-section for the inverse process must be the same as that for the forward process (Reif 1965). In other words,

$$\sigma(\mathbf{v}_1', \mathbf{v}_2'; \mathbf{v}_1, \mathbf{v}_2) = \sigma(\mathbf{v}_1, \mathbf{v}_2; \mathbf{v}_1', \mathbf{v}_2'). \tag{3.16}$$

The rate at which particles with the original velocities $\mathbf{v}_1$ and $\mathbf{v}_2$ are scattered into the range $\mathbf{v}_1'$ to $\mathbf{v}_1' + d\mathbf{v}_1'$ and $\mathbf{v_2}'$ to $\mathbf{v_2}' + d\mathbf{v}_2'$ is

$$u\, f_1(\mathbf{r}, \mathbf{v}_1, t)\, f_2(\mathbf{r}, \mathbf{v}_2, t)\, \sigma(\mathbf{v}_1, \mathbf{v}_2; \mathbf{v}_1', \mathbf{v}_2')\, d^3\mathbf{v}_1'\, d^3\mathbf{v}_2'. \tag{3.17}$$

Here, $u = |\mathbf{v}_1 - \mathbf{v}_2|$. Moreover, $f_1(\mathbf{r}, \mathbf{v}_1, t)$ and $f_2(\mathbf{r}, \mathbf{v}_2, t)$ are the ensemble-averaged distribution functions for particles of types 1 and 2, respectively. In writing the previous expression, we have assumed that the distribution functions f_1 and f_2 are uncorrelated. This assumption is reasonable provided that the mean-free-path is much longer than the effective range of the inter-particle force. (This follows because, before they encounter one another, two colliding particles originate at different points that are typically separated by a mean-free-path. However, the typical correlation length is of similar magnitude to the range of the inter-particle force.) In writing the previous expression, we have also implicitly assumed that the inter-particle force responsible for the collisions is sufficiently short-range that the particle position vectors do not change appreciably (on a macroscopic lengthscale) during a collision. (Both of the previous assumptions are valid in a conventional weakly coupled plasma, because the range of the inter-particle force is of order the Debye length, which is assumed to be much smaller than any macroscopic lengthscale. Moreover, the mean-free-path is much longer than the Debye length—see Section 1.7.) By analogy with Equation (3.17), the rate at which particles with the original velocities $\mathbf{v}_1'$ and $\mathbf{v}_2'$ are scattered into the range $\mathbf{v}_1$ to $\mathbf{v}_1 + d\mathbf{v}_1$ and $\mathbf{v_2}$ to $\mathbf{v_2} + d\mathbf{v}_2$ is

$$u'\, f_1(\mathbf{r}, \mathbf{v}_1', t)\, f_2(\mathbf{r}, \mathbf{v}_2', t)\, \sigma(\mathbf{v}_1', \mathbf{v}_2'; \mathbf{v}_1, \mathbf{v}_2)\, d^3\mathbf{v}_1\, d^3\mathbf{v}_2, \tag{3.18}$$

where $\mathbf{u}' = \mathbf{v}_1' - \mathbf{v}_2'$. Now, it is easily demonstrated from Equations (3.12) and (3.13) that

$$d^3\mathbf{v}_1\, d^3\mathbf{v}_2 = d^3\mathbf{U}\, d^3\mathbf{u} = d^3\mathbf{U}\, d^3\mathbf{u}' = d^3\mathbf{v}_1'\, d^3\mathbf{v}_2'. \tag{3.19}$$

The result $d^3\mathbf{u} = d^3\mathbf{u}'$ follows from the fact that the vectors $\mathbf{u}$ and $\mathbf{u}'$ differ only in direction. Thus, the net rate of change of the distribution function of particles of type 1 with velocities $\mathbf{v}_1$ (at position $\mathbf{r}$ and time t) due to collisions with particles of type 2 [i.e., the collision operator—see Equation (3.9)] is given by

$$\left(\frac{\partial f_1}{\partial t}\right)_2 \equiv C_{12}(f_1, f_2) = \iiint u\, \sigma(\mathbf{v}_1, \mathbf{v}_2; \mathbf{v}_1', \mathbf{v}_2')\, (f_1'\, f_2' - f_1\, f_2)\, d^3\mathbf{v}_2\, d^3\mathbf{v}_1'\, d^3\mathbf{v}_2'. \tag{3.20}$$

Here, use has been made of Equation (3.16), as well as $u' = u$. Moreover, f_1, f_2. f_1', and f_2' are short-hand for $f_1(\mathbf{r}, \mathbf{v}_1, t)$, $f_2(\mathbf{r}, \mathbf{v}_2, t)$, $f_1(\mathbf{r}, \mathbf{v}_1', t)$, and $f_2(\mathbf{r}, \mathbf{v}_2', t)$, respectively. The previous expression is known as the *Boltzmann collision operator*. By an analogous argument, the net rate of change of the distribution function of particles of type 2 with velocities $\mathbf{v}_2$ (at position $\mathbf{r}$ and time t) due to collisions with particles of type 1 is given by

$$\left(\frac{\partial f_2}{\partial t}\right)_1 \equiv C_{21}(f_1, f_2) = \int\!\!\int\!\!\int u\,\sigma(\mathbf{v}_1, \mathbf{v}_2; \mathbf{v}_1', \mathbf{v}_2')\,(f_1'\,f_2' - f_1\,f_2)\,d^3\mathbf{v}_1\,d^3\mathbf{v}_1'\,d^3\mathbf{v}_2'. \tag{3.21}$$

Expression (3.20) for the Boltzmann collision operator can be further simplified for elastic collisions because, in this case, the collision cross-section $\sigma(\mathbf{v}_1, \mathbf{v}_2; \mathbf{v}_1', \mathbf{v}_2')$ is a function only of the magnitude of the relative velocity vector, $\mathbf{u}$, and its change in direction as a result of the collision. Furthermore, the integral over the final velocities $\mathbf{v}_1'$ and $\mathbf{v}_2'$ reduces to an integral over all solid angles for the change in direction of $\mathbf{u}$. Thus, we can write

$$\sigma(\mathbf{v}_1, \mathbf{v}_2; \mathbf{v}_1', \mathbf{v}_2')\,d^3\mathbf{v}_1'\,d^3\mathbf{v}_2' = \frac{d\sigma(u, \chi, \phi)}{d\Omega}\,d\Omega, \tag{3.22}$$

where $\Omega = \sin\chi\,d\chi\,d\phi$. Here, χ is the angle through which the direction of $\mathbf{u}$ is deflected as a consequence of the collision (see Figure 3.1), and ϕ is an azimuthal angle that determines the orientation of the plane in which the vector $\mathbf{u}$ is confined during the collision. (See Section 3.7.) Moreover, $d\sigma/d\Omega$ is a conventional differential scattering cross-section (Reif 1965). Hence, we obtain

$$C_{12}(f_1, f_2) = \int\!\!\int\!\!\int u\,\frac{d\sigma(u, \chi, \phi)}{d\Omega}\,(f_1'\,f_2' - f_1\,f_2)\,d^3\mathbf{v}_2\,d\Omega. \tag{3.23}$$

Note, finally, that if we exchange the identities of particles 1 and 2 in Equation (3.22) then $\mathbf{u} \to -\mathbf{u}$, but $u \to u$, $\chi \to \chi$, and $\phi \to \phi$. Thus, we conclude that

$$\sigma(\mathbf{v}_2, \mathbf{v}_1; \mathbf{v}_2', \mathbf{v}_1') = \sigma(\mathbf{v}_1, \mathbf{v}_2; \mathbf{v}_1', \mathbf{v}_2'). \tag{3.24}$$

3.5 Collisional Conservation Laws

Consider

$$\int C_{12}\,d^3\mathbf{v}_1 = \int\!\!\int\!\!\int\!\!\int u\,\sigma(\mathbf{v}_1, \mathbf{v}_2; \mathbf{v}_1', \mathbf{v}_2')\,(f_1'\,f_2' - f_1\,f_2)\,d^3\mathbf{v}_1\,d^3\mathbf{v}_2\,d^3\mathbf{v}_1'\,d^3\mathbf{v}_2', \tag{3.25}$$

which follows from Equation (3.20). Interchanging primed and unprimed dummy variables of integration on the right-hand side, we obtain

$$\int C_{12}\,d^3\mathbf{v}_1 = \int\!\!\int\!\!\int\!\!\int u'\,\sigma(\mathbf{v}_1', \mathbf{v}_2'; \mathbf{v}_1, \mathbf{v}_2)\,(f_1\,f_2 - f_1'\,f_2')\,d^3\mathbf{v}_1'\,d^3\mathbf{v}_2'\,d^3\mathbf{v}_1\,d^3\mathbf{v}_2. \tag{3.26}$$

Hence, making use of Equation (3.16), as well as the fact that $u' = u$, we deduce that

$$\int C_{12}\, d^3\mathbf{v}_1 = -\int\!\!\int\!\!\int\!\!\int u\,\sigma(\mathbf{v}_1,\mathbf{v}_2;\mathbf{v}_1',\mathbf{v}_2')\,(f_1' f_2' - f_1 f_2)\, d^3\mathbf{v}_1\, d^3\mathbf{v}_2\, d^3\mathbf{v}_1'\, d^3\mathbf{v}_2'$$
$$= -\int C_{12}\, d^3\mathbf{v}_1, \tag{3.27}$$

which implies that

$$\int C_{12}\, d^3\mathbf{v}_1 = 0. \tag{3.28}$$

The previous expression states that collisions with particles of type 2 give rise to zero net rate of change of the number density of particles of type 1 at position $\mathbf{r}$ and time t. In other words, the collisions conserve the number of particles of type 1. Now, it is easily seen from Equations (3.20) and (3.21) that

$$C_{12}\, d^3\mathbf{v}_1 = C_{21}\, d^3\mathbf{v}_2. \tag{3.29}$$

Hence, Equation (3.28) also implies that

$$\int C_{21}\, d^3\mathbf{v}_2 = 0. \tag{3.30}$$

In other words, collisions also conserve the number of particles of type 2.

Consider

$$(m_1 + m_2)\int \mathbf{U}\, C_{12}\, d^3\mathbf{v}_1 = \mathbf{0}. \tag{3.31}$$

This integral is obviously zero, as indicated, as a consequence of the conservation law (3.28), as well as the fact that the center of mass velocity, $\mathbf{U}$, is a constant of the motion. However, making use of Equations (3.10) and (3.29), the previous expression can be rewritten in the form

$$\int m_1\, \mathbf{v}_1\, C_{12}\, d^3\mathbf{v}_1 = -\int m_2\, \mathbf{v}_2\, C_{21}\, d^3\mathbf{v}_2. \tag{3.32}$$

This equation states that the rate at which particles of type 1 gain momentum due to collisions with particles of type 2 is equal to the rate at which particles of type 2 lose momentum due to collisions with particles of type 1. In other words, the collisions conserve momentum.

Finally, consider

$$\int K\, C_{12}\, d^3\mathbf{v}_1 = 0. \tag{3.33}$$

This integral is obviously zero, as indicated, as a consequence of the conservation law (3.28), as well as the fact that the kinetic energy, K, is the same before and after an elastic collision. It follows from Equations (3.15) and (3.29) that

$$\int \frac{1}{2}\, m_1\, v_1^2\, C_{12}\, d^3\mathbf{v}_1 = -\int \frac{1}{2}\, m_2\, v_2^2\, C_{21}\, d^3\mathbf{v}_2. \tag{3.34}$$

This equation states that the rate at which particles of type 1 gain kinetic energy due to collisions with particles of type 2 is equal to the rate at which particles of type 2 lose kinetic energy due to collisions with particles of type 1. In other words, the collisions conserve energy.

3.6 Boltzmann H-Theorem

Equation (3.20) can be written

$$\left(\frac{\partial f_1}{\partial t}\right)_2 = \iiint u\,\sigma(\mathbf{v}_1, \mathbf{v}_2; \mathbf{v}_1', \mathbf{v}_2')\,(f_1' f_2' - f_1 f_2)\, d^3\mathbf{v}_2\, d^3\mathbf{v}_1'\, d^3\mathbf{v}_2'. \tag{3.35}$$

Consider the quantity

$$H = \int f_1 \ln f_1\, d^3\mathbf{v}_1. \tag{3.36}$$

It follows from Equation (3.35) that

$$\begin{aligned}\frac{dH}{dt} &= \int (1 + \ln f_1)\,\frac{\partial f_1}{\partial t}\, d^3\mathbf{v}_1 \\ &= \iiiint u\,\sigma\,(1 + \ln f_1)\,(f_1' f_2' - f_1 f_2)\, d^3\mathbf{v}_1\, d^3\mathbf{v}_2\, d^3\mathbf{v}_1'\, d^3\mathbf{v}_2',\end{aligned} \tag{3.37}$$

where σ is short-hand for $\sigma(\mathbf{v}_1, \mathbf{v}_2; \mathbf{v}_1', \mathbf{v}_2')$. Suppose that we swap the dummy labels 1 and 2. This process leaves both $u = |\mathbf{v}_1 - \mathbf{v}_2|$ and the value of the integral unchanged [assuming that there is an implicit summation over different species in Equation (3.36)]. According to Equation (3.24), it also leaves the scattering cross-section $\sigma(\mathbf{v}_1, \mathbf{v}_2; \mathbf{v}_1', \mathbf{v}_2')$ unchanged. Hence, we deduce that

$$\frac{dH}{dt} = \iiiint u\,\sigma\,(1 + \ln f_2)\,(f_1' f_2' - f_1 f_2)\, d^3\mathbf{v}_1\, d^3\mathbf{v}_2\, d^3\mathbf{v}_1'\, d^3\mathbf{v}_2'. \tag{3.38}$$

Suppose that we swap primed and unprimed dummy variables of integration in Equation (3.37). This leaves the value of the integral unchanged. Making use of Equation (3.16), as well as the fact that $u' = u$, we obtain

$$\frac{dH}{dt} = -\iiiint u\,\sigma\,(1 + \ln f_1')\,(f_1' f_2' - f_1 f_2)\, d^3\mathbf{v}_1\, d^3\mathbf{v}_2\, d^3\mathbf{v}_1'\, d^3\mathbf{v}_2'. \tag{3.39}$$

Finally, swapping primed and unprimed variables in Equation (3.38) yields

$$\frac{dH}{dt} = -\iiiint u\,\sigma\,(1 + \ln f_2')\,(f_1' f_2' - f_1 f_2)\, d^3\mathbf{v}_1\, d^3\mathbf{v}_2\, d^3\mathbf{v}_1'\, d^3\mathbf{v}_2'. \tag{3.40}$$

The previous four equations can be combined to give

$$\frac{dH}{dt} = \frac{1}{4}\iiiint u\,\sigma\,\ln\!\left(\frac{f_1 f_2}{f_1' f_2'}\right)(f_1' f_2' - f_1 f_2)\, d^3\mathbf{v}_1\, d^3\mathbf{v}_2\, d^3\mathbf{v}_1'\, d^3\mathbf{v}_2'. \tag{3.41}$$

Now, $\ln(f_1 f_2/f_1' f_2')$ is positive when $f_1' f_2' - f_1 f_2$ is negative, and vice versa. We, therefore, deduce that the integral on the right-hand side of the previous expression can never take a positive value. In other words,

$$\frac{dH}{dt} \leq 0. \tag{3.42}$$

This result is known as the *Boltzmann H-theorem*.

In fact, the quantity H is bounded below (i.e., it cannot take the value minus infinity). Hence, H cannot decrease indefinitely, but must tend to a limit in which $dH/dt = 0$. According to Equation (3.41), the distribution function associated with this limiting state is characterized by

$$f_1\, f_2 = f_1'\, f_2', \tag{3.43}$$

or, equivalently,

$$\ln f_1 + \ln f_2 - \ln f_1' - \ln f_2' = 0. \tag{3.44}$$

Consider a distribution function that satisfies

$$\ln f_i = a_i + m_i\, \mathbf{b}\cdot\mathbf{v}_i + m_i\, c\, v_i^2, \tag{3.45}$$

where i is a species label, m_i is the particle mass, and a_i, $\mathbf{b}$, and c are constants. It follows that

$$\ln f_1 + \ln f_2 - \ln f_1' - \ln f_2' = \mathbf{b}\cdot(m_1\,\mathbf{v}_1 + m_2\,\mathbf{v}_2 - m_1\,\mathbf{v}_1' - m_2\,\mathbf{v}_2') \\ + c\,(m_1\, v_1^2 + m_2\, v_2^2 - m_1\, v_1'^{\,2} - m_2\, v_2'^{\,2}). \tag{3.46}$$

However, for an elastic collision, momentum conservation implies that (see Section 3.3)

$$m_1\,\mathbf{v}_1 + m_2\,\mathbf{v}_2 = m_1\,\mathbf{v}_1' + m_2\,\mathbf{v}_2', \tag{3.47}$$

whereas energy conservation yields (see Section 3.3)

$$m_1\, v_1^2 + m_2\, v_2^2 = m_1\, v_1'^{\,2} + m_2\, v_2'^{\,2}. \tag{3.48}$$

In other words, a distribution function that satisfies Equation (3.45) automatically satisfies Equation (3.44). We, thus, conclude that collisions act to drive the distribution functions for the colliding particles towards particular distribution functions that satisfy Equation (3.45). [Incidentally, elastic collisions generally only conserve particle number, particle momentum, and particle energy. These conservation laws correspond to the three terms appearing on the right-hand side of Equation (3.45). Hence, in the absence of other conservation laws, we can be sure that Equation (3.45) is the most general expression that satisfies Equation (3.44).]

Without loss of generality, we can set

$$a_i = \ln\left[n_i\left(\frac{m_i}{2\pi\, T}\right)^{3/2}\right] - \frac{m_i\, V^2}{2\, T}, \tag{3.49}$$

$$\mathbf{b} = \frac{1}{T}\,\mathbf{V}, \tag{3.50}$$

$$c = -\frac{1}{2\, T}, \tag{3.51}$$

where n_i, $\mathbf{V}$, and T are constants. In this case, Equation (3.45) becomes

$$f_i = n_i\left(\frac{m_i}{2\pi\, T}\right)^{3/2} \exp\left[-\frac{m_i\,(\mathbf{v}_i - \mathbf{V})^2}{2\, T}\right], \tag{3.52}$$

which we recognize as a *Maxwellian distribution function* (Reif 1965). It is easily demonstrated that

$$n_i = \int f_i \, d^3\mathbf{v}_i, \tag{3.53}$$

$$n_i \, \mathbf{V} = \int \mathbf{v}_i \, f_i \, d^3\mathbf{v}_i, \tag{3.54}$$

$$\frac{3}{2} \, n_i \, T = \int \frac{1}{2} \, m_i \, v^2 \, f_i \, d^3\mathbf{v}_i. \tag{3.55}$$

These relations allow us to identify the constants n_i, $\mathbf{V}$, and T with the species-i number density, mean flow velocity, and kinetic temperature, respectively. We conclude that collisions tend to relax the distribution functions for the colliding particles toward Maxwellian distributions characterized by a common mean flow velocity and a common temperature.

3.7 Two-Body Coulomb Collisions

Consider a two-body Coulomb collision between a particle of species 1, with mass m_1 and charge e_1, and a particle of species 2, with mass m_2 and charge e_2. The equations of motion of the two particles take the form

$$m_1 \, \ddot{\mathbf{r}}_1 = k \, \frac{\mathbf{r}}{|\mathbf{r}|^3}, \tag{3.56}$$

$$m_2 \, \ddot{\mathbf{r}}_2 = -k \, \frac{\mathbf{r}}{|\mathbf{r}|^3}, \tag{3.57}$$

where

$$k = \frac{e_1 \, e_2}{4\pi \, \epsilon_0}, \tag{3.58}$$

Here, $\mathbf{r}_1$ and $\mathbf{r}_2$ are the respective position vectors, and $\mathbf{r} = \mathbf{r}_1 - \mathbf{r}_2$ is the relative position vector. It is easily demonstrated that

$$\mathbf{r}_1 = \mathbf{R} + \frac{\mu_{12}}{m_1} \, \mathbf{r}, \tag{3.59}$$

$$\mathbf{r}_2 = \mathbf{R} - \frac{\mu_{12}}{m_2} \, \mathbf{r}, \tag{3.60}$$

where

$$\mathbf{R} = \frac{m_1 \, \mathbf{r}_1 + m_2 \, \mathbf{r}_2}{m_1 + m_2} \tag{3.61}$$

is the vector position of the center of mass (which does not accelerate), and $\mu_{12} = m_1 \, m_2/(m_1 + m_2)$ the reduced mass. Equations (3.56) and (3.57) can be combined to give a single equation of relative motion,

$$\mu_{12} \, \ddot{\mathbf{r}} = k \, \frac{\mathbf{r}}{|\mathbf{r}|^3}. \tag{3.62}$$

Two relations that immediately follow from the previous equation are

$$\frac{d\mathbf{h}}{dt} = \mathbf{0}, \tag{3.63}$$

$$\frac{dE}{dt} = 0, \tag{3.64}$$

where

$$\mathbf{h} = \mathbf{r} \times \dot{\mathbf{r}} \tag{3.65}$$

is the conserved angular momentum per unit mass, and

$$E = \frac{1}{2}\mu_{12}\,|\dot{\mathbf{r}}|^2 + \frac{k}{|\mathbf{r}|} \tag{3.66}$$

the conserved energy.

Equation (3.65) implies that $\mathbf{r}\cdot\mathbf{h} = 0$. This is the equation of a plane that passes through the origin, and whose normal is parallel to the constant vector **h**. We, therefore, conclude that the relative position vector **r** is constrained to lie in this plane, which implies that the trajectories of both colliding particles are coplanar. Let the plane $\mathbf{r}\cdot\mathbf{h} = 0$ coincide with the x-y plane, so that we can write $\mathbf{r} = (x, y)$. It is convenient to define the standard plane polar coordinates $r = (x^2 + y^2)^{1/2}$ and $\theta = \tan^{-1}(y/x)$. When expressed in terms of these coordinates, the conserved angular momentum per unit mass becomes

$$\mathbf{h} = h\,\mathbf{e}_z, \tag{3.67}$$

where

$$h = r^2\,\dot{\theta}. \tag{3.68}$$

Furthermore, the conserved energy takes the form

$$E = \frac{1}{2}\mu_{12}\left(\dot{r}^2 + r^2\,\dot{\theta}^2\right) + \frac{k}{r}. \tag{3.69}$$

Suppose that $r = z^{-1}$, where $z = z(\theta)$ and $\theta = \theta(r)$. It follows that

$$\dot{r} = -\frac{\dot{z}}{z^2} = -r^2\,\frac{dz}{d\theta}\,\frac{d\theta}{dt} = -h\,\frac{dz}{d\theta}. \tag{3.70}$$

Hence, Equation (3.69) transforms to give

$$E = \frac{1}{2}\mu_{12}\,h^2\left[\left(\frac{dz}{d\theta}\right)^2 + z^2\right] + k\,z. \tag{3.71}$$

It is convenient to define the relative velocity at large distances,

$$u = \left(\frac{2\,E}{\mu_{12}}\right)^{1/2}, \tag{3.72}$$

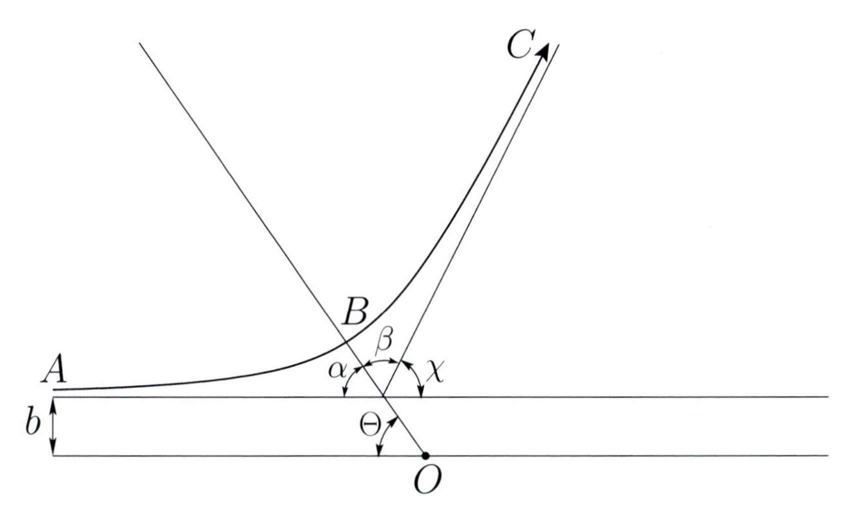

Figure 3.1
A two-body Coulomb collision.

as well as the *impact parameter*,

$$b = \frac{h}{u}. \tag{3.73}$$

The latter parameter is simply the distance of closest approach of the two particles in the situation in which there is no Coulomb force acting between them, and they, consequently, move in straight-lines. (See Figure 3.1.) The previous three equations can be combined to give

$$b^2\left(\frac{dz}{d\theta}\right)^2 = 1 - b^2 z^2 - \left(\frac{k}{E}\right)z. \tag{3.74}$$

Figure 3.1 shows the collision in a frame of reference in which particle 2 remains stationary at the origin, O, whereas particle 1 traces out the path ABC. Point B corresponds to the closest approach of the two particles. It follows, by symmetry (because Coulomb collisions are reversible), that the angles α and β shown in the figure are equal to one another. Hence, we deduce that

$$\chi = \pi - 2\,\Theta. \tag{3.75}$$

Here, χ is the angle through which the path of particle 1 (or particle 2) is deviated as a consequence of the collision, whereas Θ is the angle through which the relative position vector, $\mathbf{r}$, rotates as particle 1 moves from point A (which is assumed to be infinity far from point O) to point B. Suppose that point A corresponds to $\theta = 0$. It follows that

$$\Theta = \int_0^{z_{\rm max}} \frac{d\theta}{dz}\,dz. \tag{3.76}$$

Here, $z_{\rm max} = 1/r_{\rm min}$, where $r_{\rm min}$ is the distance of closest approach. Now, by symmetry, $(dz/d\theta)_{z_{\rm max}} = 0$, so Equation (3.74) implies that

$$1 - b^2\,z_{\rm max}^2 - \left(\frac{k}{E}\right) z_{\rm max} = 0. \tag{3.77}$$

Combining Equations (3.74) and (3.76), we obtain

$$\Theta = \int_0^{z_{\rm max}} \frac{b\,dz}{\sqrt{1-b^2\,z^2-k\,z/E}} = \int_0^{\zeta_{\rm max}} \frac{d\zeta}{1-\zeta^2-\alpha\,\zeta}, \tag{3.78}$$

where $\alpha = k/(E\,b)$, and

$$1 - \zeta_{\rm max}^2 - \alpha\,\zeta_{\rm max} = 0. \tag{3.79}$$

Integration (Speigel, Liu, and Lipschutz 1999) yields

$$\Theta = \frac{\pi}{2} - \sin^{-1}\left(\frac{\alpha}{\sqrt{4+\alpha^2}}\right). \tag{3.80}$$

Hence, from Equation (3.75), we get

$$\chi = 2\,\sin^{-1}\left(\frac{\alpha}{\sqrt{4+\alpha^2}}\right), \tag{3.81}$$

which can be rearranged to give

$$\cot\left(\frac{\chi}{2}\right) = \frac{2\,E\,b}{k} = \frac{4\pi\,\epsilon_0\,\mu_{12}\,u^2\,b}{e_1\,e_2}. \tag{3.82}$$

3.8 Rutherford Scattering Cross-Section

Consider a particle of type 1, incident with relative velocity u onto an ensemble of particles of type 2 with number density n_2. If $p_1(\Omega)\,d\Omega$ is the probability per unit time of the particle being scattered into the range of solid angle Ω to $\Omega + d\Omega$, then the differential scattering cross-section, $d\sigma/d\Omega$, is defined via (Reif 1965)

$$p_1(\Omega)\,d\Omega = n_2\,u\,\frac{d\sigma}{d\Omega}\,d\Omega. \tag{3.83}$$

Assuming that the scattering is azimuthally symmetric (i.e., symmetric in ϕ), we can write $d\Omega = 2\pi\,\sin\chi\,d\chi$. Now, the probability per unit time of a collision having an impact parameter in the range b to $b + db$ is

$$p_1(b)\,db = n_2\,u\,2\pi\,b\,db. \tag{3.84}$$

Furthermore, we can write

$$p_1(\Omega)\left|\frac{d\Omega}{db}\right| = p_1(b), \tag{3.85}$$

provided that χ and b are related according to the two-particle scattering law, Equation (3.82). It follows that

$$\frac{d\sigma}{d\Omega} = \frac{2\pi\, b}{|d\Omega/db|}. \tag{3.86}$$

Equation (3.82) yields

$$\frac{d\Omega}{db} = 2\pi\,\sin\chi\,\frac{d\chi}{db} = -2\pi\,\sin\chi\left(\frac{4\pi\,\epsilon_0\,\mu_{12}\,u^2}{e_1\,e_2}\right) 2\,\sin^2(\chi/2). \tag{3.87}$$

Finally, Equations (3.82), (3.86), and (3.87) can be combined to give the so-called *Rutherford scattering cross-section*,

$$\frac{d\sigma}{d\Omega} = \frac{1}{4}\left(\frac{e_1\,e_2}{4\pi\,\epsilon_0\,\mu_{12}\,u^2}\right)^2 \frac{1}{\sin^4(\chi/2)}. \tag{3.88}$$

It is immediately apparent, from the previous formula, that two-particle Coulomb collisions are dominated by small-angle (i.e., small χ) scattering events.

3.9 Landau Collision Operator

The fact that two-particle Coulomb collisions are dominated by small-angle scattering events allows some simplification of the Boltzmann collision operator in a plasma. According to Equations (3.23) and (3.88), the Boltzmann collision operator for two-body Coulomb collisions between particles of type 1 (with mass m_1 and charge e_1) and particles of type 2 (with mass m_2 and charge e_2) can be written

$$C_{12} = \int\!\!\int\!\!\int u\,\frac{d\sigma}{d\Omega}\,(f_1'\,f_2' - f_1\,f_2)\,d^3\mathbf{v}_2\,d\Omega, \tag{3.89}$$

where

$$\frac{d\sigma}{d\Omega} = \frac{1}{4}\left(\frac{e_1\,e_2}{4\pi\,\epsilon_0\,\mu_{12}\,u^2}\right)^2 \frac{1}{\sin^4(\chi/2)}. \tag{3.90}$$

Here, $\mathbf{u}$ is the relative velocity prior to a collision, and $d\Omega = \sin\chi\,d\chi\,d\phi$, where χ is the angle of deflection, and ϕ is an azimuthal angle that determines the orientation of the plane in which a given two-body collision occurs. Recall that f_1', f_2', f_1, and f_2 are short-hand for $f_1(\mathbf{r}, \mathbf{v}_1', t)$, $f_2(\mathbf{r}, \mathbf{v}_2', t)$, $f_1(\mathbf{r}, \mathbf{v}_1, t)$, and $f_2(\mathbf{r}, \mathbf{v}_2, t)$, respectively. Finally, $\mu_{12} = (1/m_1 + 1/m_2)^{-1}$.

The type 1 and type 2 particle velocities prior to the collision are $\mathbf{v}_1$ and $\mathbf{v}_2$, respectively, so that $\mathbf{u} = \mathbf{v_1} - \mathbf{v}_2$. Let us write the corresponding velocities after the collision as (see Section 3.3)

$$\mathbf{v}_1' = \mathbf{v}_1 + \frac{\mu_{12}}{m_1}\,\mathbf{g}, \tag{3.91}$$

$$\mathbf{v}_2' = \mathbf{v}_2 - \frac{\mu_{12}}{m_2}\,\mathbf{g}. \tag{3.92}$$

Here, $\mathbf{g} = \mathbf{u}' - \mathbf{u}$ is assumed to be small, which implies that the angle of deflection is also small. Expanding $f_1' \equiv f_1(\mathbf{r}, \mathbf{v}_1', t)$ to second order in $\mathbf{g}$, we obtain

$$f_1(\mathbf{v}_1') \simeq f_1(\mathbf{v}_1) + \frac{\mu_{12}}{m_1}\,\mathbf{g}\cdot\frac{\partial f_1(\mathbf{v}_1)}{\partial \mathbf{v}_1} + \frac{1}{2}\,\frac{\mu_{12}^2}{m_1^2}\,\mathbf{g}\mathbf{g} : \frac{\partial^2 f_1(\mathbf{v}_1)}{\partial \mathbf{v}_1 \partial \mathbf{v}_1}. \tag{3.93}$$

Likewise, expanding $f_2' \equiv f_2(\mathbf{r}, \mathbf{v}_2', t)$, we get

$$f_2(\mathbf{v}_2') \simeq f_2(\mathbf{v}_1) - \frac{\mu_{12}}{m_2}\,\mathbf{g}\cdot\frac{\partial f_2(\mathbf{v}_2)}{\partial \mathbf{v}_2} + \frac{1}{2}\,\frac{\mu_{12}^2}{m_2^2}\,\mathbf{g}\mathbf{g} : \frac{\partial^2 f_2(\mathbf{v}_2)}{\partial \mathbf{v}_2 \partial \mathbf{v}_2}. \tag{3.94}$$

Note that, in writing the previous two equations, we have neglected the $\mathbf{r}$ and t dependence of $f_1(\mathbf{r}, \mathbf{v}_1', t)$, et cetera, for ease of notation. Hence,

$$\begin{aligned} f_1' f_2' - f_1 f_2 &\simeq \mu_{12}\,\mathbf{g}\cdot\left(\frac{\partial f_1}{\partial \mathbf{v}_1}\frac{f_2}{m_1} - \frac{f_1}{m_2}\frac{\partial f_2}{\partial \mathbf{v}_2}\right) \\ &\quad + \frac{1}{2}\,\mu_{12}^2\,\mathbf{g}\mathbf{g} : \left(\frac{\partial^2 f_1}{\partial \mathbf{v}_1 \partial \mathbf{v}_1}\frac{f_2}{m_1^2} + \frac{f_1}{m_2^2}\frac{\partial^2 f_2}{\partial \mathbf{v}_2 \partial \mathbf{v}_2} - \frac{2}{m_1\,m_2}\frac{\partial f_1}{\partial \mathbf{v}_1}\frac{\partial f_2}{\partial \mathbf{v}_2}\right). \end{aligned} \tag{3.95}$$

It follows that

$$C_{12} \simeq \frac{1}{4}\left(\frac{e_1\,e_2}{4\pi\,\epsilon_0\,\mu_{12}}\right)^2 \int\!\!\int \left[\mu_{12}\,\mathbf{g}\cdot\mathbf{J} + \frac{1}{2}\,\mu_{12}^2\,\mathbf{g}\mathbf{g} : \left(\frac{1}{m_1}\frac{\partial}{\partial \mathbf{v}_1} - \frac{1}{m_2}\frac{\partial}{\partial \mathbf{v}_2}\right)\mathbf{J}\right]\frac{d^3\mathbf{v}_2\,d\Omega}{u^3\,\sin^4(\chi/2)}, \tag{3.96}$$

where

$$\mathbf{J} = \frac{\partial f_1}{\partial \mathbf{v}_1}\frac{f_2}{m_1} - \frac{f_1}{m_2}\frac{\partial f_2}{\partial \mathbf{v}_2}. \tag{3.97}$$

Let $\mathbf{l}$, $\mathbf{m}$, and $\mathbf{n}$ be a right-handed set of mutually orthogonal unit vectors. Suppose that $\mathbf{u} = u\,\mathbf{l}$. Recall that $\mathbf{u}' = \mathbf{u} + \mathbf{g}$. Now, in an elastic collision for which the angle of deviation is χ, we require $|\mathbf{u}'| = |\mathbf{u}|$, $|\mathbf{u}\times\mathbf{u}'| = u^2\,\sin\chi$, and $\mathbf{u}' = \mathbf{u}$ when $\chi = 0$. In other words, we need $|\mathbf{g} + u\,\mathbf{l}| = u$, $|\mathbf{l}\times\mathbf{g}| = u\,\sin\chi$, and $\mathbf{g} = \mathbf{0}$ when $\chi = 0$. We deduce that

$$\mathbf{g} \simeq u\,[(\cos\chi - 1)\,\mathbf{l} + \sin\chi\,\cos\phi\,\mathbf{m} + \sin\chi\,\sin\phi\,\mathbf{n}]. \tag{3.98}$$

Thus,

$$\begin{aligned} \int \frac{\mathbf{g}\,d\Omega}{\sin^4(\chi/2)} &= \int_0^\pi \oint \frac{\mathbf{g}\,\sin\chi\,d\chi\,d\phi}{\sin^4(\chi/2)} = \mathbf{u}\int\frac{(\cos\chi - 1)\,d\Omega}{\sin^4(\chi/2)} \\ &\simeq -\mathbf{u}\int\frac{2\,d\Omega}{\sin^2(\chi/2)}, \end{aligned} \tag{3.99}$$

and

$$\int\frac{\mathbf{g}\mathbf{g}\,d\Omega}{\sin^4(\chi/2)} \simeq \frac{u^2}{2}\,(\mathbf{m}\mathbf{m} + \mathbf{n}\mathbf{n})\int\frac{\sin^2\chi\,d\Omega}{\sin^4(\chi/2)} \simeq u^2\,(\mathbf{I} - \mathbf{l}\mathbf{l})\int\frac{2\,d\Omega}{\sin^2(\chi/2)}, \tag{3.100}$$

where use has again been made of the fact that χ is small.

Now,

$$\int \frac{2\,d\Omega}{\sin^2(\chi/2)} = \int \frac{4\pi\,\sin\chi\,d\chi}{\sin^2(\chi/2)} \simeq 16\pi \int \frac{d(\chi/2)}{\sin(\chi/2)} = 16\pi\,\ln\left(\frac{\chi_{\rm max}}{\chi_{\rm min}}\right), \tag{3.101}$$

where $\chi_{\rm max}$ and $\chi_{\rm min}$ are the maximum and minimum angles of deflection, respectively. However, according to Equation (3.82), small-angle two-body Coulomb collisions are characterized by

$$\chi \simeq \frac{e_1\,e_2}{2\pi\,\epsilon_0\,\mu_{12}\,u^2\,b}, \tag{3.102}$$

where b is the impact parameter. Thus, we can write

$$\int \frac{2\,d\Omega}{\sin^2(\chi/2)} = 16\pi\,\ln\Lambda_c, \tag{3.103}$$

where the quantity

$$\ln\Lambda_c = \ln\left(\frac{\chi_{\rm max}}{\chi_{\rm min}}\right) = \ln\left(\frac{b_{\rm max}}{b_{\rm min}}\right), \tag{3.104}$$

is known as the *Coulomb logarithm*.

It follows from the previous analysis that

$$C_{12} = \left(\frac{e_1\,e_2}{4\pi\,\epsilon_0\,\mu_{12}}\right)^2 4\pi\,\ln\Lambda_c \int\left[-\frac{\mu_{12}\,\mathbf{l}\cdot\mathbf{J}}{u^2} + \frac{\mu_{12}^2}{2\,u}\,(\mathbf{I}-\mathbf{ll}) : \left(\frac{1}{m_1}\frac{\partial}{\partial\mathbf{v}_1} - \frac{1}{m_2}\frac{\partial}{\partial\mathbf{v}_2}\right)\mathbf{J}\right] d^3\mathbf{v}_2. \tag{3.105}$$

If we define the tensor

$$\mathbf{w} = \frac{\mathbf{I}-\mathbf{ll}}{u} = \frac{u^2\,\mathbf{I}-\mathbf{uu}}{u^3} \tag{3.106}$$

then it is readily seen that

$$\left(\frac{\partial}{\partial\mathbf{u}}\cdot\mathbf{w}\right)_i \equiv \frac{\partial}{\partial u_j}\left(\frac{\delta_{ij}}{(u_k\,u_k)^{1/2}} - \frac{u_i\,u_j}{(u_k\,u_k)^{3/2}}\right) = -\frac{2\,u_i}{(u_k\,u_k)^{3/2}} = -\left(\frac{2\,\mathbf{l}}{u^2}\right)_i. \tag{3.107}$$

Here, i, j, et cetera, run from 1 to 3, and correspond to Cartesian components. Moreover, we have made use of the *Einstein summation convention* (that repeated indices are implicitly summed from 1 to 3) (Riley 1974). Hence, we deduce that

$$C_{12} = \left(\frac{e_1\,e_2}{4\pi\,\epsilon_0\,\mu_{12}}\right)^2 4\pi\,\ln\Lambda_c \int\left[\frac{\mu_{12}}{2}\left(\frac{\partial}{\partial\mathbf{u}}\cdot\mathbf{w}\right)\cdot\mathbf{J} + \frac{\mu_{12}^2}{2}\,\mathbf{w} : \left(\frac{1}{m_1}\frac{\partial}{\partial\mathbf{v}_1} - \frac{1}{m_2}\frac{\partial}{\partial\mathbf{v}_2}\right)\mathbf{J}\right] d^3\mathbf{v}_2. \tag{3.108}$$

Integration by parts yields

$$C_{12} = \left(\frac{e_1\,e_2}{4\pi\,\epsilon_0\,\mu_{12}}\right)^2 4\pi\,\ln\Lambda_c \int\left[\frac{\mu_{12}}{2}\left(\frac{\partial}{\partial\mathbf{u}}\cdot\mathbf{w}\right)\cdot\mathbf{J} + \frac{\mu_{12}^2}{2\,m_1}\,\mathbf{w} : \frac{\partial\mathbf{J}}{\partial\mathbf{v}_1} + \frac{\mu_{12}^2}{2\,m_2}\left(\frac{\partial}{\partial\mathbf{v}_2}\cdot\mathbf{w}\right)\cdot\mathbf{J}\right] d^3\mathbf{v}_2. \tag{3.109}$$

However,

$$\frac{\partial}{\partial \mathbf{v}_2} \cdot \mathbf{w} = -\frac{\partial}{\partial \mathbf{u}} \cdot \mathbf{w} = -\frac{\partial}{\partial \mathbf{v}_1} \cdot \mathbf{w}, \tag{3.110}$$

because $\mathbf{w}$ is a function of $\mathbf{u} = \mathbf{v}_1 - \mathbf{v}_2$. Thus, we obtain the so-called *Landau collision operator* (Laudau 1936),

$$C_{12} = \frac{\gamma_{12}}{m_1} \frac{\partial}{\partial \mathbf{v}_1} \cdot \int \mathbf{w} \cdot \mathbf{J}\, d^3\mathbf{v}_2, \tag{3.111}$$

where

$$\gamma_{12} = \left(\frac{e_1\, e_2}{4\pi\, \epsilon_0}\right)^2 2\pi\, \ln \Lambda_c. \tag{3.112}$$

It is sometimes convenient to write the Landau collision operator in the form

$$C_{12} = -\frac{1}{m_1} \frac{\partial}{\partial \mathbf{v}_1} \cdot \mathbf{A}_{12}, \tag{3.113}$$

where

$$\mathbf{A}_{12} = \mathbf{B}_{12}\, f_1 - \mathbf{D}_{12} \cdot \frac{\partial f_1}{\partial \mathbf{v}_1}, \tag{3.114}$$

and

$$\mathbf{B}_{12} = \frac{\gamma_{12}}{m_2} \int \mathbf{w} \cdot \frac{\partial f_2}{\partial \mathbf{v}_2}\, d^3\mathbf{v}_2, \tag{3.115}$$

$$\mathbf{D}_{12} = \frac{\gamma_{12}}{m_1} \int \mathbf{w}\, f_2\, d^3\mathbf{v}_2. \tag{3.116}$$

3.10 Coulomb Logarithm

According to Equation (3.103), the Coulomb logarithm can be written

$$\ln \Lambda_c = \int \frac{d\chi}{\chi}, \tag{3.117}$$

where we have made use of the fact that scattering angle χ is small, Obviously, the integral appearing in the previous expression diverges at both large and small χ.

The divergence of the integral on the right-hand side of the previous equation at large χ is a consequence of the breakdown of the small-angle approximation. The standard prescription for avoiding this divergence is to truncate the integral at some $\chi_{\max}$ above which the small-angle approximation becomes invalid. According to Equation (3.102), this truncation is equivalent to neglecting all collisions whose impact parameters fall below the value

$$b_{\min} \simeq \frac{e_1\, e_2}{2\pi\, \epsilon_0\, \mu_{12}\, u^2}. \tag{3.118}$$

The ultimate justification for the truncation of the integral appearing in Equation (3.117) at large χ is the idea that Coulomb collisions are dominated by small-angle scattering events, and that the occasional large-angle scattering events have a negligible effect on the scattering statistics. Unfortunately, this is not quite true (if it were then the integral would converge at large χ). However, the rare large-angle scattering events only make a relatively weak logarithmic contribution to the scattering statistics.

Making the estimate $(1/2)\,\mu_{12}\,u^2 \simeq T$, where T is the assumed common temperature of the two colliding species, we obtain

$$b_{\rm min} \simeq \frac{e_1\,e_2}{4\pi\,\epsilon_0\,T} = r_c, \tag{3.119}$$

where r_c is the classical distance of closest approach introduced in Section 1.6. However, as mentioned in Section 1.10, it is possible for the classical distance of closest approach to fall below the de Broglie wavelength of one or both of the colliding particles, even in the case of a weakly coupled plasma. In this situation, the most sensible thing to do is to approximate $b_{\rm min}$ as the larger de Broglie wavelength (Spitzer 1956; Braginskii 1965).

The divergence of the integral on the right-hand side of Equation (3.117) at small χ is a consequence of the infinite range of the Coulomb potential. The standard prescription for avoiding this divergence is to take the Debye shielding of the Coulomb potential into account. (See Section 1.5.) This is equivalent to neglecting all collisions whose impact parameters exceed the value

$$b_{\rm max} = \lambda_D, \tag{3.120}$$

where λ_D is the Debye length. Of course, Debye shielding is a many-particle effect. Hence, the Landau collision operator can no longer be regarded as a pure two-body collision operator. Fortunately, however, many-particle effects only make a relatively weak logarithmic contribution to the operator.

According to Equations (3.104), (3.119), and (3.120),

$$\ln\Lambda_c = \ln\left(\frac{b_{\rm max}}{b_{\rm min}}\right) = \ln\left(\frac{\lambda_D}{r_c}\right). \tag{3.121}$$

Thus, we deduce from Equation (1.20) that

$$\ln\Lambda_c \simeq \ln\Lambda. \tag{3.122}$$

In other words, the Coulomb logarithm is approximately equal to the natural logarithm of the plasma parameter. The fact that the plasma parameter is much larger than unity in a weakly coupled plasma implies that the Coulomb logarithm is large compared to unity in such a plasma. In fact, $\ln\Lambda_c$ lies in the range 10–20 for typical weakly coupled plasmas. It also follows that $b_{\rm max} \gg b_{\rm min}$ in a weakly coupled plasma, which means that there is a large range of impact parameters for which it is accurate to treat Coulomb collisions as small-angle two-body scattering events.

The official definition of the Coulomb logarithm is as follows (Huba 2000d). For a particle of type 1, with mass m_1 and charge $e_1 = Z_1\,e$, scattered by particles of type 2, with mass m_2 and charge $e_2 = Z_2\,e$, the Coulomb logarithm is defined $\ln\Lambda_c = \ln(b_{\rm max}/b_{\rm min})$. Here, $b_{\rm min}$ is the larger of $e_1\,e_2/(4\pi\,\epsilon_0\,\mu_{12}\,u^2)$ and $\hbar/(2\,\mu_{12}\,u)$, averaged over both particle distributions, where $\mu_{12} = m_1\,m_2/(m_1+m_2)$ and $\mathbf{u} = \mathbf{v}_1 - \mathbf{v}_2$. Furthermore, $b_{\rm max} = (\sum_s n_s\,e_s^2/\epsilon_0\,T_s)^{-1/2}$, where the summation extends over all species, s, for which $\bar{u}^2 \lesssim T_s/m_s$. For thermal (i.e., Maxwellian) electron-electron collisions, we obtain

$$\ln\Lambda_c = 23 - \ln\left(n_e^{1/2}\,T_e^{-3/2}\right) \qquad T_e < 10\,{\rm eV},$$
$$\ln\Lambda_c = 24 - \ln\left(n_e^{1/2}\,T_e^{-1}\right) \qquad T_e > 10\,{\rm eV}. \tag{3.123}$$

Likewise, for thermal electron-ion collisions, we get

$$\ln\Lambda_c = 30 - \ln\left(n_e^{1/2}\,T_i^{-3/2}\,Z_i^{3/2}\,\hat{m}_i\right) \qquad T_e < T_i\,m_e/m_i,$$
$$\ln\Lambda_c = 23 - \ln\left(n_e^{1/2}\,Z_i\,T_e^{-3/2}\right) \qquad T_i\,m_e/m_i < T_e < 10\,Z_i^2\,{\rm eV},$$
$$\ln\Lambda_c = 24 - \ln\left(n_e^{1/2}\,T_e^{-1}\right) \qquad T_i\,m_e/m_i < 10\,Z_i^2\,{\rm eV} < T_e. \tag{3.124}$$

Here, n_e and n_i are measured in units of cm^{-3}, whereas T_e and T_i are measured in units of electron-volts. Moreover, $\hat{m}_i = m_i/m_p$, where m_p is the proton mass.

The standard approach in plasma physics is to treat the Coulomb logarithm as a constant, with a value determined by the ambient electron number density, the ambient electron and ion temperatures, and the ion charge and mass numbers, as has just been described. This approximation ensures that the Landau collision operator, $C_{12}(f_1, f_2)$, is strictly bilinear in its two arguments.

3.11 Rosenbluth Potentials

It is convenient to define

$$G_2(\mathbf{v}_1) = \int u\,f_2\,d^3\mathbf{v}_2, \tag{3.125}$$

$$H_2(\mathbf{v}_1) = \int u^{-1}\,f_2\,d^3\mathbf{v}_2. \tag{3.126}$$

Now, from Equation (3.106),

$$w_{ij} = \frac{\delta_{ij}}{u} - \frac{u_i\,u_j}{u^3}. \tag{3.127}$$

Moreover,

$$\frac{\partial u}{\partial u_i} = \frac{u_i}{u}, \tag{3.128}$$

$$\frac{\partial u_i}{\partial u_j} = \delta_{ij}. \tag{3.129}$$

Hence, it is easily demonstrated that

$$w_{ij} = \frac{\partial^2 u}{\partial u_i\,\partial u_j}, \tag{3.130}$$

$$\frac{\partial w_{ij}}{\partial u_j} = \frac{\partial w_{jj}}{\partial u_i} = 2\,\frac{\partial}{\partial u_i}\left(\frac{1}{u}\right). \tag{3.131}$$

According to Equations (3.115) and (3.116),

$$\mathbf{B}_{12} = \frac{\gamma_{12}}{m_2}\int \mathbf{w}\cdot\frac{\partial f_2}{\partial \mathbf{v}_2}\,d^3\mathbf{v}_2 = \frac{\gamma_{12}}{m_2}\int \frac{\partial}{\partial \mathbf{u}}\cdot\mathbf{w}\,f_2\,d^3\mathbf{v}_2, \tag{3.132}$$

$$\mathbf{D}_{12} = \frac{\gamma_{12}}{m_1}\int \mathbf{w}\,f_2\,d^3\mathbf{v}_2, \tag{3.133}$$

where we have integrated the first equation by parts, making use of Equation (3.110). Thus, we deduce from Equations (3.130) and (3.131) that

$$\mathbf{B}_{12} = \frac{2\,\gamma_{12}}{m_2}\,\frac{\partial H_2}{\partial \mathbf{v}_1}, \tag{3.134}$$

$$\mathbf{D}_{12} = \frac{\gamma_{12}}{m_1}\,\frac{\partial^2 G_2}{\partial \mathbf{v}_1 \partial \mathbf{v}_1}. \tag{3.135}$$

The quantities $H_2(\mathbf{v})$ and $G_2(\mathbf{v})$ are known as *Rosenbluth potentials* (Rosenbluth, MacDonald, and Judd 1957), and can easily be seen to satisfy

$$\nabla_v^2 H_2 = -4\pi\,f_2(\mathbf{v}), \tag{3.136}$$

$$\nabla_v^2 G_2 = 2\,H_2(\mathbf{v}), \tag{3.137}$$

where ∇_v^2 denotes a velocity-space Laplacian operator. The former result follows because $\nabla_v^2(1/v) = -4\pi\,\delta(\mathbf{v})$, and the latter because $\nabla_v^2(v) = 2/v$. In particular, if $f_2(\mathbf{v})$ is isotropic in velocity space then we obtain

$$\frac{d}{dv}\left(v^2\,\frac{dH_2}{dv}\right) = -4\pi\,v^2\,f_2(v), \tag{3.138}$$

$$\frac{d}{dv}\left(v^2\,\frac{dG_2}{dv}\right) = 2\,v^2\,H_2(v). \tag{3.139}$$

Suppose that $f_2(\mathbf{v})$ is a Maxwellian distribution of characteristic number density n_2, mean flow velocity zero, and temperature T_2. In other words,

$$f_2(\mathbf{v}) = n_2\left(\frac{m_2}{2\pi\,T_2}\right)^{3/2}\exp\left(-\frac{m_2\,v^2}{2\,T_2}\right). \tag{3.140}$$

In this case, Equation (3.138) reduces to

$$\frac{d^2}{d\zeta^2}(\zeta\,H_2) = -\frac{4}{\sqrt{\pi}}\,\frac{n_2}{v_{t\,2}}\,\zeta\,e^{-\zeta^2} = \frac{2}{\sqrt{\pi}}\,\frac{n_2}{v_{t\,2}}\,\frac{d}{d\zeta}\,e^{-\zeta^2}, \tag{3.141}$$

where $\zeta = v/v_{t\,2}$, and $v_{t\,2} = \sqrt{2\,T_2/m_2}$. Hence, requiring $H_2(\zeta)$ to be finite at $\zeta = 0$, we can integrate the previous expression to give

$$H_2(\zeta) = \frac{n_2}{v_{t\,2}}\,\frac{\mathrm{erf}(\zeta)}{\zeta}, \tag{3.142}$$

where

$$\mathrm{erf}(\zeta) = \frac{2}{\sqrt{\pi}}\int_0^{\zeta} e^{-t^2}\,dt \tag{3.143}$$

is a so-called *error function* (Abramowitz and Stegun 1965b). This function is such that

$$\mathrm{erf}(\zeta) = \frac{2}{\sqrt{\pi}}\left[\zeta - \frac{\zeta^3}{3} + O\left(\zeta^5\right)\right] \tag{3.144}$$

when $0 < \zeta \ll 1$, and

$$\mathrm{erf}(\zeta) = 1 - \frac{e^{-\zeta^2}}{\sqrt{\pi}\,\zeta}\left[1 + O\left(\frac{1}{\zeta^2}\right)\right] \tag{3.145}$$

when $\zeta \gg 1$. Equation (3.139) yields

$$\frac{d^2}{d\zeta^2}(\zeta\,G_2) = 2\,n_2\,v_{t\,2}\,\mathrm{erf}(\zeta), \tag{3.146}$$

which can be integrated, subject to the constraint that G_2 be finite at $\zeta = 0$, to give

$$G_2(\zeta) = \frac{n_2\,v_{t\,2}}{2\,\zeta}\left[\zeta\,\frac{d\,\mathrm{erf}}{d\zeta} + \left(1 + 2\,\zeta^2\right)\mathrm{erf}(\zeta)\right]. \tag{3.147}$$

According to Equations (3.128), (3.129), (3.142), and (3.147),

$$\frac{\partial H_2}{\partial \mathbf{v}} = -n_2\,F_1(\zeta)\,\frac{\mathbf{v}}{v^3}, \tag{3.148}$$

$$\frac{\partial^2 G_2}{\partial \mathbf{v}\,\partial \mathbf{v}} = \frac{n_2\,v_{t\,2}^2}{2\,v^3}\left[-F_2(\zeta)\,\mathbf{I} + 3\,F_3(\zeta)\,\frac{\mathbf{v}\mathbf{v}}{v^2}\right], \tag{3.149}$$

where

$$F_1(\zeta) = \mathrm{erf}(\zeta) - \zeta\,\frac{d\,\mathrm{erf}}{d\zeta}, \tag{3.150}$$

$$F_2(\zeta) = \left(1 - 2\,\zeta^2\right)\mathrm{erf}(\zeta) - \zeta\,\frac{d\,\mathrm{erf}}{d\zeta}, \tag{3.151}$$

$$F_3(\zeta) = \left(1 - \frac{2}{3}\,\zeta^2\right)\mathrm{erf}(\zeta) - \zeta\,\frac{d\,\mathrm{erf}}{d\zeta}. \tag{3.152}$$

Finally, it follows from Equations (3.114), (3.134), (3.135), (3.148), and (3.149) that

$$\mathbf{A}_{12}(\mathbf{v}) = -\frac{\gamma_{12}\,n_2}{m_2}\left\{2\,F_1(\zeta)\,\frac{\mathbf{v}}{v^3}\,f_1(\mathbf{v}) + \frac{m_2}{m_1}\,\frac{v_{t2}^2}{2\,v^3}\left[-F_2(\zeta)\,\mathbf{I} + 3\,F_3(\zeta)\,\frac{\mathbf{v}\mathbf{v}}{v^2}\right]\cdot\frac{\partial f_1}{\partial \mathbf{v}}\right\}. \tag{3.153}$$

Suppose that $f_1(\mathbf{v})$ is a Maxwellian distribution of characteristic number density n_1, mean flow velocity zero, and temperature T_1. In other words,

$$f_1(\mathbf{v}) = n_1\left(\frac{m_1}{2\pi\,T_1}\right)^{3/2}\exp\left(-\frac{m_1\,v^2}{2\,T_1}\right). \tag{3.154}$$

It follows that

$$\frac{\partial f_1}{\partial \mathbf{v}} = -\frac{m_1}{T_1}\,\mathbf{v}\,f_1. \tag{3.155}$$

Hence, Equations (3.113) and (3.153) yield

$$C_{12}(\mathbf{v}) = -\frac{1}{m_1}\,\frac{\partial}{\partial \mathbf{v}}\cdot(\mathbf{R}_{12}\,f_1), \tag{3.156}$$

where

$$\mathbf{R}_{12} = \frac{2\,\gamma_{12}\,n_2}{m_2\,v_{t2}^3}\left(\frac{T_2 - T_1}{T_1}\right)\frac{F_1(\zeta)}{\zeta^3}\,\mathbf{v}, \tag{3.157}$$

and use has been made of the fact that $3\,F_3 - F_2 = 2\,F_1$. The ensemble-averaged kinetic equation, Equation (3.9), can thus be written in the form

$$\frac{\partial f_1}{\partial t} + \frac{\partial}{\partial \mathbf{r}}\cdot(\mathbf{v}\,f_1) + \frac{1}{m_1}\,\frac{\partial}{\partial \mathbf{v}}\cdot[(\mathbf{F}_1 + \mathbf{R}_{12})\,f_1] = \mathbf{0}, \tag{3.158}$$

where

$$\mathbf{F}_1 = e_1\,(\mathbf{E} + \mathbf{v}\times\mathbf{B}) \tag{3.159}$$

is the ensemble-averaged Lorentz force. In deriving Equation (3.158), we have made use of the easily proved result

$$\frac{\partial}{\partial \mathbf{v}}\cdot\mathbf{F}_1 = \mathbf{0}. \tag{3.160}$$

According to Equation (3.158), collisions with particles of type 2 give rise to a velocity dependent effective force, $\mathbf{R}_{12}$, acting on individual particles of type 1. As expected, this force is zero if the temperatures of the two species are equal. On the other hand, if particles of type 2 have a higher kinetic temperature than particles of type 1 (i.e., if $T_2 > T_1$) then the collisional force acts to speed up the latter particles—in other words, the force always acts in the same direction as the particle's instantaneous velocity. [This follows because $F_1(\zeta) \geq 0$.] Conversely, if particles of type 2 have a lower kinetic temperature than particles of type 1, then the collisional force acts to slow down the latter particles—in other words, the force always acts in the opposite direction to the particle's instantaneous velocity. In both cases, the collisional force is clearly acting to equalize the kinetic temperatures.

Suppose that $f_1(\mathbf{v})$ is a Maxwellian distribution of characteristic number density n_1, mean flow velocity $\mathbf{V}$, and temperature T_2. In other words,

$$f_1(\mathbf{v}) = n_1 \left(\frac{m_1}{2\pi T_2}\right)^{3/2} \exp\left[-\frac{m_1 (\mathbf{v}-\mathbf{V})^2}{2 T_2}\right]. \tag{3.161}$$

It follows that

$$\frac{\partial f_1}{\partial \mathbf{v}} = -\frac{m_1}{T_2} (\mathbf{v}-\mathbf{V}) f_1. \tag{3.162}$$

Hence, Equation (3.153) yields

$$\mathbf{A}_{12} = -\frac{\gamma_{12} n_2}{m_2}\left[-\frac{F_2(\zeta)}{v^3}\mathbf{V} + \frac{3 F_3(\zeta)}{v^5} (\mathbf{v}\cdot\mathbf{V})\mathbf{v}\right] f_1, \tag{3.163}$$

which implies that

$$C_{12} = -\frac{1}{m_1}\frac{\partial}{\partial \mathbf{v}}\cdot (\mathbf{R}_{12} f_1), \tag{3.164}$$

where

$$\mathbf{R}_{12} = -\frac{\gamma_{12} n_2}{m_2}\left[-\frac{F_2(\zeta)}{v^3}\mathbf{V} + \frac{3 F_3(\zeta)}{v^5} (\mathbf{v}\cdot\mathbf{V})\mathbf{v}\right]. \tag{3.165}$$

As before, collisions with particles of type 2 give rise to a velocity dependent effective force, $\mathbf{R}_{12}$, acting on individual particles of type 1. In particular, if $\mathbf{v}$ is parallel to $\mathbf{V}$, then

$$\mathbf{R}_{12} = -\frac{2\gamma_{12} n_2}{m_2 v_{t2}^3}\frac{F_1(\zeta)}{\zeta^3}\mathbf{V}. \tag{3.166}$$

We conclude that particles of type 1 moving parallel to the mean drift velocity $\mathbf{V}$ (of particles of type 1 relative to particles of type 2) experience a velocity dependent force due to collisions with particles of type 2, which acts to reduce their speed. Of course, this has the effect of reducing the drift velocity.

It is easily demonstrated that $F_1(\zeta) \to 1$ as $\zeta \to \infty$. Hence, Equations (3.157) and (3.166) yield $R_{12} \propto v^{-2}$ and $R_{12} \propto v^{-3}$, respectively, in the limit $v \gg v_{t2}$, implying that collisions only have a relatively weak effect on high speed particles. In fact, collisions are unable to prevent an imposed electric field from accelerating super-thermal particles (whose number is generally only a very small fraction of the total number of particles) to relativistic speeds (Rose and Clark 1961). Such particles are known as *runaway particles*.

3.12 Collision Times

Consider collisions between particles of type 1 (with number density n_1 and mass m_1), possessing the Maxwellian distribution function

$$f_1(\mathbf{v}) = n_1 \left(\frac{m_1}{2\pi T}\right)^{3/2} \exp\left[-\frac{m_1 (\mathbf{v}-\mathbf{V})^2}{2 T}\right], \tag{3.167}$$

and particles of type 2 (with number density n_2 and mass m_2), possessing the Maxwellian distribution function

$$f_2(\mathbf{v}) = n_2 \left(\frac{m_2}{2\pi T}\right)^{3/2} \exp\left(-\frac{m_2 v^2}{2T}\right). \tag{3.168}$$

Here, T is the common temperature of the two species, and $\mathbf{V}$ is the mean drift velocity of species 1 relative to species 2. As we saw in the previous section, collisions with particles of type 2 give rise to a velocity-dependent force acting on individual particles of type 1. This force takes the form [see Equation (3.165)]

$$\mathbf{R}_{12} = -\frac{\gamma_{12} n_2}{m_2} \left[-\frac{F_2(\zeta)}{v^3} \mathbf{V} + \frac{3 F_3(\zeta)}{v^5} (\mathbf{v} \cdot \mathbf{V}) \mathbf{v}\right], \tag{3.169}$$

where $\zeta = v/v_{t2}$ and $v_{t2} = \sqrt{2T/m_2}$. The net force per unit volume acting on type 1 particles due to collisions with type 2 particles is thus

$$\mathbf{F}_{12} = \int m_1 \mathbf{v} C_{12} d^3\mathbf{v} = \int \mathbf{R}_{12} f_1 d^3\mathbf{v}. \tag{3.170}$$

Suppose that the drift velocity, $\mathbf{V}$, is much smaller than the thermal velocity, $v_{t1} = \sqrt{2T/m_1}$, of type 1 particles. In this case, we can write

$$f_1(\mathbf{v}) \simeq \frac{n_1}{\pi^{3/2} v_{t1}^3} \exp\left(-\frac{v^2}{v_{t1}^2}\right)\left(1 - \frac{2\mathbf{v} \cdot \mathbf{V}}{v^2}\right). \tag{3.171}$$

Hence, Equations (3.169) and (3.170) yield

$$(\mathbf{F}_{12})_i = -\frac{\gamma_{12} n_1 n_2}{\pi^{3/2} m_2 v_{t1}^3} \int \exp\left(-\frac{v^2}{v_{t1}^2}\right)\left(1 - \frac{2 v_j V_j}{v^2}\right)\left[-\frac{F_2(\zeta)}{v^3} V_i + \frac{3 F_3(\zeta)}{v^5} v_k V_k v_i\right] d^3\mathbf{v}. \tag{3.172}$$

However, it follows from symmetry that

$$\int H(v) v_i d^3\mathbf{v} = 0, \tag{3.173}$$

$$\int H(v) v_i v_j d^3\mathbf{v} = \frac{\delta_{ij}}{3} \int H(v) v^2 d^3\mathbf{v}, \tag{3.174}$$

$$\int H(v) v_i v_j v_k d^3\mathbf{v} = 0, \tag{3.175}$$

where $H(v)$ is a general function. Hence, Equation (3.172) reduces to

$$\mathbf{F}_{12} = -\frac{\gamma_{12} n_1 n_2 \mathbf{V}}{\pi^{3/2} m_2 v_{t1}^3} \int \exp\left(-\frac{v^2}{v_{t1}^2}\right)\left[\frac{F_3(\zeta) - F_2(\zeta)}{v^3}\right] d^3\mathbf{v}. \tag{3.176}$$

It follows from Equations (3.151) and (3.152) that

$$\mathbf{F}_{12} = -\frac{16 \gamma_{12} n_1 n_2 \mathbf{V}}{3 \pi^{1/2} m_2 v_{t1}^3} \int_0^\infty \zeta \operatorname{erf}(\zeta) \exp\left(-\frac{v_{t2}^2}{v_{t1}^2} \zeta^2\right) d\zeta. \tag{3.177}$$

Integration by parts gives

$$\mathbf{F}_{12} = -\frac{16\,\gamma_{12}\,n_1\,n_2\,\mathbf{V}}{3\,\pi\,m_2\,v_{t\,1}\,v_{t\,2}^2}\int_0^\infty \exp\left[-\left(1+\frac{v_{t\,2}^2}{v_{t\,1}^2}\right)\zeta^2\right]d\zeta, \tag{3.178}$$

which reduces to

$$\mathbf{F}_{12} = -\left[\frac{8\,\gamma_{12}\,n_1\,n_2}{3\,\pi^{1/2}\,m_2\,v_{t\,2}^2\,(v_{t\,1}^2+v_{t\,2}^2)^{1/2}}\right]\mathbf{V}. \tag{3.179}$$

The *collision time*, τ_{12}, associated with collisions of particles of type 1 with particles of type 2, is conventionally defined via the following equation,

$$\mathbf{F}_{12} = -\frac{m_1\,n_1}{\tau_{12}}\,\mathbf{V}. \tag{3.180}$$

According to this definition, the collision time is the time required for collisions with particles of type 2 to decelerate particles of type 1 to such an extent that the mean drift velocity of the latter particles with respect to the former is eliminated. At the individual particle level, the collision time is the mean time required for the direction of motion of an individual type 1 particle to deviate through approximately 90° as a consequence of collisions with particles of type 2. According to Equations (3.112) and (3.179), we can write

$$\tau_{12} = \frac{3\pi^{1/2}\,m_1\,T^{3/2}}{2\,\sqrt{2}\,\mu_{12}^{1/2}\,\gamma_{12}\,n_2} = \frac{6\,\sqrt{2}\,\pi^{3/2}\,\epsilon_0^2\,m_1\,T^{3/2}}{\ln\Lambda_c\,\mu_{12}^{1/2}\,e_1^2\,e_2^2\,n_2}. \tag{3.181}$$

Consider a quasi-neutral plasma consisting of electrons of mass m_e, charge $-e$, and number density n_e, and ions of mass m_i, charge $+e$, and number density $n_i = n_e$. Let the two species both have Maxwellian distributions characterized by a common temperature T, and a small relative drift velocity. It follows, from the previous analysis, that we can identify four different collision times. First, the *electron-electron collision time*,

$$\tau_{ee} = \frac{12\,\sqrt{2}\,\pi^{3/2}\,\epsilon_0^2\,m_e^{1/2}\,T^{3/2}}{\ln\Lambda_c\,e^4\,n_e}, \tag{3.182}$$

which is the mean time required for the direction of motion of an individual electron to deviate through approximately 90° as a consequence of collisions with other electrons. Second, the *electron-ion collision time*,

$$\tau_{ei} = \frac{6\,\sqrt{2}\,\pi^{3/2}\,\epsilon_0^2\,m_e^{1/2}\,T^{3/2}}{\ln\Lambda_c\,e^4\,n_e}, \tag{3.183}$$

which is the mean time required for the direction of motion of an individual electron to deviate through approximately 90° as a consequence of collisions with ions. Third, the *ion-ion collision time*,

$$\tau_{ii} = \frac{12\,\sqrt{2}\,\pi^{3/2}\,\epsilon_0^2\,m_i^{1/2}\,T^{3/2}}{\ln\Lambda_c\,e^4\,n_e}, \tag{3.184}$$

which is the mean time required for the direction of motion of an individual ion to deviate through approximately 90° as a consequence of collisions with other ions. Finally, the *ion-electron collision time*,

$$\tau_{ie} = \frac{6\sqrt{2}\,\pi^{3/2}\,\epsilon_0^2\,m_i\,T^{3/2}}{\ln\Lambda_c\,e^4\,n_e\,m_e^{1/2}}, \tag{3.185}$$

which is the mean time required for the direction of motion of an individual ion to deviate through approximately 90° as a consequence of collisions with electrons. Note that these collision times are not all of the same magnitude, as a consequence of the large difference between the electron and ion masses. In fact,

$$\tau_{ee} \sim \tau_{ei} \sim (m_e/m_i)^{1/2}\,\tau_{ii} \sim (m_i/m_e)\,\tau_{ie}, \tag{3.186}$$

which implies that electrons scatter electrons (through 90°) at about the same rate that ions scatter electrons, but that ions scatter ions at a significantly lower rate than ions scatter electrons, and, finally, that electrons scatter ions at a significantly lower rate than ions scatter ions.

The *collision frequency* is simply the inverse of the collision time. Thus, the electron-electron collision frequency is written

$$\nu_{ee} \equiv \frac{1}{\tau_{ei}} = \frac{\ln\Lambda_c\,e^4\,n_e}{12\sqrt{2}\,\pi^{3/2}\,\epsilon_0^2\,m_e^{1/2}\,T^{3/2}}. \tag{3.187}$$

Given that $\ln\Lambda_c \sim \ln\Lambda$ (see Section 3.10), where $\Lambda = 4\pi\,\epsilon_0^{3/2}\,T^{3/2}/(e^3\,n_e^{1/2})$ is the plasma parameter (see Section 1.6), we obtain the estimate (see Section 1.7)

$$\nu_{ee} \sim \frac{\ln\Lambda}{\Lambda}\,\Pi_e \tag{3.188}$$

where $\Pi_e = (n_e\,e^2/\epsilon_0\,m_e)^{1/2}$ is the electron plasma frequency (see Section 1.4). Likewise, the ion-ion collision frequency is such that

$$\nu_{ii} \equiv \frac{1}{\tau_{ii}} \sim \frac{\ln\Lambda}{\Lambda}\,\Pi_i, \tag{3.189}$$

where $\Pi_i = (n_i\,e^2/\epsilon_0\,m_i)^{1/2}$ is the ion plasma frequency.

3.13 Exercises

3.1 Consider the Maxwellian distribution

$$f(\mathbf{v}) = n\left(\frac{m}{2\pi\,T}\right)^{3/2}\,\exp\left[-\frac{m\,(\mathbf{v}-\mathbf{V})^2}{2\,T}\right].$$

Demonstrate that

$$n = \int f\, d^3\mathbf{v},$$

$$n\,\mathbf{V} = \int \mathbf{v}\, f\, d^3\mathbf{v},$$

$$\frac{3}{2}\,n\,T = \int \frac{1}{2}\,m\,v^2\, f\, d^3\mathbf{v}.$$

3.2 The species-s entropy per unit volume is conventionally defined as

$$s_s = -\int f_s\, \ln f_s\, d^3\mathbf{v}_s.$$

The Boltzmann H-theorem thus states that collisions drive the system toward a maximum entropy state characterized by Maxwellian distribution functions with common mean velocities and common temperatures. Demonstrate that for a Maxwellian distribution,

$$f_s = n_s\left(\frac{m_s}{2\pi\,T_s}\right)^{3/2}\exp\left(-\frac{m\,v_s^2}{2\,T_s}\right),$$

the entropy per unit volume takes the form

$$s_s = n_s\left[\ln\left(\frac{T_s^{3/2}}{n_s}\right) + \frac{3}{2}\,\ln\left(\frac{2\pi}{m_s}\right) + \frac{3}{2}\right].$$

3.3 The Landau collision operator is written

$$C_{12}(f_1, f_2) = \frac{\gamma_{12}}{m_1}\,\frac{\partial}{\partial\mathbf{v}_1}\cdot\int \mathbf{w}_{12}\cdot\mathbf{J}_{12}\, d^3\mathbf{v}_2,$$

where

$$\gamma_{12} = \left(\frac{e_1\, e_2}{4\pi\,\epsilon_0}\right)^2 2\pi\,\ln\Lambda_c,$$

$$\mathbf{w}_{12} = \frac{u_{12}^2\,\mathbf{I} - \mathbf{u}_{12}\mathbf{u}_{12}}{u_{12}^3},$$

$$u_{12} = |\mathbf{v}_1 - \mathbf{v}_2|,$$

$$\mathbf{J}_{12} = \frac{\partial f_1}{\partial\mathbf{v}_1}\,\frac{f_2}{m_1} - \frac{f_1}{m_2}\,\frac{\partial f_2}{\partial\mathbf{v}_2}.$$

Demonstrate directly that this collision operator satisfies the same conservation laws as the Boltzmann collision operator. Namely,

$$\int C_{12}\, d^3\mathbf{v}_1 = 0,$$

$$\int m_1\, \mathbf{v}_1\, C_{12}\, d^3\mathbf{v}_1 = -\int m_2\, \mathbf{v}_2\, C_{21}\, d^3\mathbf{v}_2,$$

$$\int \frac{1}{2}\, m_1\, v_1^2\, C_{12}\, d^3\mathbf{v}_1 = -\int \frac{1}{2}\, m_2\, v_2^2\, C_{21}\, d^3\mathbf{v}_2.$$

3.4 The net heating rate per unit volume of type 1 particles due to Coulomb collisions with type 2 particles is

$$W_{12} = \int \frac{1}{2}\, m_1\, v^2\, C_{12}\, d^3\mathbf{v},$$

where C_{12} is the Landau collision operator. Suppose that both species have Maxwellian distribution functions with zero mean velocities:

$$f_1 = n_1 \left(\frac{m_1}{2\pi\, T_1}\right)^{3/2} \exp\left(-\frac{m_1\, v_1^2}{2\, T_1}\right),$$

$$f_2 = n_2 \left(\frac{m_2}{2\pi\, T_2}\right)^{3/2} \exp\left(-\frac{m_1\, v_2^2}{2\, T_2}\right).$$

Suppose, further, that the kinetic temperatures of the two species are almost the same (i.e., $T_1 \simeq T_2$). Demonstrate that

$$W_{12} = 3\, \frac{\mu_{12}}{m_2}\, \frac{n_1}{\tau_{12}}\, (T_2 - T_1),$$

where the collision time, τ_{12}, is defined in Equation (3.181). In particular, show that in an electron-ion plasma

$$W_{ei} \simeq 3\, \frac{m_e}{m_i}\, \frac{n_e}{\tau_{ei}}\, (T_i - T_e).$$

4

Plasma Fluid Theory

4.1 Introduction

In plasma fluid theory, a plasma is characterized by a few local parameters—such as the particle density, the kinetic temperature, and the flow velocity—the time evolutions of which are determined by means of fluid equations. These equations are analogous to, but generally more complicated than, the equations of gas dynamics.

Fluid equations are conventionally obtained by taking velocity space moments of the kinetic equation (see Section 3.2),

$$\frac{\partial f_s}{\partial t} + \mathbf{v}\cdot\nabla f_s + \mathbf{a}_s\cdot\nabla_v f_s = C_s(f). \tag{4.1}$$

Here, $\nabla \equiv \partial/\partial\mathbf{r}$, $\nabla_v \equiv \partial/\partial\mathbf{v}$, and

$$\mathbf{a}_s = \frac{e_s}{m_s}\,(\mathbf{E} + \mathbf{v}\times\mathbf{B}). \tag{4.2}$$

Furthermore, e_s and m_s are the species-s electrical charge and mass, respectively, whereas $\mathbf{E}$ and $\mathbf{B}$ are the ensemble-averaged electromagnetic fields.

In general, it is extremely difficult to solve the kinetic equation directly, because of the complexity of the collision operator. However, there are some situations in which collisions can be completely neglected. In such cases, the kinetic equation simplifies to give the so-called *Vlasov equation*,

$$\frac{\partial f_s}{\partial t} + \mathbf{v}\cdot\nabla f_s + \mathbf{a}_s\cdot\nabla_v f_s = 0. \tag{4.3}$$

The Vlasov equation is tractable in sufficiently simple geometry. (See Chapter 8.) Nevertheless, the fluid approach possesses significant advantages, even in the Vlasov limit. These advantages are as follows.

First, fluid equations involve fewer dimensions than the Vlasov equation. That is, three spatial dimensions instead of six phase-space dimensions. This advantage is especially important in computer simulations.

Second, the fluid description is intuitively appealing. We immediately understand the significance of fluid quantities such as density and temperature, whereas the significance of distribution functions is far less obvious. Moreover, fluid variables are relatively easy to measure in experiments, whereas, in most cases, it is extraordinarily difficult to measure a distribution function accurately. There seems remarkably little point in centering our theoretical description of plasmas on something that we cannot generally measure.

Finally, the kinetic approach to plasma physics is spectacularly inefficient. The species distribution functions f_s provide vastly more information than is needed to obtain the constitutive relations [i.e., Equations (3.1) and (3.2)] that close Maxwell's equations. (See Section 3.2.) After all, these relations only depend on the two lowest moments of the species distribution functions.

4.2 Moments of Distribution Function

The kth velocity space moment of the (ensemble-averaged) distribution function $f_s(\mathbf{r}, \mathbf{v}, t)$ is written

$$\mathbf{M}_k(\mathbf{r}, t) = \int \mathbf{v}\mathbf{v}\cdots\mathbf{v}\, f_s(\mathbf{r}, \mathbf{v}, t)\, d^3\mathbf{v}, \tag{4.4}$$

with k factors of $\mathbf{v}$. Clearly, $\mathbf{M}_k$ is a tensor of rank k (Riley 1974).

The set $\mathbf{M}_k$, for $k = 0, 1, 2, \cdots$, can be viewed as an alternative description of the distribution function that uniquely specifies f_s when the latter is sufficiently smooth. For example, a (displaced) Gaussian distribution function is uniquely specified by three moments: M_0, the vector $\mathbf{M}_1$, and the scalar formed by contracting $\mathbf{M}_2$.

The low-order moments all have simple physical interpretations. First, we have the particle *number density*,

$$n_s(\mathbf{r}, t) = \int f_s(\mathbf{r}, \mathbf{v}, t)\, d^3\mathbf{v}, \tag{4.5}$$

and the particle *flux density*,

$$n_s\, \mathbf{V}_s(\mathbf{r}, t) = \int \mathbf{v}\, f_s(\mathbf{r}, \mathbf{v}, t)\, d^3\mathbf{v}. \tag{4.6}$$

The quantity $\mathbf{V}_s$ is, of course, the *flow velocity*. The constitutive relations, (3.1) and (3.2), are determined by these lowest moments. In fact,

$$\rho = \sum_s e_s\, n_s, \tag{4.7}$$

$$\mathbf{j} = \sum_s e_s\, n_s\, \mathbf{V}_s. \tag{4.8}$$

The second-order moment, describing the flow of momentum in the laboratory frame, is called the *stress tensor*, and takes the form

$$\mathbf{P}_s(\mathbf{r}, t) = \int m_s\, \mathbf{v}\mathbf{v}\, f_s(\mathbf{r}, \mathbf{v}, t)\, d^3\mathbf{v}. \tag{4.9}$$

Finally, there is an important third-order moment measuring the *energy flux density*,

$$\mathbf{Q}_s(\mathbf{r}, t) = \int \frac{1}{2} m_s v^2 \mathbf{v} f_s(\mathbf{r}, \mathbf{v}, t) d^3\mathbf{v}. \tag{4.10}$$

It is often convenient to measure the second- and third-order moments in the rest-frame of the species under consideration. In this case, the moments have different names. The stress tensor measured in the rest-frame is called the *pressure tensor*, $\mathbf{p}_s$, whereas the energy flux density becomes the *heat flux density*, $\mathbf{q}_s$. We introduce the relative velocity,

$$\mathbf{w}_s \equiv \mathbf{v} - \mathbf{V}_s, \tag{4.11}$$

in order to write

$$\mathbf{p}_s(\mathbf{r}, t) = \int m_s \mathbf{w}_s \mathbf{w}_s f_s(\mathbf{r}, \mathbf{v}, t) d^3\mathbf{v}, \tag{4.12}$$

and

$$\mathbf{q}_s(\mathbf{r}, t) = \int \frac{1}{2} m_s w_s^2 \mathbf{w}_s f_s(\mathbf{r}, \mathbf{v}, t) d^3\mathbf{v}. \tag{4.13}$$

The trace of the pressure tensor measures the ordinary (or scalar) pressure,

$$p_s \equiv \frac{1}{3} \mathrm{Tr}(\mathbf{p}_s). \tag{4.14}$$

In fact, $(3/2) p_s$ is the kinetic energy density of species s: that is,

$$\frac{3}{2} p_s = \int \frac{1}{2} m_s w_s^2 f_s d^3\mathbf{v}. \tag{4.15}$$

In thermodynamic equilibrium, the distribution function becomes a Maxwellian characterized by some temperature T, and Equation (4.15) yields $p = n T$. It is, therefore, natural to define the (kinetic) temperature as

$$T_s \equiv \frac{p_s}{n_s}. \tag{4.16}$$

Of course, the moments measured in the two different frames are related. By direct substitution, it is easily verified that

$$\mathbf{P}_s = \mathbf{p}_s + m_s n_s \mathbf{V}_s \mathbf{V}_s, \tag{4.17}$$

$$\mathbf{Q}_s = \mathbf{q}_s + \mathbf{p}_s \cdot \mathbf{V}_s + \frac{3}{2} p_s \mathbf{V}_s + \frac{1}{2} m_s n_s V_s^2 \mathbf{V}_s. \tag{4.18}$$

4.3 Moments of Collision Operator

Boltzmann's collision operator for a neutral gas considers only binary collisions, and is, therefore, bilinear in the distribution functions of the two colliding species. (See Section 3.4.) In other words,

$$C_s(f) = \sum_{s'} C_{ss'}(f_s, f_{s'}), \tag{4.19}$$

where $C_{ss'}$ is linear in each of its arguments. Unfortunately, such bilinearity is not strictly valid for the case of Coulomb collisions in a plasma. Because of the long-range nature of the Coulomb interaction, the closest analogue to ordinary two-particle interaction is modified by Debye shielding, which is an intrinsically many-body effect. Fortunately, the departure from bilinearity is logarithmic in a weakly coupled plasma, and can, therefore, be neglected to a fairly good approximation (because a logarithm is a comparatively weakly varying function). (See Section 3.10.) Thus, from now on, $C_{ss'}$ is presumed to be bilinear.

It is important to realize that there is no simple relationship between the quantity $C_{ss'}$, which describes the effect on species s of collisions with species s', and the quantity $C_{s's}$. The two operators can have quite distinct mathematical forms (for example, where the masses m_s and $m_{s'}$ are significantly different), and they do not appear in the same equations.

Neutral particle collisions are characterized by Boltzmann's collisional conservation laws. (See Section 3.5.) In fact, the collisional process conserves particles, momentum, and energy at each point in space. We expect the same local conservation laws to hold for Coulomb collisions in a plasma, because the maximum range of the Coulomb force in a plasma is the Debye length, which is assumed to be vanishingly small.

Collisional particle conservation is expressed as

$$\int C_{ss'}\, d^3\mathbf{v} = 0. \tag{4.20}$$

Collisional momentum conservation requires that

$$\int m_s\, \mathbf{v}\, C_{ss'}\, d^3\mathbf{v} = -\int m_{s'}\, \mathbf{v}\, C_{s's}\, d^3\mathbf{v}. \tag{4.21}$$

In other words, there is zero net momentum exchanged between species s and s'. It is useful to introduce the rate of collisional momentum exchange, which is called the collisional friction force, or simply the *friction force*:

$$\mathbf{F}_{ss'} \equiv \int m_s\, \mathbf{v}\, C_{ss'}\, d^3\mathbf{v}. \tag{4.22}$$

Clearly, $\mathbf{F}_{ss'}$ is the momentum-moment of the collision operator. The total friction force experienced by species s is

$$\mathbf{F}_s \equiv \sum_{s'} \mathbf{F}_{ss'}. \tag{4.23}$$

Momentum conservation is expressed in detailed form as

$$\mathbf{F}_{ss'} = -\mathbf{F}_{s's}, \tag{4.24}$$

and in non-detailed form as

$$\sum_s \mathbf{F}_s = \mathbf{0}. \tag{4.25}$$

Collisional energy conservation requires the quantity

$$W_{Lss'} \equiv \int \frac{1}{2} m_s v^2 C_{ss'} d^3\mathbf{v} \tag{4.26}$$

to be conserved in collisions. In other words,

$$W_{Lss'} + W_{Ls's} = 0. \tag{4.27}$$

Here, the L-subscript indicates that the kinetic energy of both species is measured in the same laboratory frame. Because of Galilean invariance, the choice of this common reference frame does not matter.

An alternative collisional energy-moment is

$$W_{ss'} \equiv \int \frac{1}{2} m_s w_s^2 C_{ss'} d^3\mathbf{v}. \tag{4.28}$$

This is the kinetic energy change experienced by species s, due to collisions with species s', measured in the rest frame of species s. The total energy change for species s is

$$W_s \equiv \sum_{s'} W_{ss'}. \tag{4.29}$$

It is easily verified that

$$W_{Lss'} = W_{ss'} + \mathbf{V}_s \cdot \mathbf{F}_{ss'}. \tag{4.30}$$

Thus, the collisional energy conservation law can be written in detailed form as

$$W_{ss'} + W_{s's} + (\mathbf{V}_s - \mathbf{V}_{s'}) \cdot \mathbf{F}_{ss'} = 0, \tag{4.31}$$

or in non-detailed form as

$$\sum_s (W_s + \mathbf{V}_s \cdot \mathbf{F}_s) = 0. \tag{4.32}$$

4.4 Moments of Kinetic Equation

We obtain fluid equations by taking appropriate moments of the kinetic equation, Equation (4.1). It is convenient to rearrange the acceleration term as follows:

$$\mathbf{a}_s \cdot \nabla_v f_s = \nabla_v \cdot (\mathbf{a}_s f_s). \tag{4.33}$$

The two forms are equivalent because flow in velocity space under the Lorentz force is incompressible: that is,

$$\nabla_v \cdot \mathbf{a}_s = 0. \tag{4.34}$$

Thus, Equation (4.1) becomes

$$\frac{\partial f_s}{\partial t} + \nabla \cdot (\mathbf{v} f_s) + \nabla_v \cdot (\mathbf{a}_s f_s) = C_s(f). \tag{4.35}$$

The rearrangement of the flow term is, of course, trivial, because **v** is independent of **r**.

The kth moment of the kinetic equation is obtained by multiplying the previous equation by k powers of **v**, and integrating over velocity space. The flow term is simplified by pulling the divergence outside the velocity integral. The acceleration term is treated by partial integration. These two terms couple the kth moment to the $(k+1)$ th and $(k-1)$ th moments, respectively.

Making use of the collisional conservation laws, the zeroth moment of Equation (4.35) yields the *continuity equation* for species s:

$$\frac{\partial n_s}{\partial t} + \nabla\cdot(n_s\,\mathbf{V}_s) = 0. \tag{4.36}$$

Likewise, the first moment gives the *momentum conservation equation* for species s:

$$\frac{\partial(m_s\,n_s\,\mathbf{V}_s)}{\partial t} + \nabla\cdot\mathbf{P}_s - e_s\,n_s\,(\mathbf{E}+\mathbf{V}_s\times\mathbf{B}) = \mathbf{F}_s. \tag{4.37}$$

Finally, the contracted second moment yields the *energy conservation equation* for species s:

$$\frac{\partial}{\partial t}\left(\frac{3}{2}\,p_s + \frac{1}{2}\,m_s\,n_s\,V_s^2\right) + \nabla\cdot\mathbf{Q}_s - e_s\,n_s\,\mathbf{E}\cdot\mathbf{V}_s = W_s + \mathbf{V}_s\cdot\mathbf{F}_s. \tag{4.38}$$

The interpretation of Equations (4.36)–(4.38) as conservation laws is straightforward. Suppose that G is some physical quantity (for instance, the total number of particles, the total energy, and so on), and $g(\mathbf{r}, t)$ is its density:

$$G = \int g\,d^3\mathbf{r}. \tag{4.39}$$

If G is conserved then g must evolve according to

$$\frac{\partial g}{\partial t} + \nabla\cdot\mathbf{g} = \Delta g, \tag{4.40}$$

where **g** is the flux density of G, and Δg is the local rate per unit volume at which G is created, or exchanged with other entities in the fluid. According to the previous equation, the density of G at some point changes because there is net flow of G towards or away from that point (characterized by the divergence term), or because of local sources or sinks of G (characterized by the right-hand side).

Applying this reasoning to Equation (4.36), we see that $n_s\,\mathbf{V}_s$ is indeed the species-s particle flux density, and that there are no local sources or sinks of species-s particles.[1] From Equation (4.37), it is apparent that the stress tensor, $\mathbf{P}_s$, is the species-s momentum flux density, and that the species-s momentum is changed locally by the

[1] In general, this is not true. Atomic or nuclear processes operating in a plasma can give rise to local sources and sinks of particles of various species. However, if a plasma is sufficiently hot to be completely ionized, but still cold enough to prevent nuclear reactions from occurring, then such sources and sinks are usually negligible.

Lorentz force, and by collisional friction with other species. Finally, from Equation (4.38), we see that $\mathbf{Q}_s$ is indeed the species-s energy flux density, and that the species-s energy is changed locally by electrical work, energy exchange with other species, and frictional heating.

4.5 Fluid Equations

It is conventional to rewrite our fluid equations in terms of the pressure tensor, $\mathbf{p}_s$, and the heat flux density, $\mathbf{q}_s$. Substituting from Equations (4.17) and (4.18), and performing a little tensor algebra, Equations (4.36)–(4.38) reduce to:

$$\frac{dn_s}{dt} + n_s \nabla\cdot\mathbf{V}_s = 0, \tag{4.41}$$

$$m_s n_s \frac{d\mathbf{V}_s}{dt} + \nabla\cdot\mathbf{p}_s - e_s n_s (\mathbf{E} + \mathbf{V}_s\times\mathbf{B}) = \mathbf{F}_s, \tag{4.42}$$

$$\frac{3}{2}\frac{dp_s}{dt} + \frac{3}{2} p_s \nabla\cdot\mathbf{V}_s + \mathbf{p}_s : \nabla\mathbf{V}_s + \nabla\cdot\mathbf{q}_s = W_s. \tag{4.43}$$

Here,

$$\frac{d}{dt} \equiv \frac{\partial}{\partial t} + \mathbf{V_s}\cdot\nabla \tag{4.44}$$

is the well-known convective derivative, and

$$\mathbf{p} : \nabla\mathbf{V}_s \equiv (p_s)_{\alpha\beta}\frac{\partial(V_s)_\beta}{\partial r_\alpha}. \tag{4.45}$$

In the previous expression, α and β refer to Cartesian components, and repeated indices are summed (in accordance with the Einstein summation convention) (Riley 1974). The convective derivative, of course, measures time variation in the local rest frame of the species-s fluid. Strictly speaking, we should include an s subscript with each convective derivative, because this operator is clearly different for different plasma species.

There is one additional refinement to our fluid equations that is worth carrying out. We introduce the *generalized viscosity tensor*, $\boldsymbol{\pi}_s$, by writing

$$\mathbf{p}_s = p_s\,\mathbf{I} + \boldsymbol{\pi}_s, \tag{4.46}$$

where $\mathbf{I}$ is the unit (identity) tensor. We expect the scalar pressure term to dominate if the plasma is relatively close to thermal equilibrium. We also expect, by analogy with conventional fluid theory, the second term to describe viscous stresses. Indeed, this is generally the case in plasmas, although the generalized viscosity tensor can also include terms that are quite unrelated to conventional viscosity. Equations (4.41)–

(4.43) can, thus, be rewritten:

$$\frac{dn_s}{dt} + n_s \nabla\cdot\mathbf{V}_s = 0, \tag{4.47}$$

$$m_s n_s \frac{d\mathbf{V}_s}{dt} + \nabla p_s + \nabla\cdot\boldsymbol{\pi}_s - e_s n_s (\mathbf{E} + \mathbf{V}_s \times \mathbf{B}) = \mathbf{F}_s, \tag{4.48}$$

$$\frac{3}{2}\frac{dp_s}{dt} + \frac{5}{2} p_s \nabla\cdot\mathbf{V}_s + \boldsymbol{\pi}_s : \nabla\mathbf{V}_s + \nabla\cdot\mathbf{q}_s = W_s. \tag{4.49}$$

According to Equation (4.47), the species-s density is constant along a fluid trajectory unless the species-s flow is non-solenoidal. For this reason, the condition

$$\nabla\cdot\mathbf{V}_s = 0 \tag{4.50}$$

is said to describe incompressible species-s flow. According to Equation (4.48), the species-s flow accelerates along a fluid trajectory under the influence of the scalar pressure gradient, the viscous stresses, the Lorentz force, and the frictional force due to collisions with other species. Finally, according to Equation (4.49), the species-s energy density (that is, p_s) changes along a fluid trajectory because of the work done in compressing the fluid, viscous heating, heat flow, and the local energy gain due to collisions with other species. The electrical contribution to plasma heating, which was explicit in Equation (4.38), has now become entirely implicit.

4.6 Entropy Production

It is instructive to rewrite the species-s energy evolution equation, Equation (4.49), as an entropy evolution equation (Hazeltine and Waelbroeck 2004). The fluid definition of *entropy density*, which coincides with the thermodynamic entropy density in the limit that the distribution function approaches a Maxwellian, is (Reif 1965)

$$s_s = n_s \ln\left(\frac{T_s^{3/2}}{n_s}\right) + c, \tag{4.51}$$

where c is a constant. The corresponding *entropy flux density* is written

$$\mathbf{s}_s = s_s\,\mathbf{V}_s + \frac{\mathbf{q}_s}{T_s}. \tag{4.52}$$

Clearly, entropy is convected by the fluid flow, but is also carried by the flow of heat, in accordance with the second law of thermodynamics (Reif 1965). After some algebra, Equation (4.49) can be rearranged to give

$$\frac{\partial s_s}{\partial t} + \nabla\cdot\mathbf{s}_s = \Theta_s, \tag{4.53}$$

where the right-hand side is given by

$$\Theta_s = \frac{W_s}{T_s} - \frac{\boldsymbol{\pi}_s : \nabla \mathbf{V}_s}{T_s} - \frac{\mathbf{q}_s}{T_s} \cdot \frac{\nabla T_s}{T_s}. \tag{4.54}$$

It follows, from our previous discussion of conservation laws, that the quantity Θ_s can be regarded as the *entropy production rate* per unit volume for species s. Evidently, entropy is produced by collisional heating, viscous heating, and heat flow down temperature gradients.

4.7 Fluid Closure

No amount of manipulation, or rearrangement, can cure our fluid equations of their most serious defect—the fact that they are incomplete. In their present form, which is specified in Equations (4.47)–(4.49), our equations relate interesting fluid quantities, such as the particle number density, n_s, the flow velocity, $\mathbf{V}_s$, and the scalar pressure, p_s, to unknown quantities, such as the viscosity tensor, $\boldsymbol{\pi}_s$, the heat flux density, $\mathbf{q}_s$, and the moments of the collision operator, $\mathbf{F}_s$ and W_s. In order to complete our set of equations, we need to use some additional information to express the latter quantities in terms of the former. This process is known as *closure*.

Lack of closure is an endemic problem in fluid theory. Because each moment is coupled to the next higher moment (for instance, the density evolution depends on the flow velocity, the flow velocity evolution depends on the viscosity tensor, and so on), any finite set of exact moment equations is bound to contain more unknowns than equations.

There are two basic types of fluid closure schemes. In *truncation schemes*, higher order moments are arbitrarily assumed to vanish, or simply prescribed in terms of lower moments. Truncation schemes can often provide quick insight into fluid systems, but always involve uncontrolled approximation. *Asymptotic schemes*, on the other hand, depend on the rigorous exploitation of some small parameter. Asymptotic closure schemes have the advantage of being systematic, and providing some estimate of the error involved in the closure. On the other hand, the asymptotic approach to closure is mathematically demanding, because it inevitably involves working with the kinetic equation.

4.8 Chapman-Enskog Closure

The classic example of an asymptotic closure scheme is the Chapman-Enskog theory of a neutral gas dominated by collisions. In this theory, the small parameter is the ratio of the mean-free-path between collisions to the macroscopic variation lengthscale. It is instructive to briefly examine this theory, which is very well described in a classic monograph by Chapman and Cowling (Chapman and Cowling 1953).

Consider a neutral gas consisting of identical hard-sphere molecules of mass m and diameter σ. Admittedly, this is not a particularly physical model of a neutral gas, but we are only considering it for illustrative purposes. The fluid equations for such a gas are similar to Equations (4.47)–(4.49):

$$\frac{dn}{dt} + n\,\nabla\cdot\mathbf{V} = 0, \tag{4.55}$$

$$m\,n\,\frac{d\mathbf{V}}{dt} + \nabla p + \nabla\cdot\boldsymbol{\pi} + m\,n\,\mathbf{g} = \mathbf{0}, \tag{4.56}$$

$$\frac{3}{2}\frac{dp}{dt} + \frac{5}{2}\,p\,\nabla\cdot\mathbf{V} + \boldsymbol{\pi} : \nabla\mathbf{V} + \nabla\cdot\mathbf{q} = 0. \tag{4.57}$$

Here, n is the particle number density, $\mathbf{V}$ the flow velocity, p the scalar pressure, and $\mathbf{g}$ the acceleration due to gravity. We have dropped the subscript s because, in this case, there is only a single species. There is no collisional friction or heating in a single species system. Of course, there are no electrical or magnetic forces in a neutral gas, so we have included gravitational forces instead. The purpose of the closure scheme is to express the viscosity tensor, $\boldsymbol{\pi}$, and the heat flux density, $\mathbf{q}$, in terms of n, $\mathbf{V}$, or p, and, thereby, complete the set of equations.

The mean-free-path, l, for hard-sphere molecules is given by

$$l = \frac{1}{\sqrt{2}\,\pi\,n\,\sigma^2}. \tag{4.58}$$

This formula is fairly easy to understand. The volume swept out by a given molecule in moving a mean-free-path must contain, on average, approximately one other molecule. Observe that l is completely independent of the speed or mass of the molecules. The mean-free-path is assumed to be much smaller than the variation lengthscale, L, of macroscopic quantities, so that

$$\epsilon = \frac{l}{L} \ll 1. \tag{4.59}$$

In the Chapman-Enskog scheme, the distribution function is expanded, order by order, in the small parameter ϵ:

$$f(\mathbf{r},\mathbf{v},t) = f_0(\mathbf{r},\mathbf{v},t) + \epsilon\,f_1(\mathbf{r},\mathbf{v},t) + \epsilon^2\,f_2(\mathbf{r},\mathbf{v},t) + \cdots. \tag{4.60}$$

Here, f_0, f_1, f_2, and so on, are all assumed to be of the same order of magnitude. In fact, only the first two terms in this expansion are ever calculated. To zeroth order in ϵ, the kinetic equation requires that f_0 be a Maxwellian:

$$f_0(\mathbf{r},\mathbf{v},t) = n(\mathbf{r})\left[\frac{m}{2\pi\,T(\mathbf{r})}\right]^{3/2}\exp\left[-\frac{m\,(\mathbf{v}-\mathbf{V})^2}{2\,T(\mathbf{r})}\right]. \tag{4.61}$$

Recall that $p = n\,T$. As is well known, there is zero heat flow or viscous stress associated with a Maxwellian distribution function (Reif 1965). Thus, both the heat flux density, $\mathbf{q}$, and the viscosity tensor, $\boldsymbol{\pi}$, depend on the first-order non-Maxwellian correction to the distribution function, f_1.

It is possible to linearize the kinetic equation, and then rearrange it so as to obtain an integral equation for f_1 in terms of f_0. This rearrangement crucially depends on the bilinearity of the collision operator. Incidentally, the equation is integral because the collision operator is an integral operator. The integral equation is solved by expanding f_1 in velocity space using Laguerre polynomials (sometimes called Sonine polynomials). It is possible to reduce the integral equation to an infinite set of simultaneous algebraic equations for the coefficients in this expansion. If the expansion is truncated, after N terms, say, then these algebraic equations can be solved for the coefficients. It turns out that the Laguerre polynomial expansion converges very rapidly. Thus, it is conventional to keep only the first two terms in this expansion, which is usually sufficient to ensure an accuracy of about 1 percent in the final result. Finally, the appropriate moments of f_1 are taken, so as to obtain expression for the heat flux density and the viscosity tensor. Strictly speaking, after evaluating f_1, we should then go on to evaluate f_2, so as to ensure that f_2 really is negligible compared to f_1. In reality, this is never done because the mathematical difficulties involved in such a calculation are prohibitive.

The Chapman-Enskog method outlined previously can be applied to any assumed force law between molecules, provided that the force is sufficiently short-range (i.e., provided that it falls off faster with increasing separation than the Coulomb force). For all sensible force laws, the viscosity tensor is given by

$$\pi_{\alpha\beta} = -\eta\left(\frac{\partial V_\alpha}{\partial r_\beta} + \frac{\partial V_\beta}{\partial r_\alpha} - \frac{2}{3}\,\nabla\cdot\mathbf{V}\,\delta_{\alpha\beta}\right), \tag{4.62}$$

whereas the heat flux density takes the form

$$\mathbf{q} = -\kappa\,\nabla T. \tag{4.63}$$

Here, η is the *coefficient of viscosity*, and κ is the *coefficient of thermal conductivity*. It is convenient to write

$$\eta = m\,n\,\chi_v, \tag{4.64}$$

$$\kappa = n\,\chi_t, \tag{4.65}$$

where χ_v is the *viscous diffusivity* and χ_t is the *thermal diffusivity*. Both χ_v and χ_t have the dimensions of length squared over time, and are, effectively, diffusion coefficients. For the special case of hard-sphere molecules, Chapman-Enskog theory yields (Chapman and Cowling 1953):

$$\chi_v = \frac{75\pi^{1/2}}{64}\left(1 + \frac{3}{202} + \cdots\right)\nu\,l^2 = A_v\,\nu\,l^2, \tag{4.66}$$

$$\chi_t = \frac{5\pi^{1/2}}{16}\left(1 + \frac{1}{44} + \cdots\right)\nu\,l^2 = A_t\,\nu\,l^2. \tag{4.67}$$

Here,

$$\nu = \frac{v_t}{l} \tag{4.68}$$

is the collision frequency, and

$$v_t = \sqrt{\frac{2\,T}{m}} \tag{4.69}$$

is the thermal velocity. The first two terms in the Laguerre polynomial expansion are shown explicitly (in the round brackets) in Equations (4.66) and (4.67).

Equations (4.66) and (4.67) have a simple physical interpretation. The viscous and thermal diffusivities of a neutral gas can be accounted for in terms of the random-walk diffusion of molecules with excess momentum and energy, respectively. Recall the standard result in stochastic theory that if particles jump an average distance l, in a random direction, ν times a second, then the diffusivity associated with such motion is $\chi \sim \nu\, l^2$ (Reif 1965). Chapman-Enskog theory basically allows us to calculate the numerical constants A_v and A_t, multiplying $\nu\, l^2$ in the expressions for χ_v and χ_t, for a given force law between molecules. Obviously, these coefficients are different for different force laws. The expression for the mean-free-path, l, is also different for different force laws.

4.9 Normalization of Neutral Gas Equations

Let $\bar{n}$, $\bar{v}_t$, and $\bar{l}$ be typical values of the particle density, the thermal velocity, and the mean-free-path, respectively. Suppose that the typical flow velocity is $\lambda\,\bar{v}_t$, and the typical variation lengthscale of macroscopic quantities is L. Let us define the following normalized quantities:

$$\hat{n} = \frac{n}{\bar{n}}, \qquad \hat{v}_t = \frac{v_t}{\bar{v}_t}, \qquad \hat{l} = \frac{l}{\bar{l}}, \qquad \hat{\mathbf{r}} = \frac{\mathbf{r}}{L},$$

$$\widehat{\nabla} = L\,\nabla, \qquad \hat{t} = \frac{\lambda\,\bar{v}_t\,t}{L}, \qquad \widehat{\mathbf{V}} = \frac{\mathbf{V}}{\lambda\,\bar{v}_t}, \qquad \widehat{T} = \frac{T}{m\,\bar{v}_t^{\,2}},$$

$$\hat{\mathbf{g}} = \frac{L\,\mathbf{g}}{(1+\lambda^2)\,\bar{v}_t^{\,2}}, \qquad \hat{p} = \frac{p}{m\,\bar{n}\,\bar{v}_t^{\,2}}, \qquad \widehat{\boldsymbol{\pi}} = \frac{\boldsymbol{\pi}}{\lambda\,\epsilon\,m\,\bar{n}\,\bar{v}_t^{\,2}}, \qquad \widehat{\mathbf{q}} = \frac{\mathbf{q}}{\epsilon\,m\,\bar{n}\,\bar{v}_t^{\,3}}.$$

Here,

$$\epsilon = \frac{\bar{l}}{L} \ll 1. \tag{4.70}$$

Note that

$$\widehat{\pi}_{\alpha\beta} = -A_v\,\hat{n}\;\hat{v}_t\,\hat{l}\left(\frac{\partial \widehat{V}_\alpha}{\partial\,\hat{r}_\beta} + \frac{\partial \widehat{V}_\beta}{\partial\,\hat{r}_\alpha} - \frac{2}{3}\,\widehat{\nabla}\cdot\widehat{\mathbf{V}}\,\delta_{\alpha\beta}\right), \tag{4.71}$$

$$\widehat{\mathbf{q}} = -A_t\,\hat{n}\;\hat{v}_t\,\hat{l}\;\widehat{\nabla}\widehat{T}. \tag{4.72}$$

All hatted quantities are designed to be $O(1)$. The normalized fluid equations are written:

$$\frac{d\hat{n}}{d\hat{t}} + \hat{n}\,\widehat{\nabla}\cdot\widehat{\mathbf{V}} = 0, \tag{4.73}$$

$$\lambda^2\,\hat{n}\,\frac{d\widehat{\mathbf{V}}}{d\hat{t}} + \widehat{\nabla}\hat{p} + \lambda\,\epsilon\,\widehat{\nabla}\cdot\widehat{\boldsymbol{\pi}} + (1+\lambda^2)\,\hat{n}\,\hat{\mathbf{g}} = \mathbf{0}, \tag{4.74}$$

$$\lambda\,\frac{3}{2}\frac{d\hat{p}}{d\hat{t}} + \lambda\,\frac{5}{2}\,\hat{p}\,\widehat{\nabla}\cdot\widehat{\mathbf{V}} + \lambda^2\,\epsilon\,\widehat{\boldsymbol{\pi}} : \widehat{\nabla\mathbf{V}} + \epsilon\,\widehat{\nabla}\cdot\widehat{\mathbf{q}} = 0, \tag{4.75}$$

where

$$\frac{d}{d\hat{t}} \equiv \frac{\partial}{\partial\hat{t}} + \widehat{\mathbf{V}}\cdot\widehat{\nabla}. \tag{4.76}$$

The only large or small quantities remaining in the previous equations are the parameters λ and ϵ.

Suppose that $\lambda \gg 1$. In other words, suppose that the flow velocity is much greater than the thermal speed. Retaining only the largest terms in Equations (4.73)–(4.75), our system of fluid equations reduces to (in unnormalized form):

$$\frac{dn}{dt} + n\,\nabla\cdot\mathbf{V} = 0, \tag{4.77}$$

$$\frac{d\mathbf{V}}{dt} + \mathbf{g} \simeq \mathbf{0}. \tag{4.78}$$

These are called the *cold-gas equations*, because they can also be obtained by formally taking the limit $T \to 0$. The cold-gas equations describe externally driven, highly supersonic, gas dynamics. The gas pressure (that is, the thermal energy density) can be neglected in the cold-gas limit, because the thermal velocity is much smaller than the flow velocity. Consequently, there is no need for an energy evolution equation. Furthermore, the viscosity can also be neglected, because the viscous diffusion velocity is also far smaller than the flow velocity.

Suppose that $\lambda \sim O(1)$. In other words, suppose the flow velocity is of similar magnitude to the thermal speed. Again, retaining only the largest terms in Equations (4.73)–(4.75), our system of fluid equations reduces to (in unnormalized form):

$$\frac{dn}{dt} + n\,\nabla\cdot\mathbf{V} = 0, \tag{4.79}$$

$$m\,n\,\frac{d\mathbf{V}}{dt} + \nabla p + m\,n\,\mathbf{g} \simeq \mathbf{0}, \tag{4.80}$$

$$\frac{3}{2}\frac{dp}{dt} + \frac{5}{2}\,p\,\nabla\cdot\mathbf{V} \simeq 0. \tag{4.81}$$

The previous equations can be rearranged to give:

$$\frac{dn}{dt} + n\,\nabla\cdot\mathbf{V} = 0, \tag{4.82}$$

$$m\,n\,\frac{d\mathbf{V}}{dt} + \nabla p + m\,n\,\mathbf{g} \simeq \mathbf{0}, \tag{4.83}$$

$$\frac{d}{dt}\left(\frac{p}{n^{5/3}}\right) \simeq 0. \tag{4.84}$$

These are called the *hydrodynamic equations*, because they are similar to the equations governing the dynamics of water. The hydrodynamic equations govern relatively fast, internally driven, gas dynamics—in particular, the dynamics of sound waves. The gas pressure is non-negligible in the hydrodynamic limit, because the thermal velocity is similar in magnitude to the flow speed. Consequently, an energy evolution equation is needed. However, the energy equation takes a particularly simple form, as Equation (4.84) is immediately recognizable as the *adiabatic equation of state* for a monatomic gas. This is not surprising, because the flow velocity is still much faster than the viscous and thermal diffusion velocities (which accounts for the absence of viscosity and thermal conductivity in the hydrodynamic equations), in which case the gas acts effectively like a perfect thermal insulator.

Suppose, finally, that $\lambda \sim \epsilon$. In other words, suppose the flow velocity is of similar magnitude to the viscous and thermal diffusion velocities. Our system of fluid equations now reduces to a force balance criterion,

$$\nabla p + m\,n\,\mathbf{g} \simeq \mathbf{0}, \tag{4.85}$$

to lowest order. To next order, we obtain a set of equations describing the relatively slow viscous and thermal evolution of the gas:

$$\frac{dn}{dt} + n\,\nabla\cdot\mathbf{V} = 0, \tag{4.86}$$

$$m\,n\,\frac{d\mathbf{V}}{dt} + \nabla\cdot\boldsymbol{\pi} \simeq \mathbf{0}, \tag{4.87}$$

$$\frac{3}{2}\frac{dp}{dt} + \frac{5}{2}\,p\,\nabla\cdot\mathbf{V} + \nabla\cdot\mathbf{q} \simeq 0. \tag{4.88}$$

Clearly, this set of equations is only appropriate to relatively quiescent, quasi-equilibrium, gas dynamics. Virtually all of the terms in our original fluid equations, (4.55)–(4.57), must be retained in this limit.

The previous investigation reveals an important truth in gas dynamics, which also applies to plasma dynamics. Namely, the form of the fluid equations crucially depends on the typical fluid velocity associated with the type of dynamics under investigation. As a general rule, the equations get simpler as the typical velocity gets faster, and vice versa.

4.10 Braginskii Equations

Let now consider the problem of closure in plasma fluid equations. There are, in fact, two possible small parameters in plasmas upon which we could base an asymptotic closure scheme. The first is the ratio of the mean-free-path, l, to the macroscopic lengthscale, L. This is only appropriate to collisional plasmas. The second is the ratio of the Larmor radius, ρ, to the macroscopic lengthscale, L. This is only appropriate to magnetized plasmas. There is, of course, no small parameter upon which to base an asymptotic closure scheme in a collisionless, unmagnetized plasma. However, such systems occur predominately in accelerator physics contexts, and are not really plasmas at all, because they exhibit virtually no collective effects. Let us investigate Chapman-Enskog-like closure schemes in a collisional, quasi-neutral plasma consisting of equal numbers of electrons and ions. We shall treat the unmagnetized and magnetized cases separately.

The first step in our closure scheme is to approximate the actual collision operator for Coulomb interactions by an operator that is strictly bilinear in its arguments. (See Section 3.10.) Once this has been achieved, the closure problem is formally of the type that can be solved using the Chapman-Enskog method.

The electron-ion and ion-ion collision times are written

$$\tau_e = \frac{6\sqrt{2}\,\pi^{3/2}\,\epsilon_0^2\,\sqrt{m_e}\,T_e^{3/2}}{\ln\Lambda_c\,e^4\,n}, \tag{4.89}$$

and

$$\tau_i = \frac{12\,\pi^{3/2}\,\epsilon_0^2\,\sqrt{m_i}\,T_i^{3/2}}{\ln\Lambda_c\,e^4\,n}, \tag{4.90}$$

respectively. (See Section 3.12.) Here, $n = n_e = n_i$ is the number density of particles, and $\ln\Lambda_c$ is the Coulomb logarithm, whose origin is the slight modification to the collision operator mentioned previously. (See Section 3.10.)

The basic forms of Equations (4.89) and (4.90) are not hard to understand. From Equation (4.58), we expect

$$\tau \sim \frac{l}{v_t} \sim \frac{1}{n\,\sigma^2\,v_t}, \tag{4.91}$$

where σ^2 is the typical "cross-section" of the electrons or ions for Coulomb "collisions" (i.e., large angle scattering events). Of course, this cross-section is simply the square of the distance of closest approach, r_c, defined in Equation (1.17). Thus,

$$\tau \sim \frac{1}{n\,r_c^2\,v_t} \sim \frac{\epsilon_0^2\,\sqrt{m}\,T^{3/2}}{e^4\,n}. \tag{4.92}$$

The most significant feature of Equations (4.89) and (4.90) is the strong variation of the collision times with temperature. As the plasma gets hotter, the distance of closest approach gets smaller, so that both electrons and ions offer much smaller cross-sections for Coulomb collisions. The net result is that such collisions become far less frequent, and the collision times (i.e., the mean times between 90° degree scattering events) get much longer. It follows that as plasmas are heated they become less collisional very rapidly.

The electron and ion fluid equations in a collisional plasma take the form [see Equations (4.47)–(4.49)]:

$$\frac{dn}{dt} + n\,\nabla\cdot\mathbf{V}_e = 0, \tag{4.93}$$

$$m_e\,n\,\frac{d\mathbf{V}_e}{dt} + \nabla p_e + \nabla\cdot\boldsymbol{\pi}_e + e\,n\,(\mathbf{E} + \mathbf{V}_e\times\mathbf{B}) = \mathbf{F}, \tag{4.94}$$

$$\frac{3}{2}\frac{dp_e}{dt} + \frac{5}{2}\,p_e\,\nabla\cdot\mathbf{V}_e + \boldsymbol{\pi}_e : \nabla\mathbf{V}_e + \nabla\cdot\mathbf{q}_e = W_e, \tag{4.95}$$

and

$$\frac{dn}{dt} + n\,\nabla\cdot\mathbf{V}_i = 0, \tag{4.96}$$

$$m_i\,n\,\frac{d\mathbf{V}_i}{dt} + \nabla p_i + \nabla\cdot\boldsymbol{\pi}_i - e\,n\,(\mathbf{E} + \mathbf{V}_i\times\mathbf{B}) = -\mathbf{F}, \tag{4.97}$$

$$\frac{3}{2}\frac{dp_i}{dt} + \frac{5}{2}\,p_i\,\nabla\cdot\mathbf{V}_i + \boldsymbol{\pi}_i : \nabla\mathbf{V}_i + \nabla\cdot\mathbf{q}_i = W_i, \tag{4.98}$$

respectively. Here, use has been made of the momentum conservation law, Equation (4.25). Equations (4.93)–(4.95) and (4.96)–(4.98) are called the *Braginskii equations*, because they were first obtained in a celebrated article by S.I. Braginskii (Braginskii 1965).

In the unmagnetized limit, which actually corresponds to

$$\Omega_i\,\tau_i,\quad \Omega_e\,\tau_e \ll 1, \tag{4.99}$$

the standard two-Laguerre-polynomial Chapman-Enskog closure scheme yields

$$\mathbf{F} = \frac{n\,e}{\sigma_\parallel}\,\mathbf{j} - 0.71\,n\,\nabla T_e, \tag{4.100}$$

$$W_i = \frac{3\,m_e}{m_i}\frac{n\,(T_e - T_i)}{\tau_e}, \tag{4.101}$$

$$W_e = -W_i + \frac{\mathbf{j}\cdot\mathbf{F}}{n\,e} = -W_i + \frac{j^2}{\sigma_\parallel} - 0.71\,\frac{\mathbf{j}\cdot\nabla T_e}{e}. \tag{4.102}$$

Here, $\mathbf{j} = -n\,e\,(\mathbf{V}_e - \mathbf{V}_i)$ is the net plasma current, and the *electrical conductivity*, $\sigma_\parallel$, is given by

$$\sigma_\parallel = 1.96\,\frac{n\,e^2\,\tau_e}{m_e}. \tag{4.103}$$

Moreover, use has been made of the conservation law, Equation (4.32).

Let us examine each of the previous collisional terms, one by one. The first term on the right-hand side of Equation (4.100) is a friction force caused by the relative motion of electrons and ions, and obviously controls the electrical conductivity of the plasma. The form of this term is fairly easy to understand. The electrons lose their ordered velocity with respect to the ions, $\mathbf{U} = \mathbf{V}_e - \mathbf{V}_i$, in an electron-ion collision time, τ_e, and consequently lose momentum $m_e\,\mathbf{U}$ per electron (which is given to the ions) in this time. This means that a frictional force $(m_e\,n/\tau_e)\,\mathbf{U} \sim n\,e\,\mathbf{j}/(n\,e^2\,\tau_e/m_e)$ is exerted on the electrons. An equal and opposite force is exerted on the ions. Because the Coulomb cross-section diminishes with increasing electron energy (i.e., $\tau_e \sim T_e^{3/2}$), the conductivity of the fast electrons in the distribution function is higher than that of the slow electrons (because $\sigma_\parallel \sim \tau_e$). Hence, electrical current in plasmas is carried predominately by the fast electrons. This effect has some important and interesting consequences.

One immediate consequence is the second term on the right-hand side of Equation (4.100), which is called the *thermal force*. To understand the origin of a frictional force proportional to minus the gradient of the electron temperature, let us assume that the electron and ion fluids are at rest (i.e., $V_e = V_i = 0$). It follows that the number of electrons moving from left to right (along the x-axis, say) and from right to left per unit time is exactly the same at a given point (coordinate x_0, say) in the plasma. As a result of electron-ion collisions, these fluxes experience frictional forces, $\mathbf{F}_-$ and $\mathbf{F}_+$, respectively, of approximate magnitude $m_e\,n\,v_e/\tau_e$, where v_e is the electron thermal velocity. In a completely homogeneous plasma, these forces balance exactly, and so there is zero net frictional force. Suppose, however, that the electrons coming from the right are, on average, hotter than those coming from the left. It follows that the frictional force $\mathbf{F}_+$ acting on the fast electrons coming from the right is less than the force $\mathbf{F}_-$ acting on the slow electrons coming from the left, because τ_e increases with electron temperature. As a result, there is a net frictional force acting to the left: that is, in the direction of $-\nabla T_e$.

Let us estimate the magnitude of the frictional force. At point x_0, collisions are experienced by electrons that have traversed distances of similar magnitude to a mean-free-path, $l_e \sim v_e\,\tau_e$. Thus, the electrons coming from the right originate from regions in which the temperature is approximately $l_e\,\partial T_e/\partial x$ greater than the regions from which the electrons coming from the left originate. Because the friction force is proportional to T_e^{-1}, the net force $\mathbf{F}_+ - \mathbf{F}_-$ is approximately

$$\mathbf{F}_T \sim -\frac{l_e}{T_e}\frac{\partial T_e}{\partial x}\frac{m_e\,n\,v_e}{\tau_e} \sim -\frac{m_e\,v_e^2}{T_e}\,n\,\frac{\partial T_e}{\partial x} \sim -n\,\frac{\partial T_e}{\partial x}. \tag{4.104}$$

It must be emphasized that the thermal force is a direct consequence of collisions, despite the fact that the expression for the thermal force does not contain τ_e explicitly.

The term W_i, specified in Equation (4.101), represents the rate at which energy is acquired by the ions due to collisions with the electrons. The most striking aspect of this term is its smallness (note that it is proportional to an inverse mass ratio, m_e/m_i). The smallness of W_i is a direct consequence of the fact that electrons are considerably lighter than ions. Consider the limit in which the ion mass is infinite, and the ions are at rest on average: that is, $V_i = 0$. In this case, collisions of electrons with ions take place without any exchange of energy. The electron velocities are randomized by the collisions, so that the energy associated with their ordered velocity, $\mathbf{U} = \mathbf{V}_e - \mathbf{V}_i$, is converted into heat energy in the electron fluid [this is represented by the second term on the extreme right-hand side of Equation (4.102)]. However, the ion energy remains unchanged. Let us now assume that the ratio m_i/m_e is large, but finite, and that $U = 0$. If $T_e = T_i$ then the ions and electrons are in thermal equilibrium, so no heat is exchanged between them. However, if $T_e > T_i$ then heat is transferred from the electrons to the ions. As is well known, when a light particle collides with a heavy particle, the order of magnitude of the transferred energy is given by the mass ratio m_1/m_2, where m_1 is the mass of the lighter particle. For example, the mean fractional energy transferred in isotropic scattering is $2\,m_1/m_2$. Thus, we would expect the energy per unit time transferred from the electrons to the ions to be roughly

$$W_i \sim \frac{n}{\tau_e}\,\frac{2\,m_e}{m_i}\,\frac{3}{2}\,(T_e - T_i). \tag{4.105}$$

In fact, τ_e is defined so as to make the previous estimate exact.

The term W_e, specified in Equation (4.102), represents the rate at which energy is acquired by the electrons because of collisions with the ions, and consists of three terms. Not surprisingly, the first term is simply minus the rate at which energy is acquired by the ions due to collisions with the electrons. The second term represents the conversion of the ordered motion of the electrons, relative to the ions, into random motion (i.e., heat) via collisions with the ions. This term is positive definite, indicating that the randomization of the electron ordered motion gives rise to irreversible heat generation. Incidentally, this term is usually called the *ohmic heating* term. Finally, the third term represents the work done against the thermal force. This term can be either positive or negative, depending on the direction of the current flow relative to the electron temperature gradient, which indicates that work done against the thermal force gives rise to reversible heat generation. There is an analogous effect in metals called the *Thomson effect* (Doolittle 1959).

The electron and ion heat flux densities are given by

$$\mathbf{q}_e = -\kappa_\parallel^e\,\nabla T_e - 0.71\,\frac{T_e}{e}\,\mathbf{j}, \tag{4.106}$$

$$\mathbf{q}_i = -\kappa_\parallel^i\,\nabla T_i, \tag{4.107}$$

respectively. The electron and ion *thermal conductivities* are written

$$\kappa_\parallel^e = 3.2\,\frac{n\,\tau_e\,T_e}{m_e}, \tag{4.108}$$

$$\kappa_\parallel^i = 3.9\,\frac{n\,\tau_i\,T_i}{m_i}, \tag{4.109}$$

respectively.

It follows, by comparison with Equations (4.63)–(4.68), that the first term on the right-hand side of Equation (4.106), as well as the expression on the right-hand side of Equation (4.107), represent straightforward random-walk heat diffusion, with frequency ν, and step-length l. Recall, that $\nu = \tau^{-1}$ is the collision frequency, and

$l = \tau\, v_t$ is the mean-free-path. The electron heat diffusivity is generally much greater than that of the ions, because $\kappa_\parallel^e/\kappa_\parallel^i \sim \sqrt{m_i/m_e}$, assuming that $T_e \sim T_i$.

The second term on the right-hand side of Equation (4.106) describes a convective heat flux due to the motion of the electrons relative to the ions. To understand the origin of this flux, we need to recall that electric current in plasmas is carried predominately by the fast electrons in the distribution function. Suppose that U is non-zero. In the coordinate system in which V_e is zero, more fast electrons move in the direction of $\mathbf{U}$, and more slow electrons move in the opposite direction. Although the electron fluxes are balanced in this frame of reference, the energy fluxes are not (because a fast electron possesses more energy than a slow electron), and heat flows in the direction of $\mathbf{U}$: that is, in the opposite direction to the electric current. The net heat flux density is of approximate magnitude $n\, T_e\, U$, because there is no near cancellation of the fluxes due to the fast and slow electrons. Like the thermal force, this effect depends on collisions, despite the fact that the expression for the convective heat flux does not contain τ_e explicitly.

Finally, the electron and ion viscosity tensors take the form

$$(\pi_e)_{\alpha\beta} = -\eta_0^e \left(\frac{\partial V_\alpha}{\partial r_\beta} + \frac{\partial V_\beta}{\partial r_\alpha} - \frac{2}{3}\nabla\cdot\mathbf{V}\,\delta_{\alpha\beta}\right), \tag{4.110}$$

$$(\pi_i)_{\alpha\beta} = -\eta_0^i \left(\frac{\partial V_\alpha}{\partial r_\beta} + \frac{\partial V_\beta}{\partial r_\alpha} - \frac{2}{3}\nabla\cdot\mathbf{V}\,\delta_{\alpha\beta}\right), \tag{4.111}$$

respectively. Obviously, V_α refers to a Cartesian component of the electron fluid velocity in Equation (4.110) and the ion fluid velocity in Equation (4.111). Here, the electron and ion *viscosities* are given by

$$\eta_0^e = 0.73\, n\, \tau_e\, T_e, \tag{4.112}$$

$$\eta_0^i = 0.96\, n\, \tau_i\, T_i, \tag{4.113}$$

respectively. It follows, by comparison with Equations (4.62)–(4.68), that the previous expressions correspond to straightforward random-walk diffusion of momentum, with frequency ν, and step-length l. Again, the electron diffusivity exceeds the ion diffusivity by the square root of a mass ratio (assuming $T_e \sim T_i$). However, the ion viscosity exceeds the electron viscosity by the same factor (recall that $\eta \sim n\, m\, \chi_v$): that is, $\eta_0^i/\eta_0^e \sim \sqrt{m_i/m_e}$. For this reason, the viscosity of a plasma is determined essentially by the ions. This is not surprising, because viscosity is the diffusion of momentum, and the ions possess nearly all of the momentum in a plasma by virtue of their large masses.

Let us now examine the magnetized limit,

$$\Omega_i\, \tau_i,\;\; \Omega_e\, \tau_e \gg 1, \tag{4.114}$$

in which the electron and ion gyroradii are much smaller than the corresponding mean-free-paths. In this limit, the two-Laguerre-polynomial Chapman-Enskog closure scheme yields

$$\mathbf{F} = n\,e\left(\frac{\mathbf{j}_\parallel}{\sigma_\parallel} + \frac{\mathbf{j}_\perp}{\sigma_\perp}\right) - 0.71\,n\,\nabla_\parallel T_e - \frac{3\,n}{2\,|\Omega_e|\,\tau_e}\,\mathbf{b}\times\nabla_\perp T_e, \tag{4.115}$$

$$W_i = \frac{3\,m_e}{m_i}\frac{n\,(T_e - T_i)}{\tau_e}, \tag{4.116}$$

$$W_e = -W_i + \frac{\mathbf{j}\cdot\mathbf{F}}{n\,e}. \tag{4.117}$$

Here, the *parallel electrical conductivity*, $\sigma_\parallel$, is given by Equation (4.103), whereas the *perpendicular electrical conductivity*, $\sigma_\perp$, takes the form

$$\sigma_\perp = 0.51\,\sigma_\parallel = \frac{n\,e^2\,\tau_e}{m_e}. \tag{4.118}$$

Note that $\nabla_\parallel(\cdots) \equiv [\mathbf{b}\cdot\nabla(\cdots)]\,\mathbf{b}$ denotes a gradient parallel to the magnetic field, whereas $\nabla_\perp \equiv \nabla - \nabla_\parallel$ denotes a gradient perpendicular to the magnetic field. Likewise, $\mathbf{j}_\parallel \equiv (\mathbf{b}\cdot\mathbf{j})\,\mathbf{b}$ represents the component of the plasma current flowing parallel to the magnetic field, whereas $\mathbf{j}_\perp \equiv \mathbf{j} - \mathbf{j}_\parallel$ represents the perpendicular component of the plasma current.

We expect the presence of a strong magnetic field to give rise to a marked anisotropy in plasma properties between directions parallel and perpendicular to **B**, because of the completely different motions of the constituent ions and electrons parallel and perpendicular to the field. Thus, not surprisingly, we find that the electrical conductivity perpendicular to the field is approximately half that parallel to the field [see Equations (4.115) and (4.118)]. The thermal force is unchanged (relative to the unmagnetized case) in the parallel direction, but is radically modified in the perpendicular direction. In order to understand the origin of the last term in Equation (4.115), let us consider a situation in which there is a strong magnetic field along the z-axis, and an electron temperature gradient along the x-axis. (See Figure 4.1.) The electrons gyrate in the x-y plane in circles of radius $\rho_e \sim v_e/|\Omega_e|$. At a given point, coordinate x_0, say, on the x-axis, the electrons that come from the right and the left have traversed distances of approximate magnitude ρ_e. Thus, the electrons from the right originate from regions where the electron temperature is approximately $\rho_e\,\partial T_e/\partial x$ greater than the regions from which the electrons from the left originate. Because the friction force is proportional to T_e^{-1}, an unbalanced friction force arises, directed along the $-y$-axis. (See Figure 4.1.) This direction corresponds to the direction of $-\mathbf{b}\times\nabla T_e$. There is no friction force along the x-axis, because the x-directed fluxes are associated with electrons that originate from regions where $x = x_0$. By analogy with Equation (4.104), the magnitude of the perpendicular thermal force is

$$\mathbf{F}_{T\perp} \sim \frac{\rho_e}{T_e}\frac{\partial T_e}{\partial x}\frac{m_e\,n\,v_e}{\tau_e} \sim \frac{n}{|\Omega_e|\,\tau_e}\frac{\partial T_e}{\partial x}. \tag{4.119}$$

The effect of a strong magnetic field on the perpendicular component of the thermal force is directly analogous to a well-known phenomenon in metals called the *Nernst effect* (Rowe 2006).

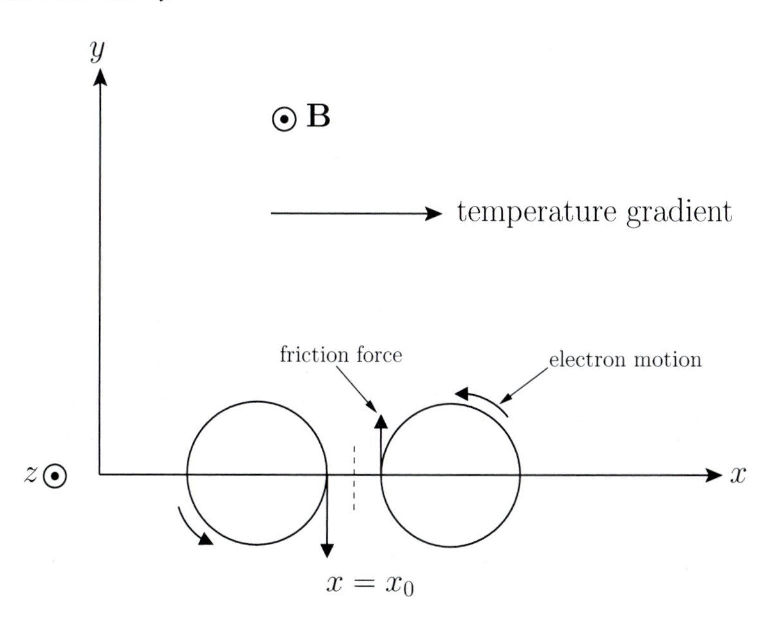

Figure 4.1
Origin of the perpendicular thermal force in a magnetized plasma.

In the magnetized limit, the electron and ion heat flux densities become

$$\mathbf{q}_e = -\kappa_\parallel^e \nabla_\parallel T_e - \kappa_\perp^e \nabla_\perp T_e - \kappa_\times^e \mathbf{b}\times\nabla_\perp T_e - 0.71\,\frac{T_e}{e}\,\mathbf{j}_\parallel - \frac{3\,T_e}{2\,|\Omega_e|\,\tau_e\,e}\,\mathbf{b}\times\mathbf{j}_\perp, \tag{4.120}$$

$$\mathbf{q}_i = -\kappa_\parallel^i \nabla_\parallel T_i - \kappa_\perp^i \nabla_\perp T_i + \kappa_\times^i \mathbf{b}\times\nabla_\perp T_i, \tag{4.121}$$

respectively. Here, the *parallel thermal conductivities* are given by Equations (4.108)–(4.109), and the *perpendicular thermal conductivities* take the form

$$\kappa_\perp^e = 4.7\,\frac{n\,T_e}{m_e\,\Omega_e^2\,\tau_e}, \tag{4.122}$$

$$\kappa_\perp^i = 2\,\frac{n\,T_i}{m_i\,\Omega_i^2\,\tau_i}. \tag{4.123}$$

Finally, the *cross thermal conductivities* are written

$$\kappa_\times^e = \frac{5\,n\,T_e}{2\,m_e\,|\Omega_e|}, \tag{4.124}$$

$$\kappa_\times^i = \frac{5\,n\,T_i}{2\,m_i\,\Omega_i}. \tag{4.125}$$

The first two terms on the right-hand sides of Equations (4.120) and (4.121) correspond to diffusive heat transport by the electron and ion fluids, respectively. According to the first terms, the diffusive transport in the direction parallel to the magnetic field is exactly the same as that in the unmagnetized case: that is, it corresponds to collision-induced random-walk diffusion of the ions and electrons, with frequency ν, and step-length l. According to the second terms, the diffusive transport in the direction perpendicular to the magnetic field is far smaller than that in the parallel direction. To be more exact, it is smaller by a factor $(\rho/l)^2$, where ρ is the gyroradius, and l the mean-free-path. In fact, the perpendicular heat transport also corresponds to collision-induced random-walk diffusion of charged particles, but with frequency ν, and step-length ρ. Thus, it is the greatly reduced step-length in the perpendicular direction, relative to the parallel direction, that ultimately gives rise to the strong reduction in the perpendicular heat transport. If $T_e \sim T_i$ then the ion perpendicular heat diffusivity actually exceeds that of the electrons by the square root of a mass ratio: that is, $\kappa_\perp^i/\kappa_\perp^e \sim \sqrt{m_i/m_e}$.

The third terms on the right-hand sides of Equations (4.120) and (4.121) correspond to heat fluxes that are perpendicular to both the magnetic field and the direction of the temperature gradient. In order to understand the origin of these terms, let us consider the ion flux. Suppose that there is a strong magnetic field along the z-axis, and an ion temperature gradient along the x-axis. (See Figure 4.2.) The ions gyrate in the x-y plane in circles of radius $\rho_i \sim v_i/\Omega_i$, where v_i is the ion thermal velocity. At a given point, coordinate x_0, say, on the x-axis, the ions that come from the right and the left have traversed distances of approximate magnitude ρ_i. The ions from the right are clearly somewhat hotter than those from the left. If the unidirectional particle fluxes, of approximate magnitude $n\,v_i$, are balanced, then the unidirectional heat fluxes, of approximate magnitude $n\,T_i\,v_i$, will have an unbalanced component of relative magnitude $(\rho_i/T_i)\,\partial T_i/\partial x$. As a result, there is a net heat flux in the $+y$-direction (i.e., the direction of $\mathbf{b}\times\nabla T_i$). The magnitude of this flux is

$$q_\times^i \sim n\,v_i\,\rho_i\,\frac{\partial T_i}{\partial x} \sim \frac{n\,T_i}{m_i\,|\Omega_i|}\,\frac{\partial T_i}{\partial x}. \tag{4.126}$$

There is an analogous expression for the electron flux, except that the electron flux is in the opposite direction to the ion flux (because the electrons gyrate in the opposite direction to the ions). Both the ion and electron fluxes transport heat along isotherms, and do not, therefore, give rise to any change in plasma temperature.

The fourth and fifth terms on the right-hand side of Equation (4.120) correspond to the convective component of the electron heat flux density, driven by motion of the electrons relative to the ions. It is clear from the fourth term that the convective flux parallel to the magnetic field is exactly the same as in the unmagnetized case [see Equation (4.106)]. However, according to the fifth term, the convective flux is radically modified in the perpendicular direction. Probably the easiest method of explaining the fifth term is via an examination of Equations (4.100), (4.106), (4.115), and (4.120). There is clearly a very close connection between the electron thermal force and the convective heat flux. In fact, starting from general principles of the thermodynamics of irreversible processes—the so-called *Onsager principles* (Reif

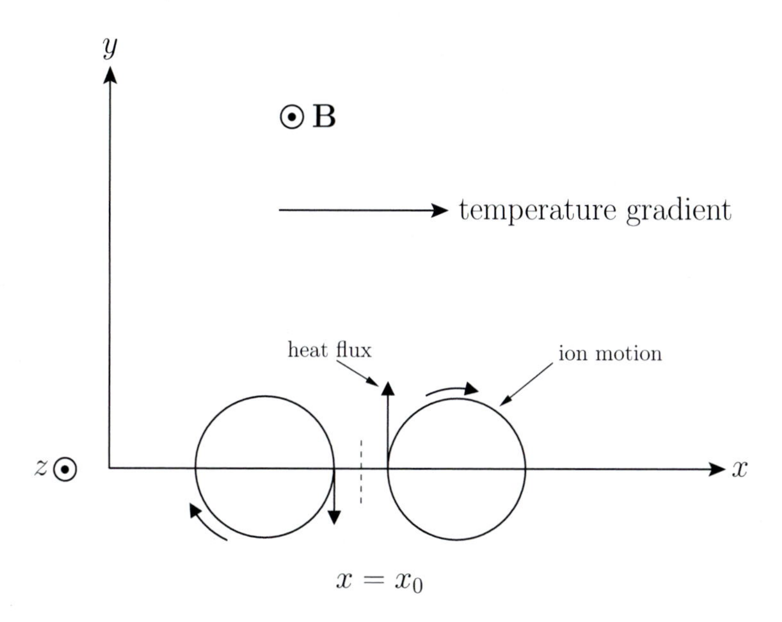

Figure 4.2
Origin of the convective perpendicular heat flux in a magnetized plasma.

1965)—it is possible to demonstrate that an electron frictional force of the form $\alpha\,(\nabla\,T_e)_\beta\,\mathbf{i}$ necessarily gives rise to an electron heat flux of the form $\alpha\,(T_e\,j_\beta/n\,e)\,\mathbf{i}$, where the subscript β corresponds to a general Cartesian component, and $\mathbf{i}$ is a unit vector. Thus, the fifth term on the right-hand side of Equation (4.120) follows by Onsager symmetry from the third term on the right-hand side of Equation (4.115). This is one of many Onsager symmetries that occur in plasma transport theory.

In order to describe the viscosity tensor in a magnetized plasma, it is helpful to define the *rate-of-strain tensor*

$$W_{\alpha\beta} = \frac{\partial V_\alpha}{\partial r_\beta} + \frac{\partial V_\beta}{\partial r_\alpha} - \frac{2}{3}\,\nabla\cdot\mathbf{V}\,\delta_{\alpha\beta}. \tag{4.127}$$

Obviously, there is a separate rate-of-strain tensor for the electron and ion fluids. It is easily demonstrated that this tensor is zero if the plasma translates, or rotates as a rigid body, or if it undergoes isotropic compression. Thus, the rate-of-strain tensor measures the deformation of plasma volume elements.

In a magnetized plasma, the viscosity tensor is best described as the sum of five component tensors,

$$\boldsymbol{\pi} = \sum_{n=0,4} \boldsymbol{\pi}_n, \tag{4.128}$$

where

$$\boldsymbol{\pi}_0 = -3\,\eta_0\left(\mathbf{b}\mathbf{b} - \frac{1}{3}\,\mathbf{I}\right)\left(\mathbf{b}\mathbf{b} - \frac{1}{3}\,\mathbf{I}\right) : \nabla\mathbf{V}, \tag{4.129}$$

with

$$\boldsymbol{\pi}_1 = -\eta_1 \left[\mathbf{I}_\perp \cdot \mathbf{W} \cdot \mathbf{I}_\perp + \frac{1}{2} \mathbf{I}_\perp (\mathbf{b} \cdot \mathbf{W} \cdot \mathbf{b}) \right], \tag{4.130}$$

and

$$\boldsymbol{\pi}_2 = -4\,\eta_1\, (\mathbf{I}_\perp \cdot \mathbf{W} \cdot \mathbf{bb} + \mathbf{bb} \cdot \mathbf{W} \cdot \mathbf{I}_\perp). \tag{4.131}$$

plus

$$\boldsymbol{\pi}_3 = \frac{\eta_3}{2}\, (\mathbf{b} \times \mathbf{W} \cdot \mathbf{I}_\perp - \mathbf{I}_\perp \cdot \mathbf{W} \times \mathbf{b}), \tag{4.132}$$

and

$$\boldsymbol{\pi}_4 = 2\,\eta_3\, (\mathbf{b} \times \mathbf{W} \cdot \mathbf{bb} - \mathbf{bb} \cdot \mathbf{W} \times \mathbf{b}). \tag{4.133}$$

Here, $\mathbf{I}$ is the identity tensor, and $\mathbf{I}_\perp = \mathbf{I} - \mathbf{bb}$. The previous expressions are valid for both electrons and ions.

The tensor $\boldsymbol{\pi}_0$ describes what is known as *parallel viscosity*. This is a viscosity that controls the variation along magnetic field-lines of the velocity component parallel to field-lines. The parallel viscosity coefficients, η_0^e and η_0^i, are specified in Equations (4.112)–(4.113). The parallel viscosity is unchanged from the unmagnetized case, and is caused by the collision-induced random-walk diffusion of particles, with frequency ν, and step-length l.

The tensors $\boldsymbol{\pi}_1$ and $\boldsymbol{\pi}_2$ describe what is known as *perpendicular viscosity*. This is a viscosity that controls the variation perpendicular to magnetic field-lines of the velocity components perpendicular to field-lines. The perpendicular viscosity coefficients are given by

$$\eta_1^e = 0.51\, \frac{n\, T_e}{\Omega_e^2\, \tau_e}, \tag{4.134}$$

$$\eta_1^i = \frac{3\, n\, T_i}{10\, \Omega_i^2\, \tau_i}. \tag{4.135}$$

The perpendicular viscosity is far smaller than the parallel viscosity. In fact, it is smaller by a factor $(\rho/l)^2$. The perpendicular viscosity corresponds to collision-induced random-walk diffusion of particles, with frequency ν, and step-length ρ. Thus, it is the greatly reduced step-length in the perpendicular direction, relative to the parallel direction, that accounts for the smallness of the perpendicular viscosity compared to the parallel viscosity.

Finally, the tensors $\boldsymbol{\pi}_3$ and $\boldsymbol{\pi}_4$ describe what is known as *gyroviscosity*. This is not really viscosity at all, because the associated viscous stresses are always perpendicular to the velocity, implying that there is no dissipation (i.e., viscous heating) associated with this effect. The gyroviscosity coefficients are given by

$$\eta_3^e = -\frac{n\, T_e}{2\, |\Omega_e|}, \tag{4.136}$$

$$\eta_3^i = \frac{n\, T_i}{2\, \Omega_i}. \tag{4.137}$$

The origin of gyroviscosity is very similar to the origin of the cross thermal conductivity terms in Equations (4.120)–(4.121). Both cross thermal conductivity and gyroviscosity are independent of the collision frequency.

4.11 Normalization of Braginskii Equations

As we have just seen, the Braginskii equations contain terms that describe a very wide range of different physical phenomena. For this reason, they are extremely complicated. Fortunately, however, it is not generally necessary to retain all of the terms in these equations when investigating a particular problem in plasma physics: for example, electromagnetic wave propagation through plasmas. In this section, we shall attempt to construct a systematic normalization scheme for the Braginskii equations that will, hopefully, enable us to determine which terms to keep, and which to discard, when investigating a particular aspect of plasma physics.

Let us consider a magnetized plasma. It is convenient to split the friction force $\mathbf{F}$ into a component $\mathbf{F}_U$ corresponding to resistivity, and a component $\mathbf{F}_T$ corresponding to the thermal force. Thus,

$$\mathbf{F} = \mathbf{F}_U + \mathbf{F}_T, \tag{4.138}$$

where

$$\mathbf{F}_U = n\,e\left(\frac{\mathbf{j}_\parallel}{\sigma_\parallel} + \frac{\mathbf{j}_\perp}{\sigma_\perp}\right), \tag{4.139}$$

$$\mathbf{F}_T = -0.71\,n\,\nabla_\parallel T_e - \frac{3\,n}{2\,|\Omega_e|\,\tau_e}\,\mathbf{b}\times\nabla_\perp T_e. \tag{4.140}$$

Likewise, the electron collisional energy gain term W_e is split into a component $-W_i$ corresponding to the energy lost to the ions (in the ion rest frame), a component W_U corresponding to work done by the friction force $\mathbf{F}_U$, and a component W_T corresponding to work done by the thermal force $\mathbf{F}_T$. Thus,

$$W_e = -W_i + W_U + W_T, \tag{4.141}$$

where

$$W_U = \frac{\mathbf{j}\cdot\mathbf{F}_U}{n\,e}, \tag{4.142}$$

$$W_T = \frac{\mathbf{j}\cdot\mathbf{F}_T}{n\,e}. \tag{4.143}$$

Finally, it is helpful to split the electron heat flux density $\mathbf{q}_e$ into a diffusive component $\mathbf{q}_{Te}$ and a convective component $\mathbf{q}_{Ue}$. Thus,

$$\mathbf{q}_e = \mathbf{q}_{Te} + \mathbf{q}_{Ue}, \tag{4.144}$$

where

$$\mathbf{q}_{Te} = -\kappa_\parallel^e\,\nabla_\parallel T_e - \kappa_\perp^e\,\nabla_\perp T_e - \kappa_\times^e\,\mathbf{b}\times\nabla_\perp T_e, \tag{4.145}$$

$$\mathbf{q}_{Ue} = 0.71\,\frac{T_e}{e}\,\mathbf{j}_\parallel - \frac{3\,T_e}{2\,|\Omega_e|\,\tau_e\,e}\,\mathbf{b}\times\mathbf{j}_\perp. \tag{4.146}$$

Let us, first of all, consider the electron fluid equations, which can be written:

$$\frac{dn}{dt} + n\,\nabla\cdot\mathbf{V}_e = 0, \tag{4.147}$$

$$m_e\,n\,\frac{d\mathbf{V}_e}{dt} + \nabla p_e + \nabla\cdot\boldsymbol{\pi}_e + e\,n\,(\mathbf{E} + \mathbf{V}_e\times\mathbf{B}) = \mathbf{F}_U + \mathbf{F}_T, \tag{4.148}$$

$$\frac{3}{2}\frac{dp_e}{dt} + \frac{5}{2}\,p_e\,\nabla\cdot\mathbf{V}_e + \boldsymbol{\pi}_e : \nabla\mathbf{V}_e + \nabla\cdot\mathbf{q}_{Te} + \nabla\cdot\mathbf{q}_{Ue} = -W_i + W_U + W_T. \tag{4.149}$$

Let $\bar{n}$, $\bar{v}_e$, $\bar{l}_e$, $\bar{B}$, and $\bar{\rho}_e = \bar{v}_e/(e\bar{B}/m_e)$, be typical values of the particle density, the electron thermal velocity, the electron mean-free-path, the magnetic field-strength, and the electron gyroradius, respectively. Suppose that the typical electron flow velocity is $\lambda_e\,\bar{v}_e$, and the typical variation lengthscale is L. Let

$$\delta_e = \frac{\bar{\rho}_e}{L}, \tag{4.150}$$

$$\zeta_e = \frac{\bar{\rho}_e}{\bar{l}_e}, \tag{4.151}$$

$$\mu = \sqrt{\frac{m_e}{m_i}}. \tag{4.152}$$

All three of these parameters are assumed to be small compared to unity.

We define the following normalized quantities:

$$\hat{n} = \frac{n}{\bar{n}}, \qquad \hat{v}_e = \frac{v_e}{\bar{v}_e}, \qquad \hat{\mathbf{r}} = \frac{\mathbf{r}}{L},$$

$$\widehat{\nabla} = L\,\nabla, \qquad \hat{t} = \frac{\lambda_e\,\bar{v}_e\,t}{L}, \qquad \widehat{\mathbf{V}}_e = \frac{\mathbf{V}_e}{\lambda_e\,\bar{v}_e},$$

$$\widehat{\mathbf{B}} = \frac{\mathbf{B}}{\bar{B}}, \qquad \widehat{\mathbf{E}} = \frac{\mathbf{E}}{\lambda_e\,\bar{v}_e\,\bar{B}}, \qquad \widehat{\mathbf{U}} = \frac{\mathbf{U}}{(1+\lambda_e^2)\,\delta_e\,\bar{v}_e},$$

$$\hat{p}_e = \frac{p_e}{m_e\,\bar{n}\,\bar{v}_e^2}, \qquad \widehat{\boldsymbol{\pi}}_e = \frac{\boldsymbol{\pi}_e}{\lambda_e\,\delta_e\,\zeta_e^{-1}\,m_e\,\bar{n}\,\bar{v}_e^2}, \qquad \widehat{\mathbf{q}}_{Te} = \frac{\mathbf{q}_{Te}}{\delta_e\,\zeta_e^{-1}\,m_e\,\bar{n}\,\bar{v}_e^3},$$

$$\widehat{\mathbf{q}}_{Ue} = \frac{\mathbf{q}_{Ue}}{(1+\lambda_e^2)\,\delta_e\,m_e\,\bar{n}\,\bar{v}_e^3}, \qquad \widehat{\mathbf{F}}_U = \frac{\mathbf{F}_U}{(1+\lambda_e^2)\,\zeta_e\,m_e\,\bar{n}\,\bar{v}_e^2/L}, \qquad \widehat{\mathbf{F}}_T = \frac{\mathbf{F}_T}{m_e\,\bar{n}\,\bar{v}_e^2/L},$$

$$\widehat{W}_i = \frac{W_i}{\delta_e^{-1}\,\zeta_e\,\mu^2\,m_e\,\bar{n}\,\bar{v}_e^3/L}, \qquad \widehat{W}_U = \frac{W_U}{(1+\lambda_e^2)^2\,\delta_e\,\zeta_e\,m_e\,\bar{n}\,\bar{v}_e^3/L}, \qquad \widehat{W}_T = \frac{W_T}{(1+\lambda_e^2)\,\delta_e\,m_e\,\bar{n}\,\bar{v}_e^3/L}.$$

The normalization procedure is designed to make all hatted quantities $\mathcal{O}(1)$. The normalization of the electric field is chosen such that the $\mathbf{E}\times\mathbf{B}$ velocity is of similar magnitude to the electron fluid velocity. Note that the parallel viscosity makes an $\mathcal{O}(1)$ contribution to $\widehat{\boldsymbol{\pi}}_e$, whereas the gyroviscosity makes an $\mathcal{O}(\zeta_e)$ contribution, and the perpendicular viscosity only makes an $\mathcal{O}(\zeta_e^2)$ contribution. Likewise, the parallel thermal conductivity makes an $\mathcal{O}(1)$ contribution to $\widehat{\mathbf{q}}_{Te}$, whereas the cross conductivity makes an $\mathcal{O}(\zeta_e)$ contribution, and the perpendicular conductivity only makes an $O(\zeta_e^2)$ contribution. Similarly, the parallel components of $\mathbf{F}_T$ and $\mathbf{q}_{Ue}$ are $O(1)$, whereas the perpendicular components are $O(\zeta_e)$.

The normalized electron fluid equations take the form:

$$\frac{d\hat{n}}{d\hat{t}} + \hat{n}\,\widehat{\nabla}\cdot\widehat{\mathbf{V}}_e = 0, \tag{4.153}$$

$$\lambda_e^2\,\delta_e\,\hat{n}\,\frac{d\widehat{\mathbf{V}}_e}{d\hat{t}} + \delta_e\,\widehat{\nabla}\hat{p}_e + \lambda_e\,\delta_e^2\,\zeta_e^{-1}\,\widehat{\nabla}\cdot\widehat{\boldsymbol{\pi}}_e \tag{4.154}$$

$$+\lambda_e\,\hat{n}\,(\widehat{\mathbf{E}} + \widehat{\mathbf{V}}_e\times\widehat{\mathbf{B}}) = (1+\lambda_e^2)\,\delta_e\,\zeta_e\,\widehat{\mathbf{F}}_U + \delta_e\,\widehat{\mathbf{F}}_T,$$

$$\lambda_e\,\frac{3}{2}\frac{d\hat{p}_e}{d\hat{t}} + \lambda_e\,\frac{5}{2}\,\hat{p}_e\,\widehat{\nabla}\cdot\widehat{\mathbf{V}}_e + \lambda_e^2\,\delta_e\,\zeta_e^{-1}\,\widehat{\boldsymbol{\pi}}_e : \widehat{\nabla}\cdot\widehat{\mathbf{V}}_e$$

$$+\delta_e\,\zeta_e^{-1}\,\widehat{\nabla}\cdot\widehat{\mathbf{q}}_{Te} + (1+\lambda_e^2)\,\delta_e\,\widehat{\nabla}\cdot\widehat{\mathbf{q}}_{Ue} = -\delta_e^{-1}\,\zeta_e\,\mu^2\,\widehat{W}_i$$

$$+\,(1+\lambda_e^2)^2\,\delta_e\,\zeta_e\,\widehat{W}_U$$

$$+\,(1+\lambda_e^2)\,\delta_e\,\widehat{W}_T. \tag{4.155}$$

The only large or small (compared to unity) quantities in these equations are the parameters λ_e, δ_e, ζ_e, and μ. Here, $d/d\hat{t} \equiv \partial/\partial\hat{t} + \widehat{\mathbf{V}}_e\cdot\widehat{\nabla}$. It is assumed that $T_e \sim T_i$.

Let us now consider the ion fluid equations, which can be written:

$$\frac{dn}{dt} + n\,\nabla\cdot\mathbf{V}_i = 0, \tag{4.156}$$

$$m_i\,n\,\frac{d\mathbf{V}_i}{dt} + \nabla p_i + \nabla\cdot\boldsymbol{\pi}_i - en\,(\mathbf{E} + \mathbf{V}_i\times\mathbf{B}) = -\mathbf{F}_U - \mathbf{F}_T, \tag{4.157}$$

$$\frac{3}{2}\frac{dp_i}{dt} + \frac{5}{2}\,p_i\,\nabla\cdot\mathbf{V}_i + \boldsymbol{\pi}_i : \nabla\mathbf{V}_i + \nabla\cdot\mathbf{q}_i = W_i. \tag{4.158}$$

It is convenient to adopt a normalization scheme for the ion equations which is similar to, but independent of, that employed to normalize the electron equations. Let $\bar{n}$, $\bar{v}_i$, $\bar{l}_i$, $\bar{B}$, and $\bar{\rho}_i = \bar{v}_i/(e\bar{B}/m_i)$, be typical values of the particle density, the ion thermal velocity, the ion mean-free-path, the magnetic field-strength, and the ion gyroradius, respectively. Suppose that the typical ion flow velocity is $\lambda_i\,\bar{v}_i$, and the typical variation lengthscale is L. Let

$$\delta_i = \frac{\bar{\rho}_i}{L}, \tag{4.159}$$

$$\zeta_i = \frac{\bar{\rho}_i}{\bar{l}_i}, \tag{4.160}$$

$$\mu = \sqrt{\frac{m_e}{m_i}}. \tag{4.161}$$

All three of these parameters are assumed to be small compared to unity.

We define the following normalized quantities:

$$\hat{n} = \frac{n}{\bar{n}}, \qquad \hat{v}_i = \frac{v_i}{\bar{v}_i}, \qquad \hat{\mathbf{r}} = \frac{\mathbf{r}}{L},$$

$$\widehat{\nabla} = L\,\nabla, \qquad \hat{t} = \frac{\lambda_i\,\bar{v}_i\,t}{L}, \qquad \widehat{\mathbf{V}}_i = \frac{\mathbf{V}_i}{\lambda_i\,\bar{v}_i},$$

$$\widehat{\mathbf{B}} = \frac{\mathbf{B}}{\bar{B}}, \qquad \widehat{\mathbf{E}} = \frac{\mathbf{E}}{\lambda_i\,\bar{v}_i\,\bar{B}}, \qquad \widehat{\mathbf{U}} = \frac{\mathbf{U}}{(1+\lambda_i^2)\,\delta_i\,\bar{v}_i},$$

$$\hat{p}_i = \frac{p_i}{m_i\,\bar{n}\,\bar{v}_i^{\,2}}, \qquad \widehat{\boldsymbol{\pi}}_i = \frac{\boldsymbol{\pi}_i}{\lambda_i\,\delta_i\,\zeta_i^{-1}\,m_i\,\bar{n}\,\bar{v}_i^{\,2}}, \qquad \widehat{\mathbf{q}}_i = \frac{\mathbf{q}_i}{\delta_i\,\zeta_i^{-1}\,m_i\,\bar{n}\,\bar{v}_i^{\,3}},$$

$$\widehat{\mathbf{F}}_U = \frac{\mathbf{F}_U}{(1+\lambda_i^2)\,\zeta_i\,\mu\,m_i\,\bar{n}\,\bar{v}_i^{\,2}/L}, \qquad \widehat{\mathbf{F}}_T = \frac{\mathbf{F}_T}{m_i\,\bar{n}\,\bar{v}_i^{\,2}/L}, \qquad \widehat{W}_i = \frac{W_i}{\delta_i^{-1}\,\zeta_i\,\mu\,m_i\,\bar{n}\,\bar{v}_i^{\,3}/L}.$$

As before, the normalization procedure is designed to make all hatted quantities $O(1)$. The normalization of the electric field is chosen such that the $\mathbf{E}\times\mathbf{B}$ velocity is of similar magnitude to the ion fluid velocity. Note that the parallel viscosity makes an $O(1)$ contribution to $\widehat{\boldsymbol{\pi}}_i$, whereas the gyroviscosity makes an $O(\zeta_i)$ contribution, and the perpendicular viscosity only makes an $O(\zeta_i^{\,2})$ contribution. Likewise, the parallel thermal conductivity makes an $O(1)$ contribution to $\widehat{\mathbf{q}}_i$, whereas the cross conductivity makes an $O(\zeta_i)$ contribution, and the perpendicular conductivity only makes an $O(\zeta_i^{\,2})$ contribution. Similarly, the parallel component of $\mathbf{F}_T$ is $O(1)$, whereas the perpendicular component is $O(\zeta_i\,\mu)$.

The normalized ion fluid equations take the form:

$$\frac{d\hat{n}}{d\hat{t}} + \hat{n}\,\widehat{\nabla}\cdot\widehat{\mathbf{V}}_i = 0, \tag{4.162}$$

$$\begin{aligned}\lambda_i^2\,\delta_i\,\hat{n}\,\frac{d\widehat{\mathbf{V}}_i}{d\hat{t}} + \delta_i\,\widehat{\nabla}\hat{p}_i + \lambda_i\,\delta_i^2\,\zeta_i^{-1}\,\widehat{\nabla}\cdot\widehat{\boldsymbol{\pi}}_i& \\ -\lambda_i\,\hat{n}\,(\widehat{\mathbf{E}} + \widehat{\mathbf{V}}_i\times\widehat{\mathbf{B}}) &= -(1+\lambda_i^2)\,\delta_i\,\zeta_i\,\mu\,\widehat{\mathbf{F}}_U - \delta_i\,\widehat{\mathbf{F}}_T,\end{aligned} \tag{4.163}$$

$$\begin{aligned}\lambda_i\,\frac{3}{2}\frac{d\hat{p}_i}{d\hat{t}} + \lambda_i\,\frac{5}{2}\,\hat{p}_i\,\widehat{\nabla}\cdot\widehat{\mathbf{V}}_i + \lambda_i^2\,\delta_i\,\zeta_i^{-1}\,\widehat{\boldsymbol{\pi}}_i : \widehat{\nabla}\cdot\widehat{\mathbf{V}}_i& \\ +\delta_i\,\zeta_i^{-1}\,\widehat{\nabla}\cdot\widehat{\mathbf{q}}_i &= \delta_i^{-1}\,\zeta_i\,\mu\,\widehat{W}_i.\end{aligned} \tag{4.164}$$

The only large or small (compared to unity) quantities in these equations are the parameters λ_i, δ_i, ζ_i, and μ. Here, $d/d\hat{t} \equiv \partial/\partial\hat{t} + \widehat{\mathbf{V}}_i\cdot\widehat{\nabla}$.

Let us adopt the ordering

$$\delta_e, \delta_i \ll \zeta_e, \zeta_i, \mu \ll 1, \tag{4.165}$$

which is appropriate to a collisional, highly magnetized, plasma. In the first stage of our ordering procedure, we shall treat δ_e and δ_i as small parameters, and ζ_e, ζ_i, and μ as $O(1)$. In the second stage, we shall take note of the smallness of ζ_e, ζ_i, and μ. Note that the parameters λ_e and λ_i are "free ranging." In other words, they can be either large, small, or $O(1)$. In the initial stage of the ordering procedure, the ion and electron normalization schemes we have adopted become essentially identical [because $\mu \sim O(1)$], and it is convenient to write

$$\lambda_e \sim \lambda_i \sim \lambda, \tag{4.166}$$

$$\delta_e \sim \delta_i \sim \delta, \tag{4.167}$$

$$V_e \sim V_i \sim V, \tag{4.168}$$

$$v_e \sim v_i \sim v_t, \tag{4.169}$$

$$\Omega_e \sim \Omega_i \sim \Omega. \tag{4.170}$$

There are three fundamental orderings in plasma fluid theory. These are analogous to the three orderings in neutral gas fluid theory discussed in Section 4.9.

The first ordering is

$$\lambda \sim \delta^{-1}. \tag{4.171}$$

This corresponds to

$$V \gg v_t. \tag{4.172}$$

In other words, the fluid velocities are much greater than the respective thermal velocities. We also have

$$\frac{V}{L} \sim \Omega. \tag{4.173}$$

Here, V/L is conventionally termed the *transit frequency*, and is the frequency with which fluid elements traverse the system. It is clear that the transit frequencies are of approximately the same magnitudes as the gyrofrequencies in this ordering. Keeping only the largest terms in Equations (4.153)–(4.155) and (4.162)–(4.164), the Braginskii equations reduce to (in unnormalized form):

$$\frac{dn}{dt} + n\,\nabla\cdot\mathbf{V}_e = 0, \tag{4.174}$$

$$m_e\,n\,\frac{d\mathbf{V}_e}{dt} + e\,n\,(\mathbf{E} + \mathbf{V}_e\times\mathbf{B}) = [\zeta]\,\mathbf{F}_U, \tag{4.175}$$

and

$$\frac{dn}{dt} + n\,\nabla\cdot\mathbf{V}_i = 0, \tag{4.176}$$

$$m_i\,n\,\frac{d\mathbf{V}_i}{dt} - e\,n\,(\mathbf{E} + \mathbf{V}_i\times\mathbf{B}) = -[\zeta]\,\mathbf{F}_U. \tag{4.177}$$

The factors in square brackets are just to remind us that the terms they precede are smaller than the other terms in the equations (by the corresponding factors inside the brackets).

Equations (4.174)–(4.175) and (4.176)–(4.177) are called the *cold-plasma equations*, because they can be obtained from the Braginskii equations by formally taking the limit $T_e, T_i \to 0$. Likewise, the ordering (4.171) is called the *cold-plasma approximation*. The cold-plasma approximation applies not only to cold plasmas, but also to very fast disturbances that propagate through conventional plasmas. In particular, the cold-plasma equations provide a good description of the propagation of electromagnetic waves through plasmas. After all, electromagnetic waves generally have very high velocities (i.e., $V \sim c$), which they impart to plasma fluid elements, so there is usually no difficulty satisfying the inequality (4.172).

The electron and ion pressures can be neglected in the cold-plasma limit, because the thermal velocities are much smaller than the fluid velocities. It follows that there is no need for an electron or ion energy evolution equation. Furthermore, the motion of the plasma is so fast, in this limit, that relatively slow "transport" effects, such as viscosity and thermal conductivity, play no role in the cold-plasma fluid equations. In fact, the only collisional effect that appears in these equations is resistivity.

The second ordering is

$$\lambda \sim 1, \tag{4.178}$$

which corresponds to

$$V \sim v_t. \tag{4.179}$$

In other words, the fluid velocities are of similar magnitudes to the respective thermal velocities. Keeping only the largest terms in Equations (4.153)–(4.155) and (4.162)–(4.164), the Braginskii equations reduce to (in unnormalized form):

$$\frac{dn}{dt} + n\,\nabla\cdot\mathbf{V}_e = 0, \tag{4.180}$$

$$m_e\,n\,\frac{d\mathbf{V}_e}{dt} + \nabla p_e + [\delta^{-1}]\,e\,n\,(\mathbf{E} + \mathbf{V}_e\times\mathbf{B}) = [\zeta]\,\mathbf{F}_U + \mathbf{F}_T, \tag{4.181}$$

$$\frac{3}{2}\frac{dp_e}{dt} + \frac{5}{2}\,p_e\,\nabla\cdot\mathbf{V}_e = -[\delta^{-1}\,\zeta\,\mu^2]\,W_i, \tag{4.182}$$

and

$$\frac{dn}{dt} + n\,\nabla\cdot\mathbf{V}_i = 0, \tag{4.183}$$

$$m_i\,n\,\frac{d\mathbf{V}_i}{dt} + \nabla p_i - [\delta^{-1}]\,e\,n\,(\mathbf{E} + \mathbf{V}_i\times\mathbf{B}) = -[\zeta]\,\mathbf{F}_U - \mathbf{F}_T, \tag{4.184}$$

$$\frac{3}{2}\frac{dp_i}{dt} + \frac{5}{2}\,p_i\,\nabla\cdot\mathbf{V}_i = [\delta^{-1}\,\zeta\,\mu^2]\,W_i. \tag{4.185}$$

Again, the factors in square brackets remind us that the terms they precede are larger, or smaller, than the other terms in the equations.

Equations (4.180)–(4.182) and (4.183)–(4.184) are called the *magnetohydrodynamical equations*, or *MHD equations*, for short. Likewise, the ordering (4.178) is called the *MHD approximation*. The MHD equations are conventionally used to study macroscopic plasma instabilities possessing relatively fast growth-rates: for example, "sausage" modes and "kink" modes (Bateman 1978).

The electron and ion pressures cannot be neglected in the MHD limit, because the fluid velocities are similar in magnitude to the respective thermal velocities. Thus, electron and ion energy evolution equations are needed in this limit. However, MHD motion is sufficiently fast that "transport" effects, such as viscosity and thermal conductivity, are too slow to play a role in the MHD equations. In fact, the only collisional effects that appear in these equations are resistivity, the thermal force, and electron-ion collisional energy exchange.

The final ordering is

$$\lambda \sim \delta, \tag{4.186}$$

which corresponds to

$$V \sim \delta\, v_t \sim v_d, \tag{4.187}$$

where v_d is a typical drift (e.g., a curvature or grad-B drift—see Chapter 2) velocity. In other words, the fluid velocities are of similar magnitude to the respective drift velocities. Keeping only the largest terms in Equations (3.113) and (3.116), the Braginskii equations reduce to (in unnormalized form):

$$\frac{dn}{dt} + n\,\nabla\cdot\mathbf{V}_e = 0, \tag{4.188}$$

$$m_e\,n\,\frac{d\mathbf{V}_e}{dt} + [\delta^{-2}]\,\nabla p_e + [\zeta^{-1}]\,\nabla\cdot\boldsymbol{\pi}_e$$
$$+[\delta^{-2}]\,e\,n\,(\mathbf{E} + \mathbf{V}_e\times\mathbf{B}) = [\delta^{-2}\,\zeta]\,\mathbf{F}_U + [\delta^{-2}]\,\mathbf{F}_T, \tag{4.189}$$

$$\frac{3}{2}\frac{dp_e}{dt} + \frac{5}{2}\,p_e\,\nabla\cdot\mathbf{V}_e + [\zeta^{-1}]\,\nabla\cdot\mathbf{q}_{Te} + \nabla\cdot\mathbf{q}_{Ue} = -[\delta^{-2}\,\zeta\,\mu^2]\,W_i$$
$$+ [\zeta]\,W_U + W_T, \tag{4.190}$$

and

$$\frac{dn}{dt} + n\,\nabla\cdot\mathbf{V}_i = 0, \tag{4.191}$$

$$m_i\,n\,\frac{d\mathbf{V}_i}{dt} + [\delta^{-2}]\,\nabla p_i + [\zeta^{-1}]\,\nabla\cdot\boldsymbol{\pi}_i$$
$$-[\delta^{-2}]\,e\,n\,(\mathbf{E} + \mathbf{V}_i\times\mathbf{B}) = -[\delta^{-2}\,\zeta]\,\mathbf{F}_U - [\delta^{-2}]\,\mathbf{F}_T, \tag{4.192}$$

$$\frac{3}{2}\frac{dp_i}{dt} + \frac{5}{2}\,p_i\,\nabla\cdot\mathbf{V}_i + [\zeta^{-1}]\,\nabla\cdot\mathbf{q}_i = [\delta^{-2}\,\zeta\,\mu^2]\,W_i. \tag{4.193}$$

As before, the factors in square brackets remind us that the terms they precede are larger, or smaller, than the other terms in the equations.

Equations (4.188)–(4.190) and (4.191)–(4.193) are called the *drift equations*. Likewise, the ordering (4.186) is called the *drift approximation*. The drift equations are conventionally used to study equilibrium evolution, and the slow growing "micro-instabilities" that are responsible for turbulent transport in tokamaks. It is clear that virtually all of the original terms in the Braginskii equations must be retained in this limit.

In the following sections, we investigate the cold-plasma equations, the MHD equations, and the drift equations, in more detail.

4.12 Cold-Plasma Equations

Previously, we used the smallness of the magnetization parameter, δ, to derive the cold-plasma equations:

$$\frac{\partial n}{\partial t} + \nabla \cdot (n\,\mathbf{V}_e) = 0, \tag{4.194}$$

$$m_e\,n\,\frac{\partial \mathbf{V}_e}{\partial t} + m_e\,n\,(\mathbf{V}_e \cdot \nabla)\mathbf{V}_e + e\,n\,(\mathbf{E} + \mathbf{V}_e \times \mathbf{B}) = [\zeta]\,\mathbf{F}_U, \tag{4.195}$$

and

$$\frac{\partial n}{\partial t} + \nabla \cdot (n\,\mathbf{V}_i) = 0, \tag{4.196}$$

$$m_i\,n\,\frac{\partial \mathbf{V}_i}{\partial t} + m_i\,n\,(\mathbf{V}_i \cdot \nabla)\mathbf{V}_i - e\,n\,(\mathbf{E} + \mathbf{V}_i \times \mathbf{B}) = -[\zeta]\,\mathbf{F}_U. \tag{4.197}$$

Let us now use the smallness of the mass ratio m_e/m_i to further simplify these equations. In particular, we would like to write the electron and ion fluid velocities in terms of the *center-of-mass velocity*,

$$\mathbf{V} = \frac{m_i\,\mathbf{V}_i + m_e\,\mathbf{V}_e}{m_i + m_e}, \tag{4.198}$$

and the plasma current

$$\mathbf{j} = -n\,e\,\mathbf{U}, \tag{4.199}$$

where $\mathbf{U} = \mathbf{V}_e - \mathbf{V}_\mathbf{i}$. According to the ordering scheme adopted in the previous section, $U \sim V_e \sim V_i$ in the cold-plasma limit. We shall continue to regard the mean-free-path parameter ζ as $O(1)$.

It follows from Equations (4.198) and (4.199) that

$$\mathbf{V}_i \simeq \mathbf{V} + O\left(\frac{m_e}{m_i}\right), \tag{4.200}$$

and

$$\mathbf{V}_e \simeq \mathbf{V} - \frac{\mathbf{j}}{ne} + O\left(\frac{m_e}{m_i}\right). \tag{4.201}$$

Equations (4.194), (4.196), (4.200), and (4.201) yield the *continuity equation*:

$$\frac{dn}{dt} + n\,\nabla \cdot \mathbf{V} = 0, \tag{4.202}$$

where $d/dt \equiv \partial/\partial t + \mathbf{V} \cdot \nabla$. Here, use has been made of the fact that $\nabla \cdot \mathbf{j} = 0$ in a quasi-neutral plasma.

Equations (4.195) and (4.197) can be summed to give the *equation of motion*:

$$m_i\,n\,\frac{d\mathbf{V}}{dt} - \mathbf{j} \times \mathbf{B} \simeq \mathbf{0}. \tag{4.203}$$

Finally, Equations (4.195), (4.200), and (4.201) can be combined to give a modified *Ohm's law*:

$$\mathbf{E} + \mathbf{V}\times\mathbf{B} \simeq \frac{\mathbf{F}_U}{n\,e} + \frac{\mathbf{j}\times\mathbf{B}}{n\,e} + \frac{m_e}{n\,e^2}\frac{d\mathbf{j}}{dt} + \frac{m_e}{n\,e^2}(\mathbf{j}\cdot\nabla)\,\mathbf{V} - \frac{m_e}{n^2\,e^3}(\mathbf{j}\cdot\nabla)\,\mathbf{j}. \tag{4.204}$$

The first term on the right-hand side of the previous equation corresponds to resistivity, the second corresponds to the *Hall effect*, the third corresponds to the effect of electron inertia, and the remaining terms are usually negligible.

4.13 MHD Equations

The MHD equations take the form:

$$\frac{\partial n}{\partial t} + \nabla\cdot(n\,\mathbf{V}_e) = 0, \tag{4.205}$$

$$m_e\,n\,\frac{\partial\mathbf{V}_e}{\partial t} + m_e\,n\,(\mathbf{V}_e\cdot\nabla)\mathbf{V}_e + \nabla p_e + [\delta^{-1}]\,e\,n\,(\mathbf{E} + \mathbf{V}_e\times\mathbf{B}) = [\zeta]\,\mathbf{F}_U + \mathbf{F}_T, \tag{4.206}$$

$$\frac{3}{2}\frac{\partial p_e}{\partial t} + \frac{3}{2}(\mathbf{V}_e\cdot\nabla)\,p_e + \frac{5}{2}\,p_e\,\nabla\cdot\mathbf{V}_e = -[\delta^{-1}\,\zeta\,\mu^2]\,W_i, \tag{4.207}$$

and

$$\frac{\partial n}{\partial t} + \nabla\cdot(n\,\mathbf{V}_i) = 0, \tag{4.208}$$

$$m_i\,n\,\frac{\partial\mathbf{V}_i}{\partial t} + m_i\,n\,(\mathbf{V}_i\cdot\nabla)\mathbf{V}_i + \nabla p_i - [\delta^{-1}]\,e\,n\,(\mathbf{E} + \mathbf{V}_i\times\mathbf{B}) = -[\zeta]\,\mathbf{F}_U - \mathbf{F}_T, \tag{4.209}$$

$$\frac{3}{2}\frac{\partial p_i}{\partial t} + \frac{3}{2}(\mathbf{V}_i\cdot\nabla)\,p_i + \frac{5}{2}\,p_i\,\nabla\cdot\mathbf{V}_i = [\delta^{-1}\,\zeta\,\mu^2]\,W_i. \tag{4.210}$$

These equations can also be simplified by making use of the smallness of the mass ratio m_e/m_i. Now, according to the ordering adopted in Section 4.11, $U \sim \delta\,V_e \sim \delta\,V_i$ in the MHD limit. It follows from Equations (4.200) and (4.201) that

$$\mathbf{V}_i \simeq \mathbf{V} + O\!\left(\frac{m_e}{m_i}\right), \tag{4.211}$$

and

$$\mathbf{V}_e \simeq \mathbf{V} - [\delta]\,\frac{\mathbf{j}}{n\,e} + O\!\left(\frac{m_e}{m_i}\right). \tag{4.212}$$

The main point, here, is that in the MHD limit the velocity difference between the electron and ion fluids is relatively small.

Equations (4.205) and (4.208) yield the *continuity equation*:

$$\frac{dn}{dt} + n\,\nabla\cdot\mathbf{V} = 0, \tag{4.213}$$

where $d/dt \equiv \partial/\partial t + \mathbf{V}\cdot\nabla$.

Equations (4.206) and (4.209) can be summed to give the *equation of motion*:

$$m_i\,n\,\frac{d\mathbf{V}}{dt} + \nabla p - \mathbf{j}\times\mathbf{B} \simeq 0. \tag{4.214}$$

Here, $p = p_e + p_i$ is the total pressure. Note that all terms in the previous equation are the same order in δ.

The $O(\delta^{-1})$ components of Equations (4.206) and (4.209) yield the *Ohm's law*:

$$\mathbf{E} + \mathbf{V}\times\mathbf{B} \simeq \mathbf{0}. \tag{4.215}$$

This is sometimes called the *perfect conductivity equation*, because it is identical to the Ohm's law in a perfectly conducting liquid.

Equations (4.207) and (4.210) can be summed to give the *energy evolution equation*:

$$\frac{3}{2}\frac{dp}{dt} + \frac{5}{2}\,p\,\nabla\cdot\mathbf{V} \simeq 0. \tag{4.216}$$

Equations (4.213) and (4.216) can be combined to give the more familiar *adiabatic equation of state*:

$$\frac{d}{dt}\left(\frac{p}{n^{5/3}}\right) \simeq 0. \tag{4.217}$$

Finally, the $O(\delta^{-1})$ components of Equations (4.207) and (4.210) yield

$$W_i \simeq 0, \tag{4.218}$$

or $T_e \simeq T_i$ [see Equation (4.101)]. Thus, we expect equipartition of the thermal energy between electrons and ions in the MHD limit.

4.14 Drift Equations

The drift equations take the form:

$$\frac{\partial n}{\partial t} + \nabla\cdot(n\,\mathbf{V}_e) = 0, \tag{4.219}$$

$$m_e\,n\,\frac{\partial \mathbf{V}_e}{\partial t} + m_e\,n\,(\mathbf{V}_e\cdot\nabla)\mathbf{V}_e + [\delta^{-2}]\,\nabla p_e + [\zeta^{-1}]\,\nabla\cdot\boldsymbol{\pi}_e \tag{4.220}$$

$$+[\delta^{-2}]\,e\,n\,(\mathbf{E} + \mathbf{V}_e\times\mathbf{B}) = [\delta^{-2}\,\zeta]\,\mathbf{F}_U + [\delta^{-2}]\,\mathbf{F}_T,$$

$$\frac{3}{2}\frac{\partial p_e}{\partial t} + \frac{3}{2}\,(\mathbf{V}_e\cdot\nabla)\,p_e + \frac{5}{2}\,p_e\,\nabla\cdot\mathbf{V}_e$$

$$+[\zeta^{-1}]\,\nabla\cdot\mathbf{q}_{Te} + \nabla\cdot\mathbf{q}_{Ue} = -[\delta^{-2}\,\zeta\,\mu^2]\,W_i$$

$$+\,[\zeta]\,W_U + W_T, \tag{4.221}$$

and

$$\frac{\partial n}{\partial t} + \nabla\cdot(n\,\mathbf{V}_i) = 0, \tag{4.222}$$

$$m_i\,n\,\frac{\partial \mathbf{V}_i}{\partial t} + m_i\,n\,(\mathbf{V}_i\cdot\nabla)\mathbf{V}_i + [\delta^{-2}]\,\nabla p_i + [\zeta^{-1}]\,\nabla\cdot\boldsymbol{\pi}_i \tag{4.223}$$

$$-[\delta^{-2}]\,e\,n\,(\mathbf{E} + \mathbf{V}_i\times\mathbf{B}) = -[\delta^{-2}\,\zeta]\,\mathbf{F}_U - [\delta^{-2}]\,\mathbf{F}_T,$$

$$\frac{3}{2}\frac{\partial p_i}{\partial t} + \frac{3}{2}\,(\mathbf{V}_i\cdot\nabla)\,p_i + \frac{5}{2}\,p_i\,\nabla\cdot\mathbf{V}_i$$

$$+[\zeta^{-1}]\,\nabla\cdot\mathbf{q}_i = [\delta^{-2}\,\zeta\,\mu^2]\,W_i. \tag{4.224}$$

In the drift limit, the motions of the electron and ion fluids are sufficiently different that there is little to be gained in rewriting the drift equations in terms of the center-of-mass velocity and the plasma current. Instead, let us consider the $O(\delta^{-2})$ components of Equations (4.220) and (4.223):

$$\mathbf{E} + \mathbf{V}_e\times\mathbf{B} \simeq -\frac{\nabla p_e}{e\,n} - \frac{0.71\,\nabla_\parallel T_e}{e}, \tag{4.225}$$

$$\mathbf{E} + \mathbf{V}_i\times\mathbf{B} \simeq +\frac{\nabla p_i}{e\,n} - \frac{0.71\,\nabla_\parallel T_e}{e}. \tag{4.226}$$

In the previous equations, we have neglected all $O(\zeta)$ terms for the sake of simplicity. Equations (4.225)–(4.226) can be inverted to give

$$\mathbf{V}_{\perp\,e} \simeq \mathbf{V}_E + \mathbf{V}_{*\,e}, \tag{4.227}$$

$$\mathbf{V}_{\perp\,i} \simeq \mathbf{V}_E + \mathbf{V}_{*\,i}. \tag{4.228}$$

Here, $\mathbf{V}_E \equiv \mathbf{E}\times\mathbf{B}/B^2$ is the $\mathbf{E}\times\mathbf{B}$ velocity, whereas

$$\mathbf{V}_{*\,e} \equiv \frac{\nabla p_e\times\mathbf{B}}{e\,n\,B^2}, \tag{4.229}$$

and

$$\mathbf{V}_{*i} \equiv -\frac{\nabla p_i \times \mathbf{B}}{e\,n\,B^2}, \tag{4.230}$$

are termed the *electron diamagnetic velocity* and the *ion diamagnetic velocity*, respectively.

According to Equations (4.227)–(4.228), in the drift approximation the velocity of the electron fluid perpendicular to the magnetic field is the sum of the $\mathbf{E} \times \mathbf{B}$ velocity and the electron diamagnetic velocity. A similar statement can be made for the ion fluid. By contrast, in the MHD approximation the perpendicular velocities of the two fluids consist of the $\mathbf{E} \times \mathbf{B}$ velocity alone, and are, therefore, identical to lowest order. The main difference between the two orderings lies in the assumed magnitude of the electric field. In the MHD limit

$$\frac{E}{B} \sim v_t, \tag{4.231}$$

whereas in the drift limit

$$\frac{E}{B} \sim \delta\,v_t \sim v_d. \tag{4.232}$$

Thus, the MHD ordering can be regarded as a strong electric field ordering, whereas the drift ordering corresponds to a weak electric field ordering.

The diamagnetic velocities are so named because the *diamagnetic current*,

$$\mathbf{j}_* \equiv -e\,n\,(\mathbf{V}_{*e} - \mathbf{V}_{*i}) = -\frac{\nabla p \times \mathbf{B}}{B^2}, \tag{4.233}$$

generally acts to reduce the magnitude of the magnetic field inside the plasma.

The electron diamagnetic velocity can be written

$$\mathbf{V}_{*e} = \frac{T_e\,\nabla n \times \mathbf{b}}{e\,n\,B} + \frac{\nabla T_e \times \mathbf{b}}{e\,B}. \tag{4.234}$$

In order to account for this velocity, let us consider a simplified case in which the electron temperature is uniform, there is a uniform density gradient running along the x-direction, and the magnetic field is parallel to the z-axis. (See Figure 4.3.) The electrons gyrate in the x-y plane in circles of radius $\rho_e \sim v_e/|\Omega_e|$. At a given point, coordinate x_0, say, on the x-axis, the electrons that come from the right and the left have traversed distances of approximate magnitude ρ_e. Thus, the electrons from the right originate from regions where the particle density is approximately $\rho_e\,\partial n/\partial x$ greater than the regions from which the electrons from the left originate. It follows that the y-directed particle flux is unbalanced, with slightly more particles moving in the $-y$-direction than in the $+y$-direction. Thus, there is a net particle flux in the $-y$-direction: that is, in the direction of $\nabla n \times \mathbf{b}$. The magnitude of this flux is

$$n\,V_{*e} \sim \rho_e\,\frac{\partial n}{\partial x}\,v_e \sim \frac{T_e}{e\,B}\frac{\partial n}{\partial x}. \tag{4.235}$$

There is no unbalanced particle flux in the x-direction, because the x-directed fluxes are associated with electrons that originate from regions where $x = x_0$. We have

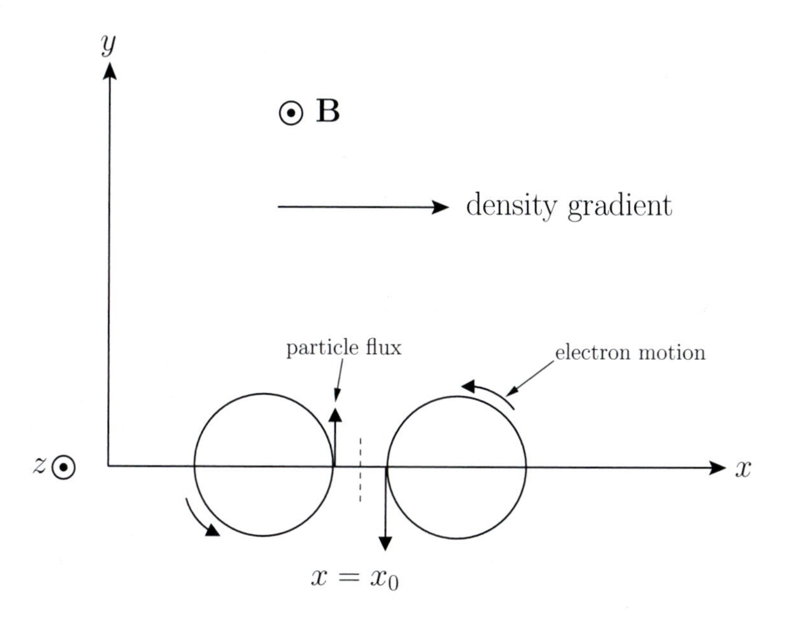

Figure 4.3
Origin of the diamagnetic velocity in a magnetized plasma.

now accounted for the first term on the right-hand side of Equation (4.234). We can account for the second term using similar arguments. The ion diamagnetic velocity is similar in magnitude to the electron diamagnetic velocity, but is oppositely directed, because ions gyrate in the opposite direction to electrons.

The most curious aspect of diamagnetic flows is that they represent fluid flows for which there is no corresponding motion of the particle guiding centers. Nevertheless, the diamagnetic velocities are real fluid velocities, and the associated diamagnetic current is a real current. For instance, the diamagnetic current contributes to force balance inside the plasma, and also gives rise to ohmic heating.

4.15 Closure in Collisionless Magnetized Plasmas

Up to now, we have only considered fluid closure in collisional magnetized plasmas. Unfortunately, the majority of magnetized plasmas encountered in nature—in particular, fusion, space, and astrophysical plasmas—are collisionless. Let us consider what happens to the cold-plasma equations, the MHD equations, and the drift equations, in the limit in which the mean-free-path goes to infinity (i.e., $\zeta \to 0$).

In the limit $\zeta \to 0$, the cold-plasma equations reduce to

$$\frac{dn}{dt} + n\,\nabla\cdot\mathbf{V} = 0, \tag{4.236}$$

$$m_i\,n\,\frac{d\mathbf{V}}{dt} - \mathbf{j}\times\mathbf{B} = \mathbf{0}, \tag{4.237}$$

$$\mathbf{E} + \mathbf{V}\times\mathbf{B} = \frac{\mathbf{j}\times\mathbf{B}}{n\,e} + \frac{m_e}{n\,e^2}\frac{d\mathbf{j}}{dt} + \frac{m_e}{n\,e^2}\,(\mathbf{j}\cdot\nabla)\,\mathbf{V} - \frac{m_e}{n^2\,e^3}\,(\mathbf{j}\cdot\nabla)\,\mathbf{j}. \tag{4.238}$$

Here, we have neglected the resistivity term, because it is $O(\zeta)$. Note that none of the remaining terms in these equations depend explicitly on collisions. Nevertheless, the absence of collisions poses a serious problem. Whereas the magnetic field effectively confines charged particles in directions perpendicular to magnetic field-lines, by forcing them to execute tight Larmor orbits, we have now lost all confinement along field-lines. But, does this matter?

The typical frequency associated with fluid motion is the transit frequency, V/L. However, according to Equation (4.173), the cold-plasma ordering implies that the transit frequency is of similar magnitude to a typical gyrofrequency:

$$\frac{V}{L} \sim \Omega. \tag{4.239}$$

So, how far is a charged particle likely to drift along a field-line in an inverse transit frequency? The answer is

$$\Delta l_{\parallel} \sim \frac{v_t\,L}{V} \sim \frac{v_t}{\Omega} \sim \rho. \tag{4.240}$$

In other words, the fluid motion in the cold-plasma limit is so fast that charged particles only have time to drift a Larmor radius along field-lines on a typical dynamical timescale. Under these circumstances, it does not really matter that the particles are not localized along field-lines—the lack of parallel confinement manifests itself too slowly to affect the plasma dynamics. We conclude, therefore, that the cold-plasma equations remain valid in the collisionless limit, provided, of course, that the plasma dynamics are sufficiently rapid for the basic cold-plasma ordering (4.239) to apply. In fact, the only difference between the collisional and collisionless cold-plasma equations is the absence of the resistivity term in Ohm's law in the latter case.

Let us now consider the MHD limit. In this case, the typical transit frequency is

$$\frac{V}{L} \sim \delta\,\Omega. \tag{4.241}$$

Thus, charged particles typically drift a distance

$$\Delta l_{\parallel} \sim \frac{v_t\,L}{V} \sim \frac{v_t}{\delta\,\Omega} \sim L \tag{4.242}$$

along field-lines in an inverse transit frequency. In other words, the fluid motion in the MHD limit is sufficiently slow that changed particles have time to drift along field-lines all the way across the system on a typical dynamical time-scale. Thus, strictly speaking, the MHD equations are invalidated by the lack of particle confinement along magnetic field-lines.

In fact, in collisionless plasmas, MHD theory is replaced by a theory known as *kinetic-MHD* (Kruskal and Oberman 1958; Rosenbluth and Rostoker 1959). The latter theory is a combination of a one-dimensional kinetic theory, describing particle motion along magnetic field-lines, and a two-dimensional fluid theory, describing perpendicular motion. Unfortunately, the equations of kinetic-MHD are considerably more complicated that the conventional MHD equations. Is there any situation in which we can salvage the simpler MHD equations in a collisionless plasma? Fortunately, there is one case in which this is possible.

It turns out that in both varieties of MHD the motion of the plasma parallel to magnetic field-lines is associated with the dynamics of sound waves, whereas the motion perpendicular to field-lines is associated with the dynamics of a new type of wave called an *Alfvén wave*. As we shall see, in Chapter 5, Alfvén waves involve the "twanging" motion of magnetic field-lines—a bit like the twanging of guitar strings. It is only the sound wave dynamics that are significantly modified when we move from a collisional to a collisionless plasma. It follows, therefore, that the MHD equations remain a reasonable approximation in a collisionless plasma in situations where the dynamics of sound waves, parallel to the magnetic field, are unimportant compared to the dynamics of Alfvén waves, perpendicular to the field. This situation arises whenever the parameter

$$\beta = \frac{2\,\mu_0\,p}{B^2} \tag{4.243}$$

(see Section 1.9) is much less than unity. In fact, it is easily demonstrated that

$$\beta \sim \left(\frac{V_S}{V_A}\right)^2, \tag{4.244}$$

where V_S is the sound speed (i.e., thermal velocity), and V_A is the speed of an Alfvén wave. Thus, the inequality

$$\beta \ll 1 \tag{4.245}$$

ensures that the collisionless parallel plasma dynamics are too slow to affect the perpendicular dynamics.

We conclude, therefore, that in a low-β, collisionless, magnetized plasma, the MHD equations,

$$\frac{dn}{dt} + n\,\nabla\cdot\mathbf{V} = 0, \tag{4.246}$$

$$m_i\,n\,\frac{d\mathbf{V}}{dt} = \mathbf{j}\times\mathbf{B} - \nabla p, \tag{4.247}$$

$$\mathbf{E} + \mathbf{V}\times\mathbf{B} = \mathbf{0}, \tag{4.248}$$

$$\frac{d}{dt}\left(\frac{p}{n^{5/3}}\right) = 0, \tag{4.249}$$

describe fairly well plasma dynamics that satisfy the basic MHD ordering (4.241).

Let us, finally, consider the drift limit. In this case, the typical transit frequency is

$$\frac{V}{L} \sim \delta^2 \Omega. \tag{4.250}$$

Thus, charged particles typically drift a distance

$$\Delta l_{\parallel} \sim \frac{v_t L}{V} \sim \frac{L}{\delta} \tag{4.251}$$

along field-lines in an inverse transit frequency. In other words, the fluid motion in the drift limit is so slow that charged particles drifting along field-lines have time to traverse the system very many times on a typical dynamical time-scale. In fact, in this limit we have to draw a distinction between those particles that always drift along field-lines in the same direction, and those particles that are trapped between magnetic mirror points and, therefore, continually reverse their direction of motion along field-lines. The former are termed *passing particles*, whereas the latter are termed *trapped particles*.

Now, in the drift limit, the perpendicular drift velocity of charged particles, which is a combination of $\mathbf{E} \times \mathbf{B}$ drift, grad-B drift, and curvature drift (see Chapter 2), is approximately

$$v_d \sim \delta\, v_t. \tag{4.252}$$

Thus, charged particles typically drift a distance

$$\Delta l_{\perp} \sim \frac{v_d L}{V} \sim L \tag{4.253}$$

across field-lines in an inverse transit time. In other words, the fluid motion in the drift limit is so slow that charged particles have time to drift perpendicular to field-lines all the way across the system on a typical dynamical time-scale. It is, thus, clear that in the drift limit the absence of collisions implies lack of confinement both parallel and perpendicular to the magnetic field. This means that the collisional drift equations, (4.219)–(4.221) and (4.222)–(4.224), are completely invalid in the long mean-free-path limit.

In fact, in collisionless plasmas, Braginskii-type transport theory—conventionally known as *classical transport theory*—is replaced by a new theory—known as *neoclassical transport theory*—which is a combination of a two-dimensional kinetic theory, describing particle motion on drift surfaces, and a one-dimensional fluid theory, describing motion perpendicular to the drift surfaces (Bernstein 1974; Hinton and Hazeltine 1976). Here, a drift surface is a closed surface formed by the locus of a charged particle's drift orbit (including drifts parallel and perpendicular to the magnetic field). Of course, the orbits only form closed surfaces if the plasma is confined, but there is little point in examining transport in an unconfined plasma. Unlike classical transport theory, which is strictly local in nature, neoclassical transport theory is nonlocal, in the sense that the transport coefficients depend on the average values of plasma properties taken over drift surfaces.

4.16 Langmuir Sheaths

Virtually all terrestrial plasmas are contained within solid vacuum vessels. But, what happens to plasma in the immediate vicinity of a vessel wall? Actually, to a first approximation, when ions and electrons hit a solid surface they recombine and are lost to the plasma. Hence, we can treat the wall as a perfect sink of particles. Now, given that the electrons in a plasma generally move much faster than the ions, the initial electron flux into the wall greatly exceeds the ion flux, assuming that the wall starts off unbiased with respect to the plasma. Of course, this flux imbalance causes the wall to charge up negatively, and so generates a potential barrier that repels the electrons, and thereby reduces the electron flux. Debye shielding confines this barrier to a thin layer of plasma, whose thickness is a few Debye lengths, coating the inside surface of the wall. This layer is known as a *plasma sheath* or a *Langmuir sheath*. The height of the potential barrier continues to grow as long as there is a net flux of negative charge into the wall. This process presumably comes to an end, and a steady-state is attained, when the potential barrier becomes sufficiently large to make electron flux equal to the ion flux (Hazeltine and Waelbroeck 2004).

Let us construct a one-dimensional model of an unmagnetized, steady-state, Langmuir sheath. Suppose that the wall lies at $x = 0$, and that the plasma occupies the region $x > 0$. Let us treat the ions and the electrons inside the sheath as collisionless fluids. The ion and electron equations of motion are thus written

$$m_i\,n_i\,V_i\,\frac{dV_i}{dx} = -T_i\,\frac{dn_i}{dx} - e\,n_i\,\frac{d\phi}{dx}, \tag{4.254}$$

$$m_e\,n_e\,V_e\,\frac{dV_e}{dx} = -T_e\,\frac{dn_e}{dx} + e\,n_e\,\frac{d\phi}{dx}, \tag{4.255}$$

respectively. Here, $\phi(x)$ is the electrostatic potential. Moreover, we have assumed uniform ion and electron temperatures, T_i and T_e, respectively, for the sake of simplicity. We have also neglected any off-diagonal terms in the ion and electron stress-tensors, because these terms are comparatively small. Note that quasi-neutrality does not apply inside the sheath, and so the ion and electron number densities, n_e and n_i, respectively, are not necessarily equal to one another.

Consider the ion fluid. Let us assume that the mean ion velocity, V_i, is much greater than the ion thermal velocity, $(T_i/m_i)^{1/2}$. Because, as will become apparent, $V_i \sim (T_e/m_i)^{1/2}$, this ordering necessarily implies that $T_i \ll T_e$: that is, that the ions are cold with respect to the electrons. It turns out that plasmas in the immediate vicinity of solid walls often have comparatively cold ions, so our ordering assumption is fairly reasonable. In the cold ion limit, the pressure term in Equation (4.254) is negligible, and the equation can be integrated to give

$$\frac{1}{2}\,m_i\,V_i^2(x) + e\,\phi(x) = \frac{1}{2}\,m_i\,V_s^2 + e\,\phi_s. \tag{4.256}$$

Here, V_s and ϕ_s are the mean ion velocity and electrostatic potential, respectively, at the edge of the sheath (i.e., $x \to \infty$). Now, ion fluid continuity requires that

$$n_i(x)\,V_i(x) = n_s\,V_s, \tag{4.257}$$

where n_s is the ion number density at the sheath boundary. Incidentally, because we expect quasi-neutrality to hold in the plasma outside the sheath, the electron number density at the edge of the sheath must also be n_s (assuming singly-charged ions). The previous two equations can be combined to give

$$V_i = V_s\left[1 - \frac{2\,e}{m_i\,V_s^2}\,(\phi - \phi_s)\right]^{1/2}, \tag{4.258}$$

$$n_i = n_s\left[1 - \frac{2\,e}{m_i\,V_s^2}\,(\phi - \phi_s)\right]^{-1/2}. \tag{4.259}$$

Consider the electron fluid. Let us assume that the mean electron velocity, V_e, is much less than the electron thermal velocity, $(m_e/T_e)^{1/2}$. In fact, this must be the case, otherwise, the electron flux to the wall would greatly exceed the ion flux. Now, if the electron fluid is essentially stationary then the left-hand side of Equation (4.255) is negligible, and the equation can be integrated to give

$$n_e = n_s\,\exp\left[\frac{e\,(\phi - \phi_s)}{T_e}\right]. \tag{4.260}$$

Here, we have made use of the fact that $n_e = n_s$ at the edge of the sheath.

Poisson's equation is written

$$\epsilon_0\,\frac{d^2\phi}{dx^2} = e\,(n_e - n_i). \tag{4.261}$$

It follows that

$$\epsilon_0\,\frac{d^2\phi}{dx^2} = e\,n_s\left(\exp\left[\frac{e\,(\phi - \phi_s)}{T_e}\right] - \left[1 - \frac{2\,e}{m_i\,V_s^2}\,(\phi - \phi_s)\right]^{-1/2}\right). \tag{4.262}$$

Let $\Phi = -e\,(\phi - \phi_s)/T_e$, $y = \sqrt{2}\,x/\lambda_D$, and

$$K = \frac{m_i\,V_s^2}{2\,T_e}, \tag{4.263}$$

where $\lambda_D = (\epsilon_0\,T_e/e^2\,n_s)^{1/2}$ is the Debye length. Equation (4.262) transforms to

$$2\,\frac{d^2\Phi}{dy^2} = -\mathrm{e}^{-\Phi} + \left(1 + \frac{\Phi}{K}\right)^{-1/2}, \tag{4.264}$$

subject to the boundary condition $\Phi \to 0$ as $y \to \infty$. Multiplying through by $d\Phi/dy$, integrating with respect to y, and making use of the boundary condition, we obtain

$$\left(\frac{d\Phi}{dy}\right)^2 = \mathrm{e}^{-\Phi} - 1 + 2\,K\left[\left(1 + \frac{\Phi}{K}\right)^{1/2} - 1\right]. \tag{4.265}$$

Unfortunately, the previous equation is highly nonlinear, and can only be solved numerically. However, it is not necessary to attempt this to see that a physical solution can only exist if the right-hand side of the equation is positive for all $y \geq 0$. Consider the limit $y \to \infty$. It follows from the boundary condition that $\Phi \to 0$. Expanding the right-hand side of Equation (4.265) in powers of Φ, we find that the zeroth- and first-order terms cancel, and we are left with

$$\left(\frac{d\Phi}{dy}\right)^2 \simeq \frac{\Phi^2}{2}\left(1 - \frac{1}{2\,K}\right) + \frac{\Phi^3}{3}\left(\frac{3}{8\,K^2} - 1\right) + O(\Phi^4). \tag{4.266}$$

Now, the purpose of the sheath is to shield the plasma from the wall potential. It can be seen, from the previous expression, that the physical solution with maximum possible shielding corresponds to $K = 1/2$, because this choice eliminates the first term on the right-hand side (thereby making Φ as small as possible at large y) leaving the much smaller, but positive (note that Φ is positive), second term. Hence, we conclude that

$$V_s = \left(\frac{T_e}{m_i}\right)^{1/2}. \tag{4.267}$$

This result is known as the *Bohm sheath criterion*. It is a somewhat surprising result, because it indicates that ions at the edge of the sheath are already moving toward the wall at a considerable velocity. Of course, the ions are further accelerated as they pass through the sheath. Because the ions are presumably at rest in the interior of the plasma, it is clear that there must exist a region sandwiched between the sheath and the main plasma in which the ions are accelerated from rest to the Bohm velocity, $V_s = (T_e/m_i)^{1/2}$. This region is called the *pre-sheath*, and is both quasi-neutral and much wider than the sheath (the actual width depends on the nature of the ion source).

The ion current density at the wall is

$$j_i = -e\,n_i(0)\,V_i(0) = -e\,n_s\,V_s = -e\,n_s\left(\frac{T_e}{m_i}\right)^{1/2}. \tag{4.268}$$

This current density is negative because the ions are moving in the negative x-direction. What about the electron current density? Well, the number density of electrons at the wall is $n_e(0) = n_s\,\exp[\,e\,(\phi_w - \phi_s)/T_e)]$, where $\phi_w = \phi(0)$ is the wall potential. Let us assume that the electrons have a Maxwellian velocity distribution peaked at zero velocity (because the electron fluid velocity is much less than the electron thermal velocity). It follows that half of the electrons at $x = 0$ are moving in the negative-x direction, and half in the positive-x direction. Of course, the former electrons hit the wall, and thereby constitute an electron current to the wall. This current is $j_e = (1/4)\,e\,n_e(0)\,\bar{V}_e$, where the $1/4$ comes from averaging over solid angle, and $\bar{V}_e = (8\,T_e/\pi\,m_e)^{1/2}$ is the mean electron speed corresponding to a Maxwellian velocity distribution (Reif 1965). Thus, the electron current density at the wall is

$$j_e = e\,n_s\left(\frac{T_e}{2\pi\,m_e}\right)^{1/2}\exp\left[\frac{e\,(\phi_w - \phi_s)}{T_e}\right]. \tag{4.269}$$

In order to replace the electrons lost to the wall, the electrons must have a mean velocity

$$V_{e\,s} = \frac{j_e}{e\,n_s} = \left(\frac{T_e}{2\pi\,m_e}\right)^{1/2}\exp\left[\frac{e\,(\phi_w - \phi_s)}{T_e}\right] \tag{4.270}$$

at the edge of the sheath. However, we previously assumed that any electron fluid velocity was much less than the electron thermal velocity, $(T_e/m_e)^{1/2}$. As is clear from the previous equation, this is only possible provided that

$$\exp\left[\frac{e\,(\phi_w - \phi_s)}{T_e}\right] \ll 1: \tag{4.271}$$

that is, provided that the wall potential is sufficiently negative to strongly reduce the electron number density at the wall. The net current density at the wall is

$$j = e\,n_s\left(\frac{T_e}{m_i}\right)^{1/2}\left(\left[\frac{m_i}{2\pi\,m_e}\right]^{1/2}\exp\left[\frac{e\,(\phi_w - \phi_s)}{T_e}\right] - 1\right). \tag{4.272}$$

Of course, we require $j = 0$ in a steady-state sheath, in order to prevent wall charging, and so we obtain

$$e\,(\phi_w - \phi_s) = -T_e\,\ln\left(\frac{m_i}{2\pi\,m_e}\right)^{1/2}. \tag{4.273}$$

We conclude that, in a steady-state sheath, the wall is biased negatively with respect to the sheath edge by an amount that is proportional to the electron temperature.

For a hydrogen plasma, $\ln(m_i/2\pi\,m_e) \simeq 2.8$. Thus, hydrogen ions enter the sheath with an initial energy $(1/2)\,m_i\,V_s^2 = 0.5\,T_e$ eV, fall through the sheath potential, and so impact the wall with energy $3.3\,T_e$ eV.

A *Langmuir probe* is a device used to determine the electron temperature and electron number density of a plasma. It works by inserting an electrode that is biased with respect to the vacuum vessel into the plasma. Provided that the bias voltage is not too positive, we would expect the probe current to vary as

$$I = A\,e\,n_s\left(\frac{T_e}{m_i}\right)^{1/2}\left[\left(\frac{m_i}{2\pi\,m_e}\right)^{1/2}\exp\left(\frac{e\,V}{T_e}\right) - 1\right], \tag{4.274}$$

where A is the surface area of the probe, and V its bias with respect to the vacuum vessel. [See Equation (4.272).] For strongly negative biases, the probe current saturates in the ion (negative) direction. The characteristic current that flows in this situation is called the *ion saturation current*, and is of magnitude

$$I_s = A\,e\,n_s\left(\frac{T_e}{m_i}\right)^{1/2}. \tag{4.275}$$

For less negative biases, the current-voltage relation of the probe has the general form

$$\ln I = C + \frac{e\,V}{T_e}, \tag{4.276}$$

where C is a constant. Thus, a plot of $\ln I$ versus V gives a straight-line from whose slope the electron temperature can be deduced. Note, however, that if the bias voltage becomes too positive then electrons cease to be effectively repelled from the probe surface, and the current-voltage relation (4.274) breaks down. Given the electron temperature, a measurement of the ion saturation current allows the electron number density at the sheath edge, n_s, to be calculated from Equation (4.275). Now, in order to accelerate ions to the Bohm velocity, the potential drop across the pre-sheath needs to be $e(\phi_p - \phi_s) = -T_e/2$, where ϕ_p is the electric potential in the interior of the plasma. It follows from Equation (4.260) that the relationship between the electron number density at the sheath boundary, n_s, and the number density in the interior of the plasma, n_p, is

$$n_s = n_p\,\mathrm{e}^{-0.5} \simeq 0.61\,n_p. \tag{4.277}$$

Thus, n_p can also be determined from the probe.

4.17 Exercises

4.1 Verify Equations (4.17) and (4.18).

4.2 Verify Equation (4.30).

4.3 Derive Equations (4.36)–(4.38) from Equation (4.35).

4.4 Derive Equations (4.41)–(4.43) from Equations (4.36)–(4.38).

4.5 Derive Equation (4.53) from Equation (4.49).

4.6 Consider the Maxwellian distribution

$$f(\mathbf{v}) = \frac{n}{\pi^{3/2}\,v_t^3}\,\exp\left(-\frac{v^2}{v_t^2}\right).$$

Let

$$I_n = \int \frac{f}{n}\left(\frac{v}{v_t}\right)^n d^3\mathbf{v}.$$

Demonstrate that $I_{-2} = 2$, $I_0 = 1$, $I_2 = 3/2$, and $I_4 = 15/4$.

4.7 Consider a neutral gas in a force-free steady-state equilibrium. The particle distribution function f satisfies the simplified kinetic equation

$$\mathbf{v}\cdot\nabla f = C(f).$$

We can crudely approximate the collision operator as

$$C = -\nu\,(f - f_0)$$

where ν is the effective collision frequency, and

$$f_0 = \frac{n}{\pi^{3/2}\,v_t^3}\exp\left[-\frac{(\mathbf{v}-\mathbf{V})^2}{v_t^2}\right].$$

Here, $v_t = \sqrt{2\,T/m}$. Suppose that the mean-free-path $l = v_t/\nu$ is much less than the typical variation lengthscale of equilibrium quantities (such as n, T, and $\mathbf{V}$). Demonstrate that it is a good approximation to write

$$f = f_0 - \nu^{-1}\,\mathbf{v}\cdot\nabla f_0.$$

(a) Suppose that n and T are uniform, but that $\mathbf{V} = V_y(x)\,\mathbf{e}_y$. Demonstrate that the only non-zero components of the viscosity tensor are

$$\pi_{xy} = \pi_{yx} = -\eta\,\frac{dV_y}{dx},$$

where

$$\eta = \frac{1}{2}\,m\,n\,\nu\,l^2.$$

(b) Suppose that n is uniform, and $\mathbf{V} = \mathbf{0}$, but that $T = T(x)$. Demonstrate that the only non-zero component of the heat flux density is

$$q_x = -\kappa\,\frac{dT}{dx},$$

where

$$\kappa = \frac{5}{2}\,n\,\nu\,l^2.$$

(c) Suppose that $\mathbf{V} = \mathbf{0}$, and $n = n(x)$ and $T = T(x)$, but that $p = n\,T$ is constant. Demonstrate that the only non-zero component of the heat flux density is

$$q_x = -\kappa\,\frac{dT}{dx},$$

where

$$\kappa = \frac{5}{4}\,n\,\nu\,l^2.$$

4.8 Consider a spatially uniform, unmagnetized plasma in which both species have zero mean flow velocity. Let n_e and T_e be the electron number density and temperature, respectively. Let $\mathbf{E}$ be the ambient electric field. The electron distribution function f_e satisfies the simplified kinetic equation

$$-\frac{e}{m_e}\,\mathbf{E}\cdot\nabla_v f_e = C_e.$$

We can crudely approximate the electron collision operator as

$$C_e = -\nu_e\,(f_e - f_0)$$

where ν_e is the effective electron-ion collision frequency, and

$$f_0 = \frac{n_e}{\pi^{3/2}\,v_{te}^3}\exp\left(-\frac{v^2}{v_{te}^2}\right).$$

Here, $v_{te} = \sqrt{2\,T_e/m_e}$. Suppose that $E \ll m_e\,\nu_e\,v_{te}/e$. Demonstrate that it is a good approximation to write

$$f_e = f_0 + \frac{e}{m_e\,\nu_e}\,\mathbf{E}\cdot\nabla_v f_0.$$

Hence, show that

$$\mathbf{j} = \sigma\,\mathbf{E},$$

where

$$\sigma = \frac{e^2\,n_e}{m_e\,\nu_e}.$$

5

Waves in Cold Plasmas

5.1 Introduction

The cold-plasma equations describe waves (and other perturbations) that propagate through a plasma much faster than a typical thermal velocity. (See Section 4.11.) The collective motions described by the cold-plasma model are closely related to the individual particle motions discussed in Chapter 2. In fact, in the cold-plasma model, all particles (of a given species) at a given position effectively move with the same velocity. It follows that the fluid velocity is identical to the particle velocity, and is, therefore, governed by the same equations. However, the cold-plasma model goes beyond the single-particle description because it determines the electromagnetic fields self-consistently in terms of the charge and current densities generated by the particle motions. In this chapter, we shall use the cold-plasma equations to investigate the properties of small amplitude plasma waves.

What role, if any, does the geometry of the plasma equilibrium play in determining the properties of plasma waves? Clearly, geometry plays a key role for modes whose wavelengths are comparable to the dimensions of the plasma. However, it is plausible that waves whose wavelengths are much smaller than the plasma dimensions have properties that are, in a local sense, independent of the geometry. In other words, the local properties of small wavelength plasma oscillations are universal in nature. To investigate these properties, we can, to a first approximation, represent the plasma as a homogeneous equilibrium (corresponding to the limit $kL \rightarrow 0$, where k is the magnitude of the wavevector, and L is the characteristic equilibrium length-scale).

5.2 Plane Waves in Homogeneous Plasmas

The propagation of small amplitude plasma waves is described by linearized equations that are obtained by expanding the plasma equations of motion in powers of the wave amplitude, and then neglecting terms of order higher than unity.

Consider a homogeneous, magnetized, quasi-neutral plasma, consisting of equal numbers of electrons and ions, in which the mean velocities of both plasma species are zero. It follows that $\mathbf{E}_0 = \mathbf{0}$, and $\mathbf{j}_0 = \nabla \times \mathbf{B}_0/\mu_0 = \mathbf{0}$, where the subscript 0 denotes an equilibrium quantity. In a homogeneous medium, the general solution of a system of linear equations can be constructed as a superposition of plane wave solutions of the form (Fitzpatrick 2013)

$$\mathbf{E}(\mathbf{r}, t) = \mathbf{E}_{\mathbf{k}} \exp[\,\mathrm{i}\,(\mathbf{k}\cdot\mathbf{r} - \omega\,t)], \tag{5.1}$$

with analogous expressions for $\mathbf{B}(\mathbf{r}, t)$ and $\mathbf{V}(\mathbf{r}, t)$. Here, $\mathbf{E}$, $\mathbf{B}$, and $\mathbf{V}$ are the perturbed electric field, magnetic field, and plasma center-of-mass velocity, respectively. The surfaces of constant phase,

$$\mathbf{k}\cdot\mathbf{r} - \omega\,t = \text{constant}, \tag{5.2}$$

are planes perpendicular to $\mathbf{k}$, traveling at the velocity

$$\mathbf{v}_{\rm ph} = \frac{\omega}{k}\,\hat{\mathbf{k}}, \tag{5.3}$$

where $k \equiv |\mathbf{k}|$, and $\hat{\mathbf{k}}$ is a unit vector pointing in the direction of $\mathbf{k}$. Here, $\mathbf{v}_{\rm ph}$ is termed the *phase-velocity* of the wave (Fitzpatrick 2013). Henceforth, for ease of notation, we shall omit the subscript $\mathbf{k}$ from field variables.

Substitution of the plane-wave solution (5.1) into Maxwell's equations yields

$$\mathbf{k}\times\mathbf{B} = -\mathrm{i}\,\mu_0\,\mathbf{j} - \frac{\omega}{c^2}\,\mathbf{E}, \tag{5.4}$$

$$\mathbf{k}\times\mathbf{E} = \omega\,\mathbf{B}, \tag{5.5}$$

where $\mathbf{j}(\mathbf{r}, t)$ is the perturbed current density. In linear theory, the current is related to the electric field via

$$\mathbf{j} = \boldsymbol{\sigma}\cdot\mathbf{E}, \tag{5.6}$$

where the *electrical conductivity tensor*, $\boldsymbol{\sigma}$, is a function of both $\mathbf{k}$ and ω. In the presence of a non-zero equilibrium magnetic field, this tensor is anisotropic in nature.

Substitution of Equation (5.6) into Equation (5.4) yields

$$\mathbf{k}\times\mathbf{B} = -\frac{\omega}{c^2}\,\mathbf{K}\cdot\mathbf{E}, \tag{5.7}$$

where

$$\mathbf{K} = \mathbf{I} + \frac{\mathrm{i}\,\boldsymbol{\sigma}}{\epsilon_0\,\omega} \tag{5.8}$$

is termed the *dielectric permittivity tensor*. Here, $\mathbf{I}$ is the identity tensor. Eliminating the magnetic field between Equations (5.5) and (5.7), we obtain

$$\mathbf{M}\cdot\mathbf{E} = \mathbf{0}, \tag{5.9}$$

where

$$\mathbf{M} = \left(\frac{c}{\omega}\right)^2 \mathbf{k}\mathbf{k} - \left(\frac{c\,k}{\omega}\right)^2 \mathbf{I} + \mathbf{K}. \tag{5.10}$$

The solubility condition for Equation (5.9),

$$\mathcal{M}(\omega, \mathbf{k}) \equiv \det(\mathbf{M}) = 0, \tag{5.11}$$

is called the *dispersion relation*, and relates the wave angular frequency, ω, to the wavevector, $\mathbf{k}$. Also, as the name "dispersion relation" suggests, this relation allows us to determine the rate at which the different Fourier components of a wave pulse disperse due to the variation of their phase-velocity with frequency (Fitzpatrick 2013).

5.3 Cold-Plasma Dielectric Permittivity

In a collisionless plasma, the linearized cold-plasma equations are written [see Equations (4.236)–(4.238)],

$$m_i\, n_e\, \frac{\partial \mathbf{V}}{\partial t} = \mathbf{j} \times \mathbf{B}_0, \tag{5.12}$$

$$\mathbf{E} = -\mathbf{V} \times \mathbf{B}_0 + \frac{\mathbf{j} \times \mathbf{B}_0}{n_e\, e} + \frac{m_e}{n_e\, e^2}\frac{\partial \mathbf{j}}{\partial t}, \tag{5.13}$$

where n_e is the equilibrium electron number density. Substitution of plane-wave solutions of the type (5.1) into the previous equations yields

$$-\mathrm{i}\, \omega\, m_i\, n_e\, \mathbf{V} = \mathbf{j} \times \mathbf{B}_0, \tag{5.14}$$

$$\mathbf{E} = -\mathbf{V} \times \mathbf{B}_0 + \frac{\mathbf{j} \times \mathbf{B}_0}{n_e\, e} - \mathrm{i}\, \frac{\omega\, m_e}{n_e\, e^2}\, \mathbf{j}. \tag{5.15}$$

Let

$$\Pi_e = \sqrt{\frac{n_e\, e^2}{\epsilon_0\, m_e}}, \tag{5.16}$$

$$\Pi_i = \sqrt{\frac{n_e\, e^2}{\epsilon_0\, m_i}}, \tag{5.17}$$

$$\Omega_e = -\frac{e\, B_0}{m_e}, \tag{5.18}$$

$$\Omega_i = \frac{e\, B_0}{m_i}, \tag{5.19}$$

be the *electron plasma frequency*, the *ion plasma frequency*, the *electron cyclotron frequency*, and the *ion cyclotron frequency*, respectively. Eliminating the fluid velocity, $\mathbf{V}$, between Equations (5.14) and (5.15), and making use of the previous definitions, we obtain

$$\mathrm{i}\, \omega\, \epsilon_0\, \mathbf{E} = \frac{\omega^2\, \mathbf{j} - \mathrm{i}\, \omega\, \Omega_e\, \mathbf{j} \times \mathbf{b} + \Omega_e\, \Omega_i\, \mathbf{b} \times (\mathbf{j} \times \mathbf{b})}{\Pi_e^2}, \tag{5.20}$$

where $\mathbf{b} = \mathbf{B}_0/B_0$.

The parallel component of the previous equation is readily solved to give

$$j_{\parallel} = \mathrm{i}\,\omega\,\epsilon_0\,\frac{\Pi_e^2}{\omega^2}\,E_{\parallel}, \tag{5.21}$$

where $j_{\parallel} \equiv \mathbf{j}\cdot\mathbf{b}$, et cetera. In solving for $\mathbf{j}_{\perp} \equiv \mathbf{j} - j_{\parallel}\,\mathbf{b}$, it is helpful to define the vectors

$$\mathbf{e}_{+} = \frac{\mathbf{e}_1 + \mathrm{i}\,\mathbf{e}_2}{\sqrt{2}}, \tag{5.22}$$

$$\mathbf{e}_{-} = \frac{\mathbf{e}_1 - \mathrm{i}\,\mathbf{e}_2}{\sqrt{2}}. \tag{5.23}$$

Here, $(\mathbf{e}_1, \mathbf{e}_2, \mathbf{b})$ are a set of mutually orthogonal, right-handed unit vectors. It is easily demonstrated that

$$\mathbf{e}_{\pm}\times\mathbf{b} = \pm\mathrm{i}\,\mathbf{e}_{\pm}, \tag{5.24}$$

$$\mathbf{b}\times(\mathbf{e}_{\pm}\times\mathbf{b}) = \mathbf{e}_{\pm}. \tag{5.25}$$

It follows that

$$j_{\pm} = \mathrm{i}\,\omega\,\epsilon_0\left(\frac{\Pi_e^2}{\omega^2 \pm \omega\,\Omega_e + \Omega_e\,\Omega_i}\right)E_{\pm}, \tag{5.26}$$

where $j_{\pm} = \mathbf{j}\cdot\mathbf{e}_{\pm}$, et cetera.

The conductivity tensor is diagonal in the "circular" basis $(\mathbf{e}_{+}, \mathbf{e}_{-}, \mathbf{b})$. In fact, its elements are the coefficients of $E_{\pm}$ and $E_{\parallel}$ in Equations (5.26) and (5.21), respectively. Thus, the dielectric permittivity tensor, defined in Equation (5.8), takes the form

$$\mathbf{K}_{\rm circ} = \begin{pmatrix} R, & 0, & 0\\ 0, & L, & 0\\ 0, & 0, & P\end{pmatrix}, \tag{5.27}$$

where

$$R \simeq 1 - \frac{\Pi_e^2}{\omega^2 + \omega\,\Omega_e + \Omega_e\,\Omega_i}, \tag{5.28}$$

$$L \simeq 1 - \frac{\Pi_e^2}{\omega^2 - \omega\,\Omega_e + \Omega_e\,\Omega_i}, \tag{5.29}$$

$$P \simeq 1 - \frac{\Pi_e^2}{\omega^2}. \tag{5.30}$$

Here, R and L represent the permittivities for right- and left-handed circularly polarized waves, respectively. The permittivity parallel to the magnetic field, P, is identical to that of an unmagnetized plasma.

The previous expressions are only approximate because the small mass-ratio ordering $m_e/m_i \ll 1$ has already been incorporated into the cold-plasma equations. The exact expressions, which are most easily obtained by solving the individual charged particle equations of motion, and then summing to obtain the fluid response, are

$$R = 1 - \frac{\Pi_e^2}{\omega^2}\left(\frac{\omega}{\omega + \Omega_e}\right) - \frac{\Pi_i^2}{\omega^2}\left(\frac{\omega}{\omega + \Omega_i}\right), \tag{5.31}$$

$$L = 1 - \frac{\Pi_e^2}{\omega^2}\left(\frac{\omega}{\omega - \Omega_e}\right) - \frac{\Pi_i^2}{\omega^2}\left(\frac{\omega}{\omega - \Omega_i}\right), \tag{5.32}$$

$$P = 1 - \frac{\Pi_e^2}{\omega^2} - \frac{\Pi_i^2}{\omega^2}. \tag{5.33}$$

Equations (5.28)–(5.30) and (5.31)–(5.33) are equivalent in the limit $m_e/m_i \to 0$. Furthermore, Equations (5.31)–(5.33) generalize in a fairly obvious manner to plasmas consisting of more than two particle species.

In order to obtain the standard expression for dielectric permittivity tensor, it is necessary to transform to the Cartesian basis $(\mathbf{e}_1, \mathbf{e}_2, \mathbf{b})$. Let $\mathbf{b} \equiv \mathbf{e}_3$, for ease of notation. It follows that the components of an arbitrary vector $\mathbf{a}$ in the Cartesian basis are related to the components in the "circular" basis via

$$\begin{pmatrix} a_1 \\ a_2 \\ a_3 \end{pmatrix} = \mathbf{U} \begin{pmatrix} a_+ \\ a_- \\ a_3 \end{pmatrix}, \tag{5.34}$$

where the unitary transformation matrix $\mathbf{U}$ is written

$$\mathbf{U} = \frac{1}{\sqrt{2}} \begin{pmatrix} 1, & 1, & 0 \\ \mathrm{i}, & -\mathrm{i}, & 0 \\ 0, & 0, & \sqrt{2} \end{pmatrix}. \tag{5.35}$$

The dielectric permittivity in the Cartesian basis is then

$$\mathbf{K} = \mathbf{U}\, \mathbf{K}_{\mathrm{circ}}\, \mathbf{U}^{\dagger}. \tag{5.36}$$

We obtain

$$\mathbf{K} = \begin{pmatrix} S, & -\mathrm{i}\,D, & 0 \\ \mathrm{i}\,D, & S, & 0 \\ 0, & 0, & P \end{pmatrix}, \tag{5.37}$$

where

$$S = \frac{R + L}{2}, \tag{5.38}$$

and

$$D = \frac{R - L}{2}, \tag{5.39}$$

represent the sum and difference of the right- and left-handed dielectric permittivities, respectively.

5.4 Cold-Plasma Dispersion Relation

It is convenient to define a vector

$$\mathbf{n} = \frac{c}{\omega}\,\mathbf{k} \tag{5.40}$$

that points in the same direction as the wavevector, **k**, and whose magnitude, n, is the *refractive index* (i.e., the ratio of the velocity of light in vacuum to the phase-velocity). Equation (5.9) can be rewritten

$$\mathbf{M}\cdot\mathbf{E} = (\mathbf{n}\cdot\mathbf{E})\,\mathbf{n} - n^2\,\mathbf{E} + \mathbf{K}\cdot\mathbf{E} = \mathbf{0}. \tag{5.41}$$

Without loss of generality, we can assume that the equilibrium magnetic field is directed along the z-axis, and that the wavevector, **k**, lies in the x-z plane. Let θ be the angle subtended between **k** and $\mathbf{B}_0$. The eigenmode equation (5.41) can be written

$$\begin{pmatrix} S - n^2\cos^2\theta, & -\mathrm{i}\,D, & n^2\cos\theta\,\sin\theta \\ \mathrm{i}\,D, & S - n^2, & 0 \\ n^2\cos\theta\,\sin\theta, & 0, & P - n^2\sin^2\theta \end{pmatrix}\begin{pmatrix} E_x \\ E_y \\ E_z \end{pmatrix} = \mathbf{0}. \tag{5.42}$$

The condition for a nontrivial solution is that the determinant of the square matrix be zero. With the help of the identity

$$S^2 - D^2 \equiv R\,L, \tag{5.43}$$

we find that (Hazeltine and Waelbroeck 2004)

$$\mathcal{M}(\omega,\mathbf{k}) \equiv A\,n^4 - B\,n^2 + C = 0, \tag{5.44}$$

where

$$A = S\,\sin^2\theta + P\,\cos^2\theta, \tag{5.45}$$

$$B = R\,L\,\sin^2\theta + P\,S\,(1+\cos^2\theta), \tag{5.46}$$

$$C = P\,R\,L. \tag{5.47}$$

The dispersion relation (5.44) is evidently a quadratic in n^2, with two roots. The solution can be written

$$n^2 = \frac{B \pm F}{2\,A}, \tag{5.48}$$

where

$$F^2 = (B^2 - 4\,A\,C) = (R\,L - P\,S)^2\,\sin^4\theta + 4\,P^2\,D^2\,\cos^2\theta. \tag{5.49}$$

Note that $F^2 \geq 0$. It follows that n^2 is always real, which implies that n is either purely real, or purely imaginary. In other words, the cold-plasma dispersion relation describes waves that either propagate without evanescense, or decay without spatial oscillation. The two roots of opposite sign for n, corresponding to a particular root for n^2, simply describe waves of the same type propagating, or decaying, in opposite directions.

The dispersion relation (5.44) can also be written

$$\tan^2\theta = -\frac{P\,(n^2-R)\,(n^2-L)}{(S\,n^2-R\,L)\,(n^2-P)}. \tag{5.50}$$

For the special case of wave propagation parallel to the magnetic field (i.e., $\theta = 0$), the previous expression reduces to

$$P = 0, \tag{5.51}$$

$$n^2 = R, \tag{5.52}$$

$$n^2 = L. \tag{5.53}$$

Likewise, for the special case of propagation perpendicular to the field (i.e., $\theta = \pi/2$), Equation (5.50) yields

$$n^2 = \frac{R\,L}{S}, \tag{5.54}$$

$$n^2 = P. \tag{5.55}$$

5.5 Wave Polarization

A pure right-handed circularly polarized wave propagating along the z-axis takes the form

$$E_x = A\,\cos(k\,z - \omega\,t), \tag{5.56}$$

$$E_y = -A\,\sin(k\,z - \omega\,t). \tag{5.57}$$

In terms of complex amplitudes, this becomes

$$\frac{\mathrm{i}\,E_x}{E_y} = 1. \tag{5.58}$$

Similarly, a left-handed circularly polarized wave is characterized by

$$\frac{\mathrm{i}\,E_x}{E_y} = -1. \tag{5.59}$$

The polarization of the transverse electric field is obtained from the middle line of Equation (5.42):

$$\frac{\mathrm{i}\,E_x}{E_y} = \frac{n^2-S}{D} = \frac{2\,n^2-(R+L)}{R-L}. \tag{5.60}$$

For the case of parallel propagation, with $n^2 = R$, the previous formula yields $\mathrm{i}\,E_x/E_y = 1$. Similarly, for the case of parallel propagation, with $n^2 = L$, we obtain $\mathrm{i}\,E_x/E_y = -1$. Thus, it is clear that the roots $n^2 = R$ and $n^2 = L$ in Equations (5.51)–(5.53) correspond to right- and left-handed circularly polarized waves, respectively.

5.6 Cutoff and Resonance

For certain values of n_e, B_0, and θ, the wave refractive index, n, is zero. For other values, the refractive index is infinite. In both cases (assuming that n is a slowly varying function of position), a transition is made from a region in which the wave in question propagates to a region in which the wave decays, or vice versa. It is demonstrated in Section 6.3 that wave reflection occurs at those points where n is zero, and in Section 6.4 that wave absorption occurs at those points where n is infinite. The former points are called wave *cutoffs*, whereas the latter are termed wave *resonances*.

According to Equations (5.44) and (5.45)–(5.47), cutoff occurs when

$$P = 0, \tag{5.61}$$

or

$$R = 0, \tag{5.62}$$

or

$$L = 0. \tag{5.63}$$

The cutoff points are independent of the direction of propagation of the wave relative to the magnetic field.

According to Equation (5.50), resonance takes place when

$$\tan^2\theta = -\frac{P}{S}. \tag{5.64}$$

Evidently, resonance points do depend on the direction of propagation of the wave relative to the magnetic field. For the case of parallel propagation, resonance occurs whenever $S \to \infty$. In other words, when

$$R \to \infty, \tag{5.65}$$

or

$$L \to \infty. \tag{5.66}$$

For the case of perpendicular propagation, resonance occurs when

$$S = 0. \tag{5.67}$$

5.7 Waves in Unmagnetized Plasmas

Let us now investigate the cold-plasma dispersion relation in detail. It is instructive to first consider the limit in which the equilibrium magnetic field is zero. In the absence of a magnetic field, there is no preferred direction, so we can, without loss of generality, assume that **k** is directed along the z-axis (i.e., $\theta = 0$). In the zero magnetic field limit (i.e., $\Omega_e, \Omega_i \to 0$), the eigenmode equation (5.42) reduces to

$$\begin{pmatrix} P - n^2, & 0, & 0 \\ 0, & P - n^2, & 0 \\ 0, & 0, & P \end{pmatrix} \begin{pmatrix} E_x \\ E_y \\ E_z \end{pmatrix} = \mathbf{0}, \tag{5.68}$$

where

$$P \simeq 1 - \frac{\Pi_e^2}{\omega^2}. \tag{5.69}$$

Here, we have neglected Π_i with respect to Π_e.

It is clear from Equation (5.68) that there are two types of waves. The first possesses the eigenvector $(0,\ 0,\ E_z)$, and has the dispersion relation

$$1 - \frac{\Pi_e^2}{\omega^2} = 0. \tag{5.70}$$

The second possesses the eigenvector $(E_x,\ E_y,\ 0)$, and has the dispersion relation

$$1 - \frac{\Pi_e^2}{\omega^2} - \frac{k^2\,c^2}{\omega^2} = 0. \tag{5.71}$$

Here, E_x, E_y, and E_z are arbitrary non-zero quantities.

The former wave has **k** parallel to **E**, and is, thus, a longitudinal (with respect to the electric field) wave. This wave is known as the *plasma wave*, and possesses the fixed frequency $\omega = \Pi_e$. Now, if **E** is parallel to **k** then it follows from Equation (5.5) that $\mathbf{B} = \mathbf{0}$. In other words, the plasma wave is purely electrostatic in nature. In fact, the plasma wave is an electrostatic oscillation of the type discussed in Section 1.4. Because ω is independent of **k**, the so-called *group-velocity* (Fitzpatrick 2013),

$$\mathbf{v}_g = \frac{\partial\omega}{\partial\mathbf{k}}, \tag{5.72}$$

associated with a plasma wave, is zero. As is demonstrated in Section 6.7, the group-velocity is the propagation velocity of localized wave packets. It is clear that the plasma wave is not a propagating wave, but instead has the property than an oscillation set up in one region of the plasma remains localized in that region. It should be noted, however, that in a "warm" plasma (i.e., a plasma with a finite thermal velocity) the plasma wave acquires a non-zero, albeit very small, group velocity. (See Section 8.2.)

The latter wave is a transverse wave, with **k** perpendicular to **E**. There are two independent linear polarizations of this wave, which propagate at identical velocities, just like a vacuum electromagnetic wave. The dispersion relation (5.71) can be rearranged to give

$$\omega^2 = \Pi_e^2 + k^2 c^2, \tag{5.73}$$

showing that this wave is just the conventional electromagnetic wave, whose vacuum dispersion relation is $\omega^2 = k^2 c^2$, modified by the presence of the plasma. An important conclusion, which follows immediately from the previous expression, is that this wave can only propagate if $\omega \geq \Pi_e$. Because Π_e is proportional to the square root of the electron number density, it follows that electromagnetic radiation of a given frequency can only propagate through an unmagnetized plasma when the electron number density falls below some critical value.

5.8 Low-Frequency Wave Propagation

Consider wave propagation through a magnetized plasma at frequencies far below the ion cyclotron or plasma frequencies, which are, in turn, well below the corresponding electron frequencies. In the low-frequency limit (i.e., $\omega \ll \Omega_i, \Pi_i$), we have [see Equations (5.28)–(5.30)]

$$S \simeq 1 + \frac{\Pi_i^2}{\Omega_i^2}, \tag{5.74}$$

$$D \simeq 0, \tag{5.75}$$

$$P \simeq -\frac{\Pi_e^2}{\omega^2}. \tag{5.76}$$

Here, use has been made of $\Pi_e^2/(\Omega_e\,\Omega_i) = -\Pi_i^2/\Omega_i^2$. Thus, the eigenmode equation (5.42) reduces to

$$\begin{pmatrix} 1 + \Pi_i^2/\Omega_i^2 - n^2\cos^2\theta, & 0, & n^2\cos\theta\sin\theta \\ 0, & 1 + \Pi_i^2/\Omega_i^2 - n^2, & 0 \\ n^2\cos\theta\sin\theta, & 0, & -\Pi_e^2/\omega^2 - n^2\sin^2\theta \end{pmatrix} \begin{pmatrix} E_x \\ E_y \\ E_z \end{pmatrix} = \mathbf{0}. \tag{5.77}$$

The solubility condition for Equation (5.77) yields the dispersion relation

$$\begin{vmatrix} 1 + \Pi_i^2/\Omega_i^2 - n^2\cos^2\theta, & 0, & n^2\cos\theta\sin\theta \\ 0, & 1 + \Pi_i^2/\Omega_i^2 - n^2, & 0 \\ n^2\cos\theta\sin\theta, & 0, & -\Pi_e^2/\omega^2 - n^2\sin^2\theta \end{vmatrix} = 0. \tag{5.78}$$

Now, in the low-frequency ordering, $\Pi_e^2/\omega^2 \gg \Pi_i^2/\Omega_i^2$. Thus, we can see that the bottom right-hand element of the previous determinant is far larger than any of the other elements. Hence, to a good approximation, the roots of the dispersion relation are obtained by equating the term multiplying this large factor to zero (Cairns 1985). In this manner, we obtain two roots:

$$n^2\cos^2\theta = 1 + \frac{\Pi_i^2}{\Omega_i^2}, \tag{5.79}$$

and

$$n^2 = 1 + \frac{\Pi_i^2}{\Omega_i^2}. \tag{5.80}$$

It is fairly easy to show, from the definitions of the plasma and cyclotron frequencies [see Equations (5.16)–(5.19)], that

$$\frac{\Pi_i^2}{\Omega_i^2} = \frac{c^2}{B_0^2/(\mu_0\,\rho)} = \frac{c^2}{V_A^2}. \tag{5.81}$$

Here, $\rho \simeq n_e\, m_i$ is the plasma mass density, and

$$V_A = \sqrt{\frac{B_0^2}{\mu_0\,\rho}} \tag{5.82}$$

is known as the *Alfvén velocity*. Thus, the dispersion relations (5.79) and (5.80) can be written

$$\omega = \frac{k\,V_A\,\cos\theta}{\sqrt{1 + V_A^2/c^2}} \simeq k\,V_A\,\cos\theta \equiv k_\parallel\,V_A, \tag{5.83}$$

and

$$\omega = \frac{k\,V_A}{\sqrt{1 + V_A^2/c^2}} \simeq k\,V_A, \tag{5.84}$$

respectively. Here, we have made use of the fact that $V_A \ll c$ in a conventional plasma.

The dispersion relation (5.83) corresponds to the *slow* or *shear-Alfvén* wave, whereas the dispersion relation (5.84) corresponds to the *fast* or *compressional-Alfvén* wave. The fast/slow terminology simply refers to the relative magnitudes of the phase-velocities of the two waves. The shear/compressional terminology refers to the velocity fields associated with the waves. In fact, it is clear from Equation (5.77) that $E_z = 0$ for both waves, whereas $E_y = 0$ for the shear wave, and $E_x = 0$ for the compressional wave. Both waves are, in fact, MHD modes that satisfy the linearized MHD Ohm's law [see Equation (4.215)]

$$\mathbf{E} + \mathbf{V}\times\mathbf{B}_0 = \mathbf{0}. \tag{5.85}$$

Thus, for the shear wave

$$V_y = -\frac{E_x}{B_0}, \tag{5.86}$$

and $V_x = V_z = 0$, whereas for the compressional wave

$$V_x = \frac{E_y}{B_0}, \tag{5.87}$$

and $V_y = V_z = 0$. Now, $\nabla\cdot\mathbf{V} = \mathrm{i}\,\mathbf{k}\cdot\mathbf{V} = \mathrm{i}\,k\,V_x\,\sin\theta$. Thus, the shear-Alfvén wave is a torsional wave, with zero divergence of the plasma flow, whereas the compressional wave involves a non-zero flow divergence. In fact, the former wave bends magnetic field-lines without compressing them, whereas the latter compresses magnetic field-lines without bending them (Hazeltine and Waelbroeck 2004). It is important to realize that the physical entity that resists compression in the compressional wave is the magnetic field, not the plasma, because there is negligible plasma pressure in the cold-plasma approximation.

It should be noted that the thermal velocity is not necessarily negligible compared to the Alfvén velocity in a conventional plasma. Thus, we would expect the dispersion relations (5.83) and (5.84), for the shear- and compressional-Alfvén waves, respectively, to undergo considerable modification in a "warm" plasma. (See Section 7.4.)

5.9 Parallel Wave Propagation

Consider wave propagation, at arbitrary frequencies, parallel to the equilibrium magnetic field. When $\theta = 0$, the eigenmode equation (5.42) simplifies to

$$\begin{pmatrix} S - n^2, & -\mathrm{i}\, D, & 0 \\ \mathrm{i}\, D, & S - n^2, & 0 \\ 0, & 0, & P \end{pmatrix} \begin{pmatrix} E_x \\ E_y \\ E_z \end{pmatrix} = \mathbf{0}. \tag{5.88}$$

One obvious way of solving this equation is to have

$$P \simeq 1 - \frac{\Pi_e^2}{\omega^2} = 0, \tag{5.89}$$

with the eigenvector $(0, 0, E_z)$. This is just the electrostatic plasma wave that we found previously in an unmagnetized plasma. This mode is longitudinal in nature, and, therefore, causes particles to oscillate parallel to $\mathbf{B}_0$. It follows that the particles experience zero Lorentz force due to the presence of the equilibrium magnetic field, with the result that this field has no effect on the mode dynamics.

The other two solutions to Equation (5.88) are obtained by setting the 2×2 determinant involving the x- and y-components of the electric field to zero. The first wave has the dispersion relation

$$n^2 = R \simeq 1 - \frac{\Pi_e^2}{(\omega + \Omega_e)\,(\omega + \Omega_i)}, \tag{5.90}$$

and the eigenvector $(E_x, \mathrm{i}\, E_x, 0)$. This is evidently a right-handed circularly polarized wave. The second wave has the dispersion relation

$$n^2 = L \simeq 1 - \frac{\Pi_e^2}{(\omega - \Omega_e)\,(\omega - \Omega_i)}, \tag{5.91}$$

and the eigenvector $(E_x, -\mathrm{i}\, E_x, 0)$. This is evidently a left-handed circularly polarized wave. At low frequencies (i.e., $\omega \ll \Omega_i$), both waves convert into the Alfvén wave discussed in the previous section. (The fast and slow Alfvén waves are indistinguishable for parallel propagation.) Let us now examine the high-frequency behavior of the right- and left-handed waves.

For the right-handed wave, because Ω_e is negative, it is evident that $n^2 \rightarrow \infty$ as $\omega \rightarrow |\Omega_e|$. This resonance, which corresponds to $R \rightarrow \infty$, is termed the *electron cyclotron resonance*. At the electron cyclotron resonance, the transverse electric field associated with a right-handed wave rotates at the same velocity, and in the same direction, as electrons gyrating around the equilibrium magnetic field. Thus, the electrons experience a continuous acceleration from the electric field, which tends to increase their perpendicular energy. It is, therefore, not surprising that right-handed waves, propagating parallel to the equilibrium magnetic field, and oscillating at the frequency $|\Omega_e|$, are absorbed by electrons.

When ω lies just above $|\Omega_e|$, we find that n^2 is negative, and so there is no wave propagation. However, for frequencies much greater than the electron cyclotron or plasma frequencies, the solution to Equation (5.90) is approximately $n^2 = 1$. In other words, $\omega^2 = k^2c^2$, which is the dispersion relation of a right-handed vacuum electromagnetic wave. Evidently, at some frequency above $|\Omega_e|$, the solution for n^2 must pass through zero, and become positive again. Putting $n^2 = 0$ in Equation (5.90), we find that the equation reduces to

$$\omega^2 + \Omega_e\,\omega - \Pi_e^2 \simeq 0, \tag{5.92}$$

assuming that $V_A \ll c$. The previous equation has only one positive root, at $\omega = \omega_1$, where

$$\omega_1 \simeq |\Omega_e|/2 + \sqrt{\Omega_e^2/4 + \Pi_e^2} > |\Omega_e|. \tag{5.93}$$

Above this frequency, the wave propagates once again.

The dispersion curve for a right-handed wave propagating parallel to the equilibrium magnetic field is sketched in Figure 5.1. The continuation of the Alfvén wave above the ion cyclotron frequency is called the *electron cyclotron wave*, or, sometimes, the *whistler wave*. The latter terminology is prevalent in ionospheric and space plasma physics contexts. The wave that propagates above the cutoff frequency, ω_1, is a standard right-handed circularly polarized electromagnetic wave, somewhat modified by the presence of the plasma. The low-frequency branch of the dispersion curve differs fundamentally from the high-frequency branch, because the former branch corresponds to a wave that can only propagate through the plasma in the presence of an equilibrium magnetic field, whereas the latter branch corresponds to a wave that can propagate in the absence of an equilibrium field.

The curious name "whistler wave" for the branch of the dispersion relation lying between the ion and electron cyclotron frequencies is originally derived from ionospheric physics. Whistler waves are a very characteristic type of audio-frequency radio interference, most commonly encountered at high latitudes, which take the form of brief, intermittent pulses, starting at high frequencies, and rapidly descending in pitch.

Whistlers were discovered in the early days of radio communication, but were not explained until much later (Storey 1953). Whistler waves start off as "instantaneous" radio pulses, generated by lightning flashes at high latitudes. The pulses are channeled along the Earth's dipolar magnetic field, and eventually return to ground level in the opposite hemisphere. Now, in the frequency range $\Omega_i \ll \omega \ll |\Omega_e|$, the

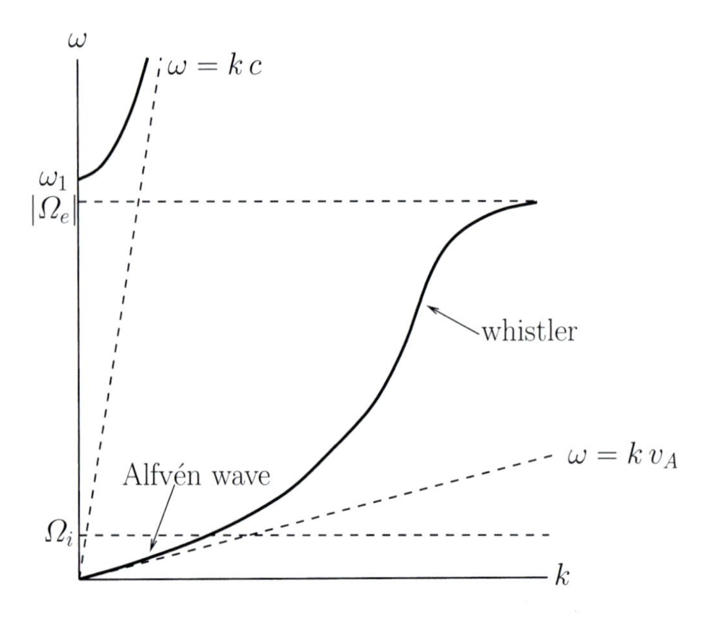

Figure 5.1
Schematic diagram showing the dispersion relation for a right-handed wave propagating parallel to the magnetic field in a magnetized plasma.

dispersion relation (5.90) reduces to

$$n^2 = \frac{k^2\,c^2}{\omega^2} \simeq \frac{\Pi_e^{\,2}}{\omega\,|\Omega_e|}. \tag{5.94}$$

As is well known, wave pulses propagate at the group-velocity,

$$v_g = \frac{d\omega}{dk} = 2\,c\,\frac{\sqrt{\omega\,|\Omega_e|}}{\Pi_e}. \tag{5.95}$$

Clearly, the low-frequency components of a pulse propagate more slowly than the high-frequency components. It follows that, by the time a pulse returns to ground level, it has been stretched out temporally, because its high-frequency components arrive slightly before its low-frequency components. This also accounts for the characteristic whistling-down effect observed at ground level.

The shape of whistler pulses, and the way in which the pulse frequency varies in time, can yield a considerable amount of information about the regions of the Earth's magnetosphere through which the pulses have passed. For this reason, many countries maintain observatories in polar regions—especially Antarctica—which monitor and collect whistler data.

For a left-handed circularly polarized wave, similar considerations to those described previously yield a dispersion curve of the form sketched in Figure 5.2. In this case, n^2 goes to infinity at the ion cyclotron frequency, Ω_i, corresponding to the so-called *ion cyclotron resonance* (at $L \to \infty$). At this resonance, the rotating electric field associated with a left-handed wave resonates with the gyromotion of the ions, allowing wave energy to be converted into perpendicular kinetic energy of the ions. There is a band of frequencies, lying above the ion cyclotron frequency, in which the left-handed wave does not propagate. At very high frequencies, a propagating mode exists, which is basically a standard left-handed circularly polarized electromagnetic wave, somewhat modified by the presence of the plasma. The cutoff frequency for this wave is

$$\omega_2 \simeq -|\Omega_e|/2 + \sqrt{\Omega_e^2/4 + \Pi_e^2}. \tag{5.96}$$

As before, the lower branch in Figure 5.2 describes a wave that can only propagate in the presence of an equilibrium magnetic field, whereas the upper branch describes a wave that can propagate in the absence an equilibrium field. The continuation of the Alfvén wave to just below the ion cyclotron frequency is generally known as the *ion cyclotron wave*.

5.10 Perpendicular Wave Propagation

Consider wave propagation, at arbitrary frequencies, perpendicular to the equilibrium magnetic field. When $\theta = \pi/2$, the eigenmode equation (5.42) simplifies to

$$\begin{pmatrix} S, & -\mathrm{i}\,D, & 0 \\ \mathrm{i}\,D, & S - n^2, & 0 \\ 0, & 0, & P - n^2 \end{pmatrix} \begin{pmatrix} E_x \\ E_y \\ E_z \end{pmatrix} = \mathbf{0}. \tag{5.97}$$

One obvious way of solving this equation is to have $P - n^2 = 0$, or

$$\omega^2 = \Pi_e^2 + k^2 c^2, \tag{5.98}$$

with the eigenvector $(0,\ 0,\ E_z)$. Because the wavevector now points in the x-direction, this is clearly a transverse wave polarized with its electric field parallel to the equilibrium magnetic field. Particle motions are along the magnetic field, so the mode dynamics are completely unaffected by this field. Thus, the wave is identical to the electromagnetic plasma wave found previously in an unmagnetized plasma. This wave is known as the *ordinary*, or *O*-, mode.

The other solution to Equation (5.97) is obtained by setting the 2×2 determinant involving the x- and y-components of the electric field to zero. The dispersion relation reduces to

$$n^2 = \frac{R\,L}{S}, \tag{5.99}$$

with the associated eigenvector $E_x\,(1,\ -\mathrm{i}\,S/D,\ 0)$.

Let us, first of all, search for the cutoff frequencies, at which n^2 goes to zero. According to Equation (5.99), these frequencies are the roots of $R = 0$ and $L = 0$.

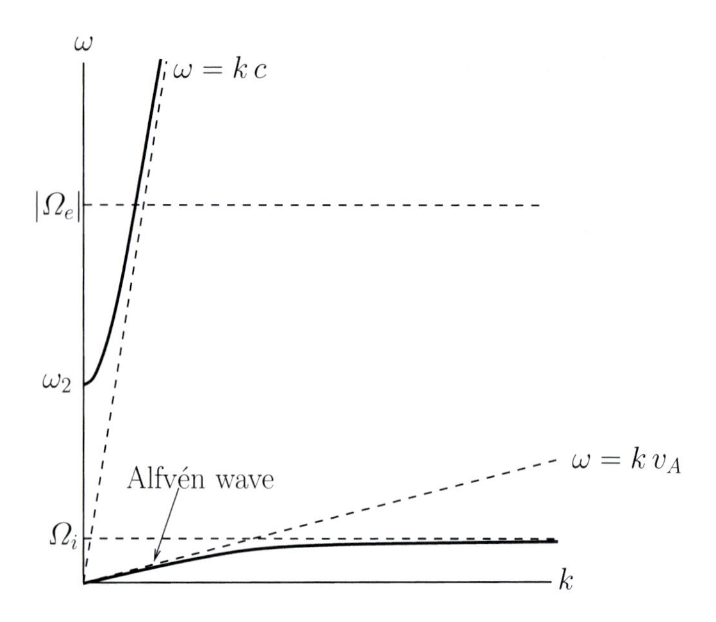

Figure 5.2
Schematic diagram showing the dispersion relation for a left-handed wave propagating parallel to the magnetic field in a magnetized plasma.

In fact, we have already solved these equations (recall that cutoff frequencies do not depend on θ). There are two cutoff frequencies, ω_1 and ω_2, which are specified by Equations (5.93) and (5.96), respectively.

Let us, next, search for the resonant frequencies, at which n^2 goes to infinity. According to Equation (5.99), the resonant frequencies are solutions of

$$S = 1 - \frac{\Pi_e^2}{\omega^2 - \Omega_e^2} - \frac{\Pi_i^2}{\omega^2 - \Omega_i^2} = 0. \tag{5.100}$$

The roots of this equation can be obtained as follows (Cairns 1985). First, we note that if the first two terms in the middle are equated to zero then we obtain $\omega = \omega_{\rm UH}$, where

$$\omega_{\rm UH} = \sqrt{\Pi_e^2 + \Omega_e^2}. \tag{5.101}$$

If this frequency is substituted into the third term in the middle then the result is far less than unity. We conclude that $\omega_{\rm UH}$ is a good approximation of one of the roots of Equation (5.100). To obtain the second root, we make use of the fact that the product of the square of the roots is

$$\Omega_e^2\,\Omega_i^2 + \Pi_e^2\,\Omega_i^2 + \Pi_i^2\,\Omega_e^2 \simeq \Omega_e^2\,\Omega_i^2 + \Pi_i^2\,\Omega_e^2. \tag{5.102}$$

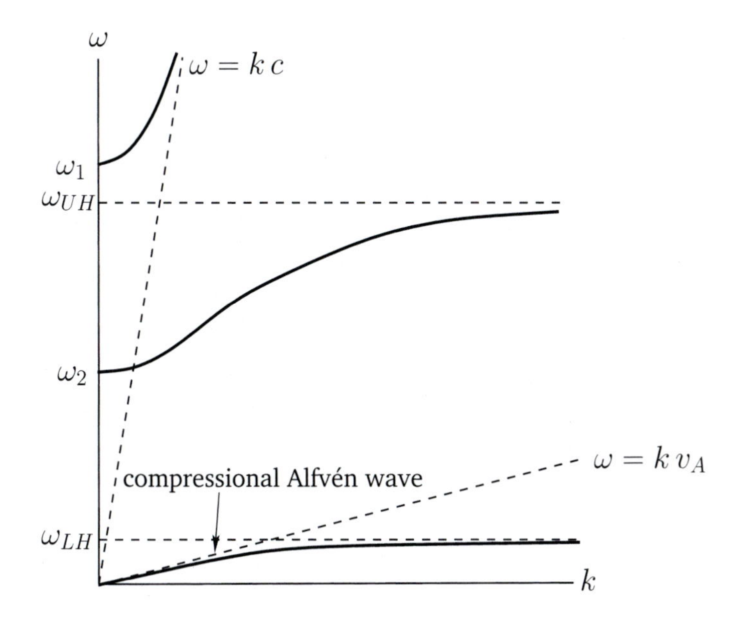

Figure 5.3
Schematic diagram showing the dispersion relation for a wave propagating perpendicular to the magnetic field in a magnetized plasma.

We, thus, obtain $\omega = \omega_{\rm LH}$, where

$$\omega_{\rm LH} = \sqrt{\frac{\Omega_e^2\,\Omega_i^2 + \Pi_i^2\,\Omega_e^2}{\Pi_e^2 + \Omega_e^2}}. \tag{5.103}$$

The first resonant frequency, $\omega_{\rm UH}$, is greater than the electron cyclotron or plasma frequencies, and is called the *upper hybrid frequency*. The second resonant frequency, $\omega_{\rm LH}$, lies between the electron and ion cyclotron frequencies, and is called the *lower hybrid frequency*. Unfortunately, there is no simple explanation of the origins of the two hybrid resonances in terms of the motions of individual particles. At low frequencies, the mode in question reverts to the compressional-Alfvén wave discussed previously. Note that the shear-Alfvén wave does not propagate perpendicular to the magnetic field.

Using the previous information, and the easily demonstrated fact that

$$\omega_{\rm LH} < \omega_2 < \omega_{\rm UH} < \omega_1, \tag{5.104}$$

we deduce that the dispersion curve for the mode in question takes the form sketched in Figure 5.3. The lowest frequency branch corresponds to the compressional-Alfvén wave. The other two branches constitute the *extraordinary*, or *X*-, mode. The upper branch is basically a linearly polarized (in the y-direction) electromagnetic wave, somewhat modified by the presence of the plasma. This branch corresponds to a wave that propagates in the absence of an equilibrium magnetic field. The lowest branch corresponds to a wave that does not propagate in the absence of an equilibrium field. Finally, the middle branch corresponds to a wave that converts into an electrostatic plasma wave in the absence of an equilibrium magnetic field.

Wave propagation at oblique angles is generally more complicated than propagation parallel or perpendicular to the equilibrium magnetic field, but does not involve any new physical effects (Stix 1992; Swanson 2003).

5.11 Exercises

5.1 Show that for fields varying as $\exp[\,\mathrm{i}\,(\mathbf{k}\cdot\mathbf{r}-\omega\,t)]$ the equations $\nabla\cdot\mathbf{E}=\rho/\epsilon_0$ and $\nabla\cdot\mathbf{B}=0$ follow from Equations (5.4) and (5.5). This explains why the former equations are not explicitly used in the study of plane waves.

5.2 Derive Equations (5.31)–(5.33) from first principles, starting from the equations of motion of individual charged particles.

5.3 Prove the identity

$$S^2-D^2=R\,L.$$

5.4 Derive the dispersion relation (5.44)–(5.47) from Equation (5.42).

5.5 Show that the square of F, defined in Equation (5.48), can be written in the positive definite form

$$F^2=(R\,L-P\,S)^2\,\sin^4\theta+4\,P^2\,D^2\,\cos^2\theta.$$

5.6 Derive the alternative dispersion relation (5.50) from (5.44).

5.7 Show that in the limit $\omega\rightarrow 0$,

$$R=L=S=1+\frac{\Pi_i^2}{\Omega_i^2}+\frac{\Pi_e^2}{\Omega_e^2},$$

$$D=0,$$

$$P=-\frac{\Pi_i^2}{\omega^2}-\frac{\Pi_e^2}{\omega^2}.$$

5.8 Show that

$$\frac{\mathrm{i}\,V_{xi}}{V_{yi}}=\frac{(\mathrm{i}\,E_x/E_y)-(\Omega_i/\omega)}{1-(\Omega_i/\omega)\,(\mathrm{i}\,E_x/E_y)},$$

$$\frac{\mathrm{i}\,V_{xe}}{V_{ye}}=\frac{(\mathrm{i}\,E_x/E_y)-(\Omega_e/\omega)}{1-(\Omega_e/\omega)\,(\mathrm{i}\,E_x/E_y)}.$$

Hence, deduce that for a right-hand/left-hand circularly polarized wave the ions and electrons execute circular orbits in the x-y plane in the electron/ion cyclotron direction.

5.9 The effect of collisions can be included in the dispersion relation for waves in cold magnetized plasmas by adding a drag force $\nu_s\, m_s\, \mathbf{V}_s$ to the equation of motion of species s. Here, ν_s is the effective collision frequency for species s, where s stands for either i or e. Thus, the species s equation of motion becomes

$$m_s\,\frac{d\mathbf{V}_s}{dt} + \nu_s\, m_s\, \mathbf{V}_s = e_s\,(\mathbf{E} + \mathbf{V}_s \times \mathbf{B}).$$

(a) Show that the effect of collisions is equivalent to the substitution

$$m_s \rightarrow m_s\left(1 + \frac{\mathrm{i}\,\nu_s}{\omega}\right).$$

(b) For high frequency transverse waves, for which $\nu_s \ll \omega$, and $\Pi_e, |\Omega_e| \ll \omega$, show that the real and imaginary parts of the wavenumber are

$$k_r \simeq \frac{\omega}{c}\left(1 - \frac{\Pi_e^{\,2}}{2\,\omega^2}\right),$$

$$k_i \simeq \frac{1}{2\,c}\sum_s \frac{\nu_s\,\Pi_s^{\,2}}{\omega^2},$$

respectively.

(c) Show that the dispersion relation for a longitudinal electron plasma oscillations is

$$\omega \simeq \Pi_e - \mathrm{i}\sum_s \frac{\nu_s\,\Pi_s^{\,2}}{2\,\Pi_e^{\,2}}.$$

Hence, demonstrate that collisions cause the oscillation to decay in time.

5.10 A cold, unmagnetized, homogeneous plasma supports oscillations at the plasma frequency, $\omega = \Pi_e$. These oscillations have the same frequency irrespective of the wavevector, $\mathbf{k}$. However, when pressure is included in the analysis, the frequency of the oscillation starts to depend on $\mathbf{k}$. We can investigate this effect by treating the (singly-charged) ions as stationary neutralizing fluid of number density n_0. The electron fluid equations are written

$$\frac{\partial n}{\partial t} + \nabla\cdot(n\,\mathbf{V}) = 0,$$

$$m_e\,n\,\frac{d\mathbf{V}}{dt} = -e\,n\,\mathbf{E} - \nabla p,$$

$$p\,n^{-\Gamma} = p_0\,n_0^{-\Gamma},$$

$$\epsilon_0\,\nabla\cdot\mathbf{E} = -e\,(n - n_0).$$

where p_0 and $\Gamma = 5/3$ are constants. Let $n = n_0 + n_1$, $p = p_0 + p_1$, $\mathbf{V} = \mathbf{V}_1$, and $\mathbf{E} = \mathbf{E}_1$, where the subscript 0 denotes an equilibrium quantity, and the subscript 1 denotes a small perturbation. Develop a set of linear equations sufficient to solve for the perturbed variables. Assuming that all perturbed quantities vary in space and time as $\exp[\,{\rm i}\,(\mathbf{k}\cdot\mathbf{r}-\omega\,t)]$, find the dispersion relation linking ω and $\mathbf{k}$. Find expressions for the phase-velocity and group-velocity of the wave as functions of ω.

6

Wave Propagation Through Inhomogeneous Plasmas

6.1 Introduction

In the previous chapter, we investigated wave propagation through homogeneous plasmas. In this chapter, we shall broaden our approach to deal with the far more interesting case of wave propagation through inhomogeneous plasmas. To be more exact, we shall consider wave propagation in the limit in which the characteristic variation lengthscale, L, of equilibrium quantities in the plasma is much longer than the wavelength of the wave. In other words, $k\,L \ll 1$, where k is the wavenumber. In this limit, we expect our wave solutions to closely resemble those found in the previous chapter (recall that the latter solutions correspond to $k\,L \rightarrow 0$). For the sake of simplicity, we shall (mostly) restrict our investigation to waves propagating through unmagnetized plasmas. However, the techniques described in this chapter can be generalized, in a fairly straightforward manner, to deal with other types of plasma wave (Budden 1985).

6.2 WKB Solutions

Let us start off by examining a very simple case. Consider a plane electromagnetic wave, of angular frequency ω, propagating along the z-axis in an unmagnetized plasma whose refractive index, n, is a function of z. Let us assume that the wave normal is initially aligned along the z-axis, and, furthermore, that the wave starts off polarized in the y-direction. It is easily demonstrated that the wave normal subsequently remains aligned along the z-axis, and also that the polarization state does not change. Thus, the wave is fully described by

$$E_y(z, t) \equiv E_y(z)\,\exp(-\mathrm{i}\,\omega\,t), \tag{6.1}$$

and

$$B_x(z, t) \equiv B_x(z)\,\exp(-\mathrm{i}\,\omega\,t). \tag{6.2}$$

It can readily be shown that $E_y(z)$ and $B_x(z)$ satisfy the differential equations

$$\frac{d^2 E_y}{dz^2} + k_0^2 n^2 E_y = 0, \tag{6.3}$$

and

$$\frac{d(c B_x)}{dz} = -\mathrm{i}\, k_0\, n^2 E_y, \tag{6.4}$$

respectively. Here, $k_0 = \omega/c$ is the wavenumber in free space. Of course, the actual wavenumber is $k = k_0\, n$.

The solution of Equation (6.3) for the case of a homogeneous plasma, for which n is constant, is simply

$$E_y(z) = A\, \mathrm{e}^{\,\mathrm{i}\,\phi(z)}, \tag{6.5}$$

where A is a constant, and

$$\phi(z) = \pm k_0\, n\, z. \tag{6.6}$$

The solution (6.5) represents a wave of constant amplitude A, and phase $\phi(z)$. According to Equation (6.6), there are two independent waves that can propagate through the plasma. The upper sign corresponds to a wave that propagates in the $+z$-direction, whereas the lower sign corresponds to a wave that propagates in the $-z$-direction. Both waves propagate at the constant phase-velocity c/n.

In general, if $n = n(z)$ then the solution of Equation (6.3) does not remotely resemble the wave-like solution (6.5). However, in the limit in which $n(z)$ is a "slowly varying" function of z (exactly how slowly varying is something that will be established later on), we expect to recover wave-like solutions. Let us suppose that $n(z)$ is indeed a "slowly varying" function, and let us try substituting the wave-like solution (6.5) into Equation (6.3). We obtain

$$\left(\frac{d\phi}{dz}\right)^2 = k_0^2 n^2 + \mathrm{i}\,\frac{d^2\phi}{dz^2}. \tag{6.7}$$

This is a non-linear differential equation which, in general, is very difficult to solve. However, we note that if n is a constant then $d^2\phi/dz^2 = 0$. It is, therefore, reasonable to suppose that if $n(z)$ is a "slowly varying" function then the last term on the right-hand side of the previous equation is relatively small. Thus, to a first approximation, Equation (6.7) yields

$$\frac{d\phi}{dz} \simeq \pm k_0\, n, \tag{6.8}$$

and

$$\frac{d^2\phi}{dz^2} \simeq \pm k_0\, \frac{dn}{dz}. \tag{6.9}$$

It is clear, from a comparison of Equations (6.7) and (6.9), that $n(z)$ can be regarded as a "slowly varying" function of z [i.e., the second term on the right-hand side of Equation (6.7) is negligible compared to the first] as long as $(dn/dz)/(k_0\, n^2) \ll 1$. In other words, the approximation holds provided that the variation lengthscale of the refractive index is far longer than the wavelength of the wave.

The second approximation to the solution is obtained by substituting Equation (6.9) into the right-hand side of Equation (6.7):

$$\frac{d\phi}{dz} \simeq \pm\left(k_0^2 n^2 \pm \mathrm{i}\, k_0\, \frac{dn}{dz}\right)^{1/2}. \tag{6.10}$$

This gives

$$\frac{d\phi}{dz} \simeq \pm k_0\, n\left(1 \pm \frac{\mathrm{i}}{k_0\, n^2}\frac{dn}{dz}\right)^{1/2} \simeq \pm k_0\, n + \frac{\mathrm{i}}{2\,n}\frac{dn}{dz}, \tag{6.11}$$

where use has been made of the binomial expansion. The previous expression can be integrated to give

$$\phi(z) \simeq \pm k_0 \int^z n\, dz' + \mathrm{i}\,\log\left(n^{1/2}\right). \tag{6.12}$$

Substitution of Equation (6.12) into Equation (6.5) yields the final result

$$E_y(z) \simeq n^{-1/2}\, \exp\left(\pm \mathrm{i}\, k_0 \int^z n\, dz'\right). \tag{6.13}$$

It follows from Equation (6.4) that

$$c\, B_x(z) \simeq \mp n^{1/2}\, \exp\left(\pm \mathrm{i}\, k_0 \int^z n\, dz'\right) - \frac{\mathrm{i}}{2\, k_0\, n^{3/2}}\frac{dn}{dz}\, \exp\left(\pm \mathrm{i}\, k_0 \int^z n\, dz'\right). \tag{6.14}$$

The second term on the right-hand side of the previous expression is small compared to the first, and is usually neglected.

We can test to what extent expression (6.13) is a good solution of Equation (6.3) by substituting this expression into the left-hand side of the equation. The result is

$$\frac{1}{n^{1/2}}\left[\frac{3}{4}\left(\frac{1}{n}\frac{dn}{dz}\right)^2 - \frac{1}{2\,n}\frac{d^2n}{dz^2}\right] \exp\left(\pm \mathrm{i}\, k_0 \int^z n\, dz'\right) = \left[\frac{3}{4}\left(\frac{1}{n}\frac{dn}{dz}\right)^2 - \frac{1}{2\,n}\frac{d^2n}{dz^2}\right] E_y. \tag{6.15}$$

This quantity needs to be small compared to $k_0^2\, n^2\, E_y$. Hence, the condition for Equation (6.13) to be a good solution of Equation (6.3) becomes

$$\frac{1}{k_0^2}\left|\frac{3}{4}\left(\frac{1}{n^2}\frac{dn}{dz}\right)^2 - \frac{1}{2\,n^3}\frac{d^2n}{dz^2}\right| \ll 1. \tag{6.16}$$

The solutions

$$E_y(z) \simeq n^{-1/2}\, \exp\left(\pm \mathrm{i}\, k_0 \int^z n\, dz'\right), \tag{6.17}$$

$$c\, B_x(z) \simeq \mp n^{1/2}\, \exp\left(\pm \mathrm{i}\, k_0 \int^z n\, dz'\right), \tag{6.18}$$

to the non-uniform wave equations (6.3) and (6.4) are usually referred to as *WKB solutions*, in honor of G. Wentzel (Wentzel 1926), H.A. Kramers (Kramers 1926), and L. Brillouin (Brilloiun 1926), who are credited with independently discovering these solutions (in a quantum mechanical context) in 1926. Actually, H. Jeffries (Jeffries 1924) wrote a paper on WKB solutions (in a wave propagation context) in 1924. Hence, these solutions are sometimes called the WKBJ solutions (or even the JWKB solutions). To be strictly accurate, the WKB solutions were first discussed by Liouville (Liouville 1837) and Green (Green 1837) in 1837, and again by Rayleigh (Rayleigh 1912) in 1912. In the following, we refer to Equations (6.17) and (6.18) as WKB solutions, because this is what they are most commonly called. However, it should be understood that, in doing so, we are not making any definitive statement as to the credit due to various scientists in discovering them. More information about WKB solutions can be found in the classic monograph of Heading (Heading 1962).

If a propagating wave is normally incident on an interface at which the refractive index suddenly changes (for instance, if a light wave propagating through air is normally incident on a glass slab) then there is generally significant reflection of the wave (Fitzpatrick 2013). However, according to the WKB solutions, (6.17) and (6.18), when a propagating wave is normally incident on a medium in which the refractive index changes slowly along the direction of propagation of the wave then the wave is not reflected at all. This is true even if the refractive index varies very substantially along the path of propagation of the wave, as long as it varies sufficiently slowly. The WKB solutions imply that, as the wave propagates through the medium, its wavelength gradually changes. In fact, the wavelength at position z is approximately $\lambda(z) = 2\pi/[k_0\, n(z)]$. Equations (6.17) and (6.18) also imply that the amplitude of the wave gradually changes as it propagates. In fact, the amplitude of the electric field component is inversely proportional to $n^{1/2}$, whereas the amplitude of the magnetic field component is directly proportional to $n^{1/2}$. Note, however, that the energy flux in the z-direction, which is given by the the Poynting vector $-(E_y\, B_x^{*} + E_y^{*}\, B_x)/(4\,\mu_0)$, remains constant (assuming that n is predominately real).

Of course, the WKB solutions (6.17) and (6.18) are only approximations. In reality, a wave propagating through a medium in which the refractive index is a slowly varying function of position is subject to a small amount of reflection. However, it is easily demonstrated that the ratio of the reflected amplitude to the incident amplitude is of order $(dn/dz)/(k_0\, n^2)$ (Budden 1985). Thus, as long as the refractive index varies on a much longer lengthscale than the wavelength of the radiation, the reflected wave is negligibly small. This conclusion remains valid as long as the inequality (6.16) is satisfied. This inequality obviously breaks down in the vicinity of a point where $n^2 = 0$. We would, therefore, expect strong reflection of the incident wave from such a point. Furthermore, the WKB solutions also break down at a point where $n^2 \to \infty$, because the amplitude of B_x becomes infinite.

6.3 Cutoffs

We have seen that electromagnetic wave propagation (in one dimension) through an inhomogeneous plasma, in the physically relevant limit in which the variation lengthscale of the plasma is much greater than the wavelength of the wave, is well described by the WKB solutions, (6.17) and (6.18). However, these solutions break down in the immediate vicinity of a *cutoff*, where $n^2 = 0$, or a *resonance*, where $n^2 \to \infty$. Let us now examine what happens to electromagnetic waves propagating through a plasma when they encounter a cutoff or a resonance.

Suppose that a cutoff is located at $z = 0$, so that

$$n^2 = a\,z + O\left(z^2\right) \tag{6.19}$$

in the immediate vicinity of this point, where $a > 0$. It is evident, from the WKB solutions, (6.17) and (6.18), that the cutoff point lies at the boundary between a region ($z > 0$) in which electromagnetic waves propagate, and a region ($z < 0$) in which the waves are evanescent. In a physically realistic solution, we would expect the wave amplitude to decay (as z decreases) in the evanescent region $z < 0$. Let us search for such a wave solution.

In the immediate vicinity of the cutoff point, $z = 0$, Equations (6.3) and (6.19) yield

$$\frac{d^2 E_y}{d\hat{z}^2} + \hat{z}\,E_y = 0, \tag{6.20}$$

where

$$\hat{z} = (k_0^2\,a)^{1/3}\,z. \tag{6.21}$$

Equation (6.20) is a standard equation, known as *Airy's equation*, and possesses two independent solutions, denoted $\mathrm{Ai}(-\hat{z})$ and $\mathrm{Bi}(-\hat{z})$ (Abramowitz and Stegun 1965d).

The second solution, $\mathrm{Bi}(-\hat{z})$, is unphysical, because it blows up as $\hat{z} \to -\infty$. The physical solution, $\mathrm{Ai}(-\hat{z})$, has the asymptotic behavior

$$\mathrm{Ai}(-\hat{z}) \simeq \frac{1}{2\sqrt{\pi}}\,|\hat{z}|^{-1/4}\,\exp\left(-\frac{2}{3}\,|\hat{z}|^{3/2}\right) \tag{6.22}$$

in the limit $\hat{z} \to -\infty$, and

$$\mathrm{Ai}(-\hat{z}) \simeq \frac{1}{\sqrt{\pi}}\,\hat{z}^{-1/4}\,\sin\left(\frac{2}{3}\,\hat{z}^{3/2} + \frac{\pi}{4}\right) \tag{6.23}$$

in the limit $\hat{z} \to +\infty$.

Suppose that a unit amplitude plane electromagnetic wave, polarized in the y-direction, is launched from an antenna, located at large positive z, toward the cutoff point at $z = 0$. It is assumed that $n = 1$ at the launch point. In the non-evanescent region, $z > 0$, the wave can be represented as a linear combination of propagating WKB solutions:

$$E_y(z) = n^{-1/2}\,\exp\left(-\mathrm{i}\,k_0\int_0^z n\,dz'\right) + R\,n^{-1/2}\,\exp\left(+\mathrm{i}\,k_0\int_0^z n\,dz'\right). \tag{6.24}$$

The first term on the right-hand side of the previous equation represents the incident wave, whereas the second term represents the reflected wave. The complex constant R is the *coefficient of reflection*. In the vicinity of the cutoff point (i.e., z small and positive, which corresponds to $\hat{z}$ large and positive), the previous expression reduces to

$$E_y(\hat{z}) = (k_0/a)^{1/6}\left[\hat{z}^{-1/4}\exp\left(-\mathrm{i}\,\frac{2}{3}\,\hat{z}^{3/2}\right) + R\,\hat{z}^{-1/4}\,\exp\left(+\mathrm{i}\,\frac{2}{3}\,\hat{z}^{3/2}\right)\right]. \tag{6.25}$$

However, we have another expression for the wave in this region:

$$E_y(\hat{z}) = C\,\mathrm{Ai}(-\hat{z}) \simeq \frac{C}{\sqrt{\pi}}\,\hat{z}^{-1/4}\,\sin\left(\frac{2}{3}\,\hat{z}^{3/2} + \frac{\pi}{4}\right), \tag{6.26}$$

where C is an arbitrary constant. The previous equation can be written

$$E_y(\hat{z}) = \frac{C}{2}\,\sqrt{\frac{\mathrm{i}}{\pi}}\left[\hat{z}^{-1/4}\,\exp\left(-\mathrm{i}\,\frac{2}{3}\,\hat{z}^{3/2}\right) - \mathrm{i}\,\hat{z}^{-1/4}\,\exp\left(+\mathrm{i}\,\frac{2}{3}\,\hat{z}^{3/2}\right)\right]. \tag{6.27}$$

A comparison of Equations (6.25) and (6.27) reveals that

$$R = -\mathrm{i}. \tag{6.28}$$

We conclude that at a cutoff point there is total reflection of the incident wave (because $|R| = 1$) with a $-\pi/2$ phase-shift.

6.4 Resonances

Suppose, now, that a resonance is located at $z = 0$, so that

$$n^2 = \frac{b}{z + \mathrm{i}\,\epsilon} + O(1) \tag{6.29}$$

in the immediate vicinity of this point, where $b > 0$. Here, ϵ is a small real constant. We introduce ϵ in our analysis principally as a mathematical artifice to ensure that E_y remains single-valued and finite. However, as will become clear later on, ϵ has a physical significance in terms of the damping or the spontaneous excitation of waves.

In the immediate vicinity of the resonance point, $z = 0$, Equations (6.3) and (6.29) yield

$$\frac{d^2E_y}{d\hat{z}^2} + \frac{E_y}{\hat{z} + \mathrm{i}\,\hat{\epsilon}} = 0, \tag{6.30}$$

where

$$\hat{z} = (k_0^2\,b)\,z, \tag{6.31}$$

and $\hat{\epsilon} = (k_0^2\,b)\,\epsilon$. This equation is singular at the point $\hat{z} = -\mathrm{i}\,\hat{\epsilon}$. Thus, it is necessary

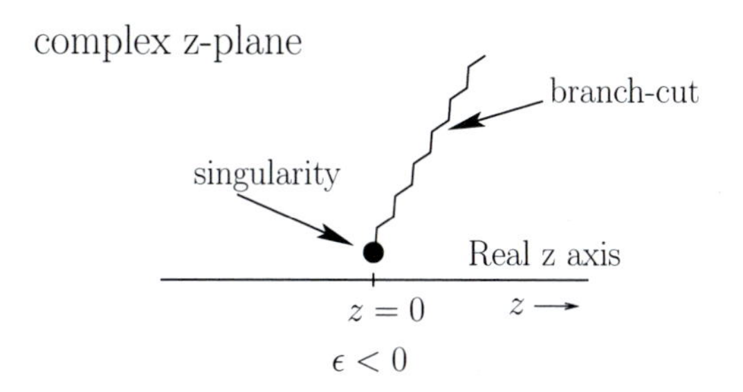

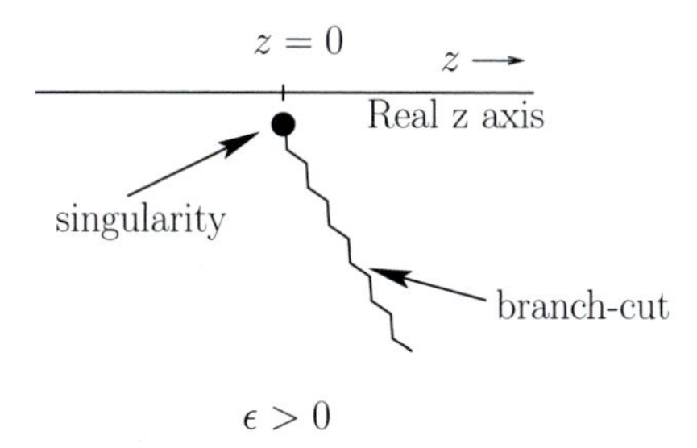

Figure 6.1
Branch-cuts in the z-plane close to a wave resonance.

to introduce a branch-cut into the complex-$\hat{z}$ plane, so as to ensure that $E_y(\hat{z})$ is single-valued. If $\epsilon > 0$ then the branch-cut lies in the lower half-plane, whereas if $\epsilon < 0$ then the branch-cut lies in the upper half-plane. (See Figure 6.1.) Suppose that the argument of $\hat{z}$ is 0 on the positive real $\hat{z}$-axis. It follows that the argument of $\hat{z}$ on the negative real $\hat{z}$-axis is $+\pi$ when $\epsilon > 0$, and $-\pi$ when $\epsilon < 0$.

Let

$$y = 2\sqrt{\hat{z}}, \tag{6.32}$$

$$E_y(y) = y\,\psi(y). \tag{6.33}$$

In the limit $\epsilon \to 0$, Equation (6.30) transforms into

$$\frac{d^2\psi}{dy^2} + \frac{1}{y}\frac{d\psi}{dy} + \left(1 - \frac{1}{y^2}\right)\psi = 0. \tag{6.34}$$

This is a standard equation, known as *Bessel's equation* of order one (Abramowitz and Stegun 1965c), and possesses two independent solutions, denoted $J_1(y)$ and $Y_1(y)$, respectively. Thus, on the positive real $\hat{z}$-axis, we can write the most general solution to Equation (6.30) in the form

$$E_y(\hat{z}) = A\sqrt{\hat{z}}\,J_1\!\left(2\sqrt{\hat{z}}\right) + B\sqrt{\hat{z}}\,Y_1\!\left(2\sqrt{\hat{z}}\right), \tag{6.35}$$

where A and B are two arbitrary constants.

Let

$$y = 2\sqrt{a\,\hat{z}}, \tag{6.36}$$

$$E_y(y) = y\,\psi(y), \tag{6.37}$$

where

$$a = \exp\left[-\mathrm{i}\,\pi\,\mathrm{sgn}(\epsilon)\right]. \tag{6.38}$$

Note that the argument of $a\,\hat{z}$ is zero on the negative real $\hat{z}$-axis. In the limit $\epsilon \to 0$, Equation (6.30) transforms into

$$\frac{d^2\psi}{dy^2} + \frac{1}{y}\frac{d\psi}{dy} - \left(1 + \frac{1}{y^2}\right)\psi = 0. \tag{6.39}$$

This is a standard equation, known as *Bessel's modified equation* of order one (Abramowitz and Stegun 1965c), and possesses two independent solutions, denoted $I_1(y)$ and $K_1(y)$, respectively. Thus, on the negative real $\hat{z}$-axis, we can write the most general solution to Equation (6.30) in the form

$$E_y(\hat{z}) = C\sqrt{a\,\hat{z}}\,I_1\!\left(2\sqrt{a\,\hat{z}}\right) + D\sqrt{a\,\hat{z}}\,K_1\!\left(2\sqrt{a\,\hat{z}}\right), \tag{6.40}$$

where C and D are two arbitrary constants.

The Bessel functions $J_1(z)$, $Y_1(z)$, $I_1(z)$, and $K_1(z)$ are all perfectly well-defined (i.e., analytic) for complex arguments, so the two expressions (6.35) and (6.40) must, in fact, be identical. In particular, the constants C and D must somehow be related to the constants A and B. In order to establish this relationship, it is convenient to investigate the behavior of the expressions (6.35) and (6.40) in the limit of small $\hat{z}$: that is, $|\hat{z}| \ll 1$. In this limit,

$$\sqrt{\hat{z}}\,J_1\!\left(2\sqrt{\hat{z}}\right) = \hat{z} + O\!\left(\hat{z}^2\right), \tag{6.41}$$

$$\sqrt{a\,\hat{z}}\,I_1\!\left(2\sqrt{a\,\hat{z}}\right) = -\hat{z} + O\!\left(\hat{z}^2\right), \tag{6.42}$$

$$\sqrt{\hat{z}}\,Y_1\!\left(2\sqrt{\hat{z}}\right) = -\frac{1}{\pi}\left[1 - (\ln|\hat{z}| + 2\,\gamma - 1)\,\hat{z}\right] + O\!\left(\hat{z}^2\right), \tag{6.43}$$

$$\sqrt{a\,\hat{z}}\,K_1\!\left(2\sqrt{a\,\hat{z}}\right) = \frac{1}{2}\left[1 - (\ln|\hat{z}| + 2\,\gamma - 1)\,\hat{z} - \mathrm{i}\arg(a)\,\hat{z}\right] + O\!\left(\hat{z}^2\right), \tag{6.44}$$

where γ is Euler's constant (Abramowitz and Stegun 1965a), and $\hat{z}$ is assumed to lie on the positive real $\hat{z}$-axis. It follows, by a comparison of Equations (6.35), (6.40), and (6.41)–(6.44), that the choice

$$C = -A + \mathrm{i}\,\frac{\pi}{2}\,\mathrm{sgn}(\epsilon)\,D = -A - \mathrm{i}\,\mathrm{sgn}(\epsilon)\,B, \tag{6.45}$$

$$D = -\frac{2}{\pi}\,B, \tag{6.46}$$

ensures that the expressions (6.35) and (6.40) are indeed identical.

In the limit $|\hat{z}| \gg 1$,

$$\sqrt{a\,\hat{z}}\,I_1\!\left(2\sqrt{a\,\hat{z}}\right) \simeq \frac{1}{2\sqrt{\pi}}\,|\hat{z}|^{\,1/4}\,\exp\!\left(+2\sqrt{|\hat{z}|}\right), \tag{6.47}$$

$$\sqrt{a\,\hat{z}}\,K_1\!\left(2\sqrt{a\,\hat{z}}\right) \simeq \frac{\sqrt{\pi}}{2}\,|\hat{z}|^{\,1/4}\,\exp\!\left(-2\sqrt{|\hat{z}|}\right), \tag{6.48}$$

where $\hat{z}$ is assumed to lie on the negative real $\hat{z}$-axis. It is clear that the I_1 solution is unphysical, because it blows up in the evanescent region ($\hat{z} < 0$). Thus, the coefficient C in expression (6.40) must be set to zero in order to prevent $E_y(\hat{z})$ from blowing up as $\hat{z} \to -\infty$. According to Equation (6.45), this constraint implies that

$$A = -\mathrm{i}\,\mathrm{sgn}(\epsilon)\,B. \tag{6.49}$$

In the limit $|\hat{z}| \gg 1$,

$$\sqrt{\hat{z}}\,J_1\!\left(2\sqrt{\hat{z}}\right) \simeq \frac{1}{\sqrt{\pi}}\,\hat{z}^{\,1/4}\,\cos\!\left(2\sqrt{\hat{z}} - \frac{3}{4}\,\pi\right), \tag{6.50}$$

$$\sqrt{\hat{z}}\,Y_1\!\left(2\sqrt{\hat{z}}\right) \simeq \frac{1}{\sqrt{\pi}}\,\hat{z}^{\,1/4}\,\sin\!\left(2\sqrt{\hat{z}} - \frac{3}{4}\,\pi\right), \tag{6.51}$$

where $\hat{z}$ is assumed to lie on the positive real $\hat{z}$-axis. It follows from Equations (6.35), (6.49), (6.50), and (6.51) that, in the non-evanescent region ($\hat{z} > 0$), the most general physical solution takes the form

$$\begin{aligned} E_y(\hat{z}) = {} & A'\,\left[\mathrm{sgn}(\epsilon) + 1\right]\,\hat{z}^{\,1/4}\,\exp\!\left[+\mathrm{i}\left(2\sqrt{\hat{z}} - \frac{3}{4}\,\pi\right)\right] \\ & + A'\,\left[\mathrm{sgn}(\epsilon) - 1\right]\,\hat{z}^{\,1/4}\,\exp\!\left[-\mathrm{i}\left(2\sqrt{\hat{z}} + \frac{3}{4}\,\pi\right)\right], \end{aligned} \tag{6.52}$$

where A' is an arbitrary constant.

Suppose that a plane electromagnetic wave, polarized in the y-direction, is launched from an antenna, located at large positive z, toward the resonance point at $z = 0$. It is assumed that $n = 1$ at the launch point. In the non-evanescent region, $z > 0$, the wave can be represented as a linear combination of propagating WKB solutions:

$$E_y(z) = E\,n^{-1/2}\,\exp\!\left(-\mathrm{i}\,k_0\int_0^z n\,dz'\right) + F\,n^{-1/2}\,\exp\!\left(+\mathrm{i}\,k_0\int_0^z n\,dz'\right). \tag{6.53}$$

The first term on the right-hand side of the previous equation represents the incident wave, whereas the second term represents the reflected wave. Here, E is the amplitude of the incident wave, and F is the amplitude of the reflected wave. In the vicinity of the resonance point (i.e., z small and positive, which corresponds to $\hat{z}$ large and positive), the previous expression reduces to

$$E_y(\hat{z}) \simeq (k_0\,b)^{-1/2}\left[E\,\hat{z}^{\,1/4}\exp\!\left(-\mathrm{i}\,2\sqrt{\hat{z}}\right) + F\,\hat{z}^{\,1/4}\exp\!\left(+\mathrm{i}\,2\sqrt{\hat{z}}\right)\right]. \tag{6.54}$$

A comparison of Equations (6.52) and (6.54) shows that if $\epsilon > 0$ then $E = 0$. In other words, there is a reflected wave, but no incident wave. This corresponds to the spontaneous excitation of waves in the vicinity of the resonance. On the other hand, if $\epsilon < 0$ then $F = 0$. In other words, there is an incident wave, but no reflected wave. This corresponds to the total absorption of incident waves in the vicinity of the resonance. It is clear that if $\epsilon > 0$ then ϵ represents some sort of spontaneous wave excitation mechanism, whereas if $\epsilon < 0$ then ϵ represents a wave absorption, or damping, mechanism. We would normally expect plasmas to absorb incident wave energy, rather than spontaneously emit waves, so we conclude that, under most circumstances, $\epsilon < 0$, and resonances absorb incident waves without reflection.

6.5 Resonant Layers

Consider the situation, studied in the previous section, in which a plane wave, polarized in the y-direction, is launched along the z-axis, from an antenna located at large positive z, and absorbed at a resonance located at $z = 0$. In the vicinity of the resonant point, the electric component of the wave satisfies

$$\frac{d^2 E_y}{dz^2} + \frac{k_0^2\, b}{z + \mathrm{i}\,\epsilon} E_y = 0, \tag{6.55}$$

where $b > 0$ and $\epsilon < 0$.

The time-averaged Poynting flux in the z-direction is written

$$P_z = -\frac{(E_y\, B_x^* + E_y^*\, B_x)}{4\,\mu_0}. \tag{6.56}$$

Now, the Faraday-Maxwell equation yields

$$\mathrm{i}\,\omega\, B_x = -\frac{dE_y}{dz}. \tag{6.57}$$

Thus, we have

$$P_z = -\frac{\mathrm{i}}{4\,\mu_0\,\omega}\left(\frac{dE_y}{dz}\, E_y^* - \frac{dE_y^*}{dz}\, E_y\right). \tag{6.58}$$

Let us ascribe any variation of P_z with z to the wave energy emitted by the plasma. We then obtain

$$\frac{dP_z}{dz} = W, \tag{6.59}$$

where W is the power emitted by the plasma per unit volume. It follows that

$$W = -\frac{\mathrm{i}}{4\,\mu_0\,\omega}\left(\frac{d^2 E_y}{dz^2}\, E_y^* - \frac{d^2 E_y^*}{dz^2}\, E_y\right). \tag{6.60}$$

Equations (6.55) and (6.60) yield

$$W = \left(\frac{k_0^2\, b}{2\,\mu_0\,\omega}\right)\left(\frac{\epsilon}{z^2+\epsilon^2}\right)|E_y|^2. \tag{6.61}$$

Note that $W < 0$, because $\epsilon < 0$, so wave energy is absorbed by the plasma. It is clear, from the previous formula, that the absorption takes place in a narrow layer, of thickness $|\epsilon|$, centered on the resonance point, $z = 0$.

6.6 Collisional Damping

Let us now consider a real-life damping mechanism. Equation (5.15) specifies the linearized Ohm's law in the collisionless cold-plasma approximation. In the presence of collisions, this expression acquires an extra term (see Section 4.12), such that

$$\mathbf{E} = -\mathbf{V}\times\mathbf{B}_0 + \frac{\mathbf{j}\times\mathbf{B}_0}{n_e\,e} - \mathrm{i}\,\frac{\omega\,m_e}{n_e\,e^2}\,\mathbf{j} + \frac{\nu_e\,m_e}{n_e\,e^2}\,\mathbf{j}, \tag{6.62}$$

where $\nu_e \equiv \tau_e^{-1}$ is the electron collision frequency. Here, for the sake of simplicity, we have neglected the small difference between the parallel and perpendicular plasma electrical conductivities. When Equation (6.62) is used to calculate the dielectric permittivity for a right-handed wave, in the limit $\omega \gg \Omega_i$, we obtain

$$R \simeq 1 - \frac{\Pi_e^2}{\omega\,(\omega + \mathrm{i}\,\nu_e - |\Omega_e|)}. \tag{6.63}$$

A right-handed circularly polarized wave, propagating parallel to the magnetic field, is governed by the dispersion relation (see Section 5.9)

$$n^2 = R \simeq 1 + \frac{\Pi_e^2}{\omega\,(|\Omega_e| - \omega - \mathrm{i}\,\nu_e)}. \tag{6.64}$$

Suppose that $n = n(z)$. Furthermore, let

$$|\Omega_e| = \omega + |\Omega_e|'\,z, \tag{6.65}$$

so that the electron cyclotron resonance is located at $z = 0$. We also assume that $|\Omega_e|' > 0$, so that the evanescent region corresponds to $z < 0$. It follows that, in the immediate vicinity of the resonance,

$$n^2 \simeq \frac{b}{z + \mathrm{i}\,\epsilon}, \tag{6.66}$$

where

$$b = \frac{\Pi_e^2}{\omega\,|\Omega_e|'}, \tag{6.67}$$

and

$$\epsilon = -\frac{\nu_e}{|\Omega_e|'}. \tag{6.68}$$

It can be seen that $\epsilon < 0$, which is consistent with the absorption of incident wave energy by the resonant layer. The approximate width of the resonant layer is

$$\delta \sim |\epsilon| = \frac{\nu_e}{|\Omega_e|'}. \tag{6.69}$$

Note that the damping mechanism—in this case collisions—controls the thickness of the resonant layer, but does not control the amount of wave energy absorbed by the layer. In fact, in the simple theory outlined previously, all of the incident wave energy is absorbed by the layer.

6.7 Pulse Propagation

Consider the situation, studied in Section 6.3, in which a plane wave, polarized in the y-direction, is launched along the z-axis, from an antenna located at large positive z, and reflected from a cutoff located at $z = 0$. Up to now, we have only considered infinite wave-trains, characterized by a discrete frequency, ω. Let us now consider the more realistic case in which the antenna emits a finite pulse of radio waves.

The pulse structure is conveniently represented as

$$E_y(t) = \int_{-\infty}^{\infty} F(\omega)\,\mathrm{e}^{-\mathrm{i}\,\omega\,t}\,d\omega, \tag{6.70}$$

where $E_y(t)$ is the electric field produced by the antenna, which is assumed to lie at $z = a$. Suppose that the pulse is a signal of roughly constant (angular) frequency ω_0, which lasts a time T, where T is long compared to $1/\omega_0$. It follows that $F(\omega)$ possesses narrow maxima around $\omega = \pm\omega_0$. In other words, only those frequencies that lie very close to the central frequency, ω_0, play a significant role in the propagation of the pulse.

Each component frequency of the pulse yields a wave that propagates independently along the z-axis, in a manner specified by the appropriate WKB solution [see Equations (6.17) and (6.18)]. Thus, if Equation (6.70) specifies the signal at the antenna (i.e., at $z = a$) then the signal at coordinate z (where $z < a$) is given by

$$E_y(z,t) = \int_{-\infty}^{\infty} \frac{F(\omega)}{n^{1/2}(\omega, z)}\,\mathrm{e}^{\,\mathrm{i}\,\phi(\omega,z,t)}\,d\omega, \tag{6.71}$$

where

$$\phi(\omega, z, t) = \frac{\omega}{c}\int_z^a n(\omega, z)\,dz' - \omega\,t. \tag{6.72}$$

Here, we have made use of the fact that $k_0 = \omega/c$.

Equation (6.71) can be regarded as a contour integral in ω-space. The quantity $F/n^{1/2}$ is a relatively slowly varying function of ω, whereas the phase, ϕ, is a large and rapidly varying function of ω. The rapid oscillations of $\exp(\mathrm{i}\,\phi)$ over most of the path of integration ensure that the integrand averages almost to zero. However, this cancellation argument does not apply to places on the integration path where the phase is stationary: that is, places where $\phi(\omega)$ has an extremum. The integral can, therefore, be estimated by finding those points where $\phi(\omega)$ has a vanishing derivative, evaluating (approximately) the integral in the neighborhood of each of these points, and summing the contributions. This procedure is called the *method of stationary phase* (Budden 1985).

Suppose that $\phi(\omega)$ has a vanishing first derivative at $\omega = \omega_s$. In the neighborhood of this point, $\phi(\omega)$ can be expanded as a Taylor series,

$$\phi(\omega) = \phi_s + \frac{1}{2}\,\phi_s''\,(\omega - \omega_s)^2 + \cdots. \tag{6.73}$$

Here, the subscript s is used to indicate ϕ or its second derivative evaluated at $\omega = \omega_s$. Because $F(\omega)/n^{1/2}(\omega, z)$ is slowly varying, the contribution to the integral from this stationary phase point is approximately

$$E_{y\,s} \simeq \frac{F(\omega_s)\,\mathrm{e}^{\,\mathrm{i}\,\phi_s}}{n^{1/2}(\omega_s, z)} \int_{-\infty}^{\infty} \exp\left[\frac{\mathrm{i}}{2}\,\phi_s''\,(\omega - \omega_s)^2\right] d\omega. \tag{6.74}$$

The previous expression can be written in the form

$$E_{y\,s} \simeq \frac{F(\omega_s)\,\mathrm{e}^{\,\mathrm{i}\,\phi_s}}{n^{1/2}(\omega_s, z)} \sqrt{\frac{4\pi}{\phi_s''}} \int_0^{\infty} \left[\cos\left(\pi\,t^2/2\right) + \mathrm{i}\,\sin\left(\pi\,t^2/2\right)\right] dt, \tag{6.75}$$

where

$$\frac{\pi}{2}\,t^2 = \frac{1}{2}\,\phi_s''\,(\omega - \omega_s)^2. \tag{6.76}$$

The integrals in the previous expression are known as *Fresnel integrals* (Abramowitz and Stegun 1965b), and can be shown to take the values

$$\int_0^{\infty} \cos\left(\pi\,t^2/2\right)\,dt = \int_0^{\infty} \sin\left(\pi\,t^2/2\right)\,dt = \frac{1}{2}. \tag{6.77}$$

It follows that

$$E_{y\,s} \simeq \sqrt{\frac{2\pi\,\mathrm{i}}{\phi_s''}}\,\frac{F(\omega_s)}{n^{1/2}(\omega_s, z)}\,\mathrm{e}^{\,\mathrm{i}\,\phi_s}. \tag{6.78}$$

If there is more than one point of stationary phase in the range of integration then the integral is approximated as a sum of terms similar to that in the previous formula.

Integrals of the form (6.71) can be calculated exactly using the *method of steepest descent* (Brillouin 1960; Budden 1985). The stationary phase approximation (6.78) agrees with the leading term of the method of steepest descent (which is far more difficult to implement than the method of stationary phase) provided that $\phi(\omega)$ is real (i.e., provided that the stationary point lies on the real axis). If ϕ is complex, however, then the stationary phase method can yield erroneous results.

It follows, from the previous discussion, that the right-hand side of Equation (6.71) averages to a very small value, expect for those special values of z and t at which one of the points of stationary phase in ω-space coincides with one of the peaks of $F(\omega)$. The locus of these special values of z and t can obviously be regarded as the equation of motion of the pulse as it propagates along the z-axis. Thus, the equation of motion is specified by

$$\left(\frac{\partial\phi}{\partial\omega}\right)_{\omega=\omega_0} = 0, \tag{6.79}$$

which yields

$$t = \frac{1}{c}\int_z^a \left[\frac{\partial(\omega\, n)}{\partial\omega}\right]_{\omega=\omega_0} dz'. \tag{6.80}$$

Suppose that the z-velocity of a pulse of central frequency ω_0 at coordinate z is given by $-u_z(\omega_0, z)$. The differential equation of motion of the pulse is then $dt = -dz/u_z$. This can be integrated, using the boundary condition $z = a$ at $t = 0$, to give the full equation of motion:

$$t = \int_z^a \frac{dz'}{u_z}. \tag{6.81}$$

A comparison of Equations (6.80) and (6.81) yields

$$u_z(\omega_0, z) = c \bigg/ \left(\frac{\partial[\omega\, n(\omega, z)]}{\partial\omega}\right)_{\omega=\omega_0}. \tag{6.82}$$

The velocity u_z is usually called the *group-velocity*. It is easily demonstrated that the previous expression for the group-velocity is entirely consistent with that given in Equation (5.72).

The dispersion relation for an electromagnetic plasma wave propagating through an unmagnetized plasma is [see Equation (6.121)]

$$n(\omega, z) = \left[1 - \frac{\Pi_e^2(z)}{\omega^2}\right]^{1/2}. \tag{6.83}$$

Here, we have assumed that equilibrium quantities are functions of z only, and that the wave propagates along the z-axis. The phase-velocity of waves of frequency ω propagating along the z-axis is given by

$$v_z(\omega, z) = \frac{c}{n(\omega, z)} = c\left[1 - \frac{\Pi_e^2(z)}{\omega^2}\right]^{-1/2}. \tag{6.84}$$

According to Equations (6.82) and (6.83), the corresponding group-velocity is

$$u_z(\omega, z) = c\left[1 - \frac{\Pi_e^2(z)}{\omega^2}\right]^{1/2}. \tag{6.85}$$

It follows that

$$v_z\, u_z = c^2. \tag{6.86}$$

Let us assume that $\Pi_e(0) = \omega$, and $\Pi_e(z) < \omega$ for $z > 0$, which implies that the reflection point corresponds to $z = 0$. It is clear from Equations (6.84) and (6.85) that the phase-velocity of the wave is always greater than the velocity of light in vacuum, whereas the group-velocity is always less than this velocity. Furthermore, as the reflection point, $z = 0$, is approached from positive z, the phase-velocity tends to infinity, whereas the group-velocity tends to zero.

Although we have only analyzed the motion of the pulse as it travels from the antenna to the reflection point, it is easily demonstrated that the speed of the reflected pulse at position z is the same as that of the incident pulse. In other words, the group velocities of pulses traveling in opposite directions are of equal magnitude.

6.8 Ray Tracing

Let us now generalize the preceding analysis so that we can deal with pulse propagation though a three-dimensional magnetized plasma.

A general wave problem can be written as a set of n coupled, linear, homogeneous, first-order, partial-differential equations, which take the form (Hazeltine and Waelbroeck 2004)

$$\mathbf{M}(\mathrm{i}\,\partial/\partial t, -\mathrm{i}\,\nabla, \mathbf{r}, t)\,\boldsymbol{\psi} = \mathbf{0}. \tag{6.87}$$

The vector-field $\boldsymbol{\psi}(\mathbf{r}, t)$ has n components (e.g., $\boldsymbol{\psi}$ might consist of $\mathbf{E}$, $\mathbf{B}$, $\mathbf{j}$, and $\mathbf{V}$) characterizing some small disturbance, and $\mathbf{M}$ is an $n \times n$ matrix characterizing the undisturbed plasma.

The lowest order WKB approximation is premised on the assumption that $\mathbf{M}$ depends so weakly on $\mathbf{r}$ and t that all of the spatial and temporal dependence of the components of $\psi(\mathbf{r}, t)$ is specified by a common factor $\exp(\mathrm{i}\,\phi)$. Thus, Equation (6.87) reduces to

$$\mathbf{M}(\omega, \mathbf{k}, \mathbf{r}, t)\,\boldsymbol{\psi} = \mathbf{0}, \tag{6.88}$$

where

$$\mathbf{k} \equiv \nabla\phi, \tag{6.89}$$

$$\omega \equiv -\frac{\partial\phi}{\partial t}. \tag{6.90}$$

In general, Equation (6.88) has many solutions, corresponding to the many different types and polarizations of waves that can propagate through the plasma in question, all of which satisfy the dispersion relation

$$\mathcal{M}(\omega, \mathbf{k}, \mathbf{r}, t) = 0, \tag{6.91}$$

where $\mathcal{M} \equiv \det(\mathbf{M})$. As is easily demonstrated (see Section 6.2), the WKB approximation is valid provided that the characteristic variation lengthscale and variation timescale of the plasma are much longer than the wavelength, $2\pi/k$, and the period, $2\pi/\omega$, respectively, of the wave in question.

Let us concentrate on one particular solution of Equation (6.88) (e.g., on one particular type of plasma wave). For this solution, the dispersion relation (6.91) yields

$$\omega = \Omega(\mathbf{k}, \mathbf{r}, t): \tag{6.92}$$

that is, the dispersion relation yields a unique frequency for a wave of a given wavevector, $\mathbf{k}$, located at a given point, ($\mathbf{r}$, t), in space and time. There is also a unique $\boldsymbol{\psi}$ associated with this frequency, which is obtained from Equation (6.88). To lowest order, we can neglect the variation of $\boldsymbol{\psi}$ with $\mathbf{r}$ and t. A general pulse solution is written

$$\boldsymbol{\psi}(\mathbf{r}, t) = \int F(\mathbf{k})\, \boldsymbol{\psi}\, e^{i\phi}\, d^3\mathbf{k}, \tag{6.93}$$

where (locally)

$$\phi = \mathbf{k}\cdot\mathbf{r} - \Omega\, t, \tag{6.94}$$

and $F(\mathbf{k})$ is a function that specifies the initial structure of the pulse in $\mathbf{k}$-space.

The integral (6.93) averages to zero, except at a point of stationary phase, where $\nabla_{\mathbf{k}}\phi = 0$. (See Section 6.7.) Here, $\nabla_{\mathbf{k}}$ is the $\mathbf{k}$-space gradient operator. It follows that the (instantaneous) trajectory of the pulse matches that of a point of stationary phase:

$$\nabla_{\mathbf{k}}\phi = \mathbf{r} - \mathbf{v}_g\, t = 0, \tag{6.95}$$

where

$$\mathbf{v}_g = \frac{\partial \Omega}{\partial \mathbf{k}} \tag{6.96}$$

is the group-velocity. Thus, the instantaneous velocity of a pulse is always equal to the local group-velocity.

Let us now determine how the wavevector, $\mathbf{k}$, and the angular frequency, ω, of a pulse evolve as the pulse propagates through the plasma. We start from the cross-differentiation rules [see Equations (6.89) and (6.90)]:

$$\frac{\partial k_i}{\partial t} + \frac{\partial \omega}{\partial r_i} = 0, \tag{6.97}$$

$$\frac{\partial k_j}{\partial r_i} - \frac{\partial k_i}{\partial r_j} = 0. \tag{6.98}$$

Equations (6.92), (6.97), and (6.98) yield [making use of the Einstein summation convention (Riley 1974)]

$$\frac{\partial k_i}{\partial t} + \frac{\partial \Omega}{\partial k_j}\frac{\partial k_j}{\partial r_i} + \frac{\partial \Omega}{\partial r_i} = \frac{\partial k_i}{\partial t} + \frac{\partial \Omega}{\partial k_j}\frac{\partial k_i}{\partial r_j} + \frac{\partial \Omega}{\partial r_i} = 0, \tag{6.99}$$

or

$$\frac{d\mathbf{k}}{dt} \equiv \frac{\partial \mathbf{k}}{\partial t} + (\mathbf{v}_g\cdot\nabla)\,\mathbf{k} = -\nabla\Omega. \tag{6.100}$$

In other words, the variation of **k**, as seen in a frame co-moving with the pulse, is determined by the spatial gradients in Ω.

Partial differentiation of Equation (6.92) with respect to t gives

$$\frac{\partial\omega}{\partial t} = \frac{\partial\Omega}{\partial k_j}\frac{\partial k_j}{\partial t} + \frac{\partial\Omega}{\partial t} = -\frac{\partial\Omega}{\partial k_j}\frac{\partial\omega}{\partial r_j} + \frac{\partial\Omega}{\partial t}, \tag{6.101}$$

which can be written

$$\frac{d\omega}{dt} \equiv \frac{\partial\omega}{\partial t} + (\mathbf{v}_g \cdot \nabla)\,\omega = \frac{\partial\Omega}{\partial t}. \tag{6.102}$$

In other words, the variation of ω, as seen in a frame co-moving with the pulse, is determined by the time variation of Ω.

According to the previous analysis, the evolution of a pulse propagating though a spatially and temporally non-uniform plasma can be determined by solving the *ray equations*:

$$\frac{d\mathbf{r}}{dt} = \frac{\partial\Omega}{\partial\mathbf{k}}, \tag{6.103}$$

$$\frac{d\mathbf{k}}{dt} = -\nabla\Omega, \tag{6.104}$$

$$\frac{d\omega}{dt} = \frac{\partial\Omega}{\partial t}. \tag{6.105}$$

The previous equations are conveniently rewritten in terms of the dispersion relation (6.91) (Hazeltine and Waelbroeck 2004):

$$\frac{d\mathbf{r}}{dt} = -\frac{\partial\mathcal{M}/\partial\mathbf{k}}{\partial\mathcal{M}/\partial\omega}, \tag{6.106}$$

$$\frac{d\mathbf{k}}{dt} = \frac{\partial\mathcal{M}/\partial\mathbf{r}}{\partial\mathcal{M}/\partial\omega}, \tag{6.107}$$

$$\frac{d\omega}{dt} = -\frac{\partial\mathcal{M}/\partial t}{\partial\mathcal{M}/\partial\omega}. \tag{6.108}$$

Incidentally, the variation in the amplitude of the pulse, as it propagates through the plasma, can only be determined by expanding the WKB solutions to higher order. (See Exercises 6.3 and 6.4.)

6.9 Ionospheric Radio Wave Propagation

To a first approximation, the Earth's ionosphere consists of an unmagnetized, horizontally stratified, partially ionized gas (Budden 1985). The dispersion relation for the electromagnetic plasma wave takes the form [see Equation (5.98)]

$$\mathcal{M} = \omega^2 - k^2c^2 - \Pi_e^2 = 0, \tag{6.109}$$

where

$$\Pi_e = \sqrt{\frac{N\,e^2}{\epsilon_0\,m_e}}. \tag{6.110}$$

Here, $N = N(z)$ is the density of free electrons in the ionosphere, and z is a coordinate that measures height above the surface of the Earth. (The curvature of the Earth, the Earth's magnetic field, and collisions, are neglected in the following analysis.)

Now,

$$\frac{\partial\mathcal{M}}{\partial\omega} = 2\,\omega, \tag{6.111}$$

$$\frac{\partial\mathcal{M}}{\partial\mathbf{k}} = -2\,c^2\,\mathbf{k}, \tag{6.112}$$

$$\frac{\partial\mathcal{M}}{\partial\mathbf{r}} = -\nabla\Pi_e^2, \tag{6.113}$$

$$\frac{\partial\mathcal{M}}{\partial t} = 0. \tag{6.114}$$

Thus, the ray equations, (6.106)–(6.108), yield

$$\frac{d\mathbf{r}}{dt} = \frac{c^2}{\omega}\,\mathbf{k}, \tag{6.115}$$

$$\frac{d\mathbf{k}}{dt} = -\frac{\nabla\Pi_e^2}{2\,\omega}, \tag{6.116}$$

$$\frac{d\omega}{dt} = 0. \tag{6.117}$$

Evidently, the frequency of a radio pulse does not change as it propagates through the ionosphere, provided that $N(z)$ does not vary in time. Furthermore, it follows from Equations (6.115)–(6.117), and the fact that $\Pi_e = \Pi_e(z)$, that a radio pulse that starts off at ground level propagating in the x-z plane, say, will continue to propagate in this plane.

For pulse propagation in the x-z plane, we have

$$\frac{dx}{dt} = \frac{c^2\,k_x}{\omega}, \tag{6.118}$$

$$\frac{dz}{dt} = \frac{c^2\,k_z}{\omega}, \tag{6.119}$$

$$\frac{dk_x}{dt} = 0. \tag{6.120}$$

The dispersion relation (6.109) yields

$$n^2 = \frac{(k_x^2 + k_z^2)\,c^2}{\omega^2} = 1 - \frac{\Pi_e^2}{\omega^2}, \tag{6.121}$$

where $n(z)$ is the refractive index.

Let us assume that $n = 1$ at $z = 0$, which is equivalent to the reasonable assumption that the atmosphere is non-ionized at ground level. It follows from Equation (6.120) that

$$k_x = k_x(z = 0) = \frac{\omega}{c} S, \tag{6.122}$$

where S is the sine of the angle of incidence of the pulse, with respect to the vertical axis, at ground level. Equations (6.121) and (6.122) yield

$$k_z = \pm\frac{\omega}{c}\sqrt{n^2 - S^2}. \tag{6.123}$$

According to Equation (6.119), the plus sign corresponds to the upward trajectory of the pulse, whereas the minus sign corresponds to the downward trajectory. Finally, Equations (6.118), (6.119), (6.122), and (6.123) yield the equations of motion of the pulse:

$$\frac{dx}{dt} = c\,S, \tag{6.124}$$

$$\frac{dz}{dt} = \pm c\sqrt{n^2 - S^2}. \tag{6.125}$$

The pulse attains its maximum altitude, $z = z_0$, when

$$n(z_0) = |S|. \tag{6.126}$$

The total distance traveled by the pulse (i.e., the distance from its launch point to the point where it intersects the Earth's surface again) is

$$x_0 = 2\,S\int_0^{z_0(S)} \frac{dz}{\sqrt{n^2(z) - S^2}}. \tag{6.127}$$

In the limit in which the radio pulse is launched vertically (i.e., $S = 0$) into the ionosphere, the turning point condition (6.126) reduces to that characteristic of a cutoff (i.e., $n = 0$). The WKB turning point described in Equation (6.126) is a generalization of the conventional turning point, which occurs when k^2 changes sign. Here, k_z^2 changes sign, while k_x^2 and k_y^2 are constrained by symmetry (i.e., k_x is constant, and k_y is zero).

According to Equations (6.115)–(6.117) and (6.121), the equation of motion of the pulse can also be written

$$\frac{d^2\mathbf{r}}{dt^2} = \frac{c^2}{2}\nabla n^2. \tag{6.128}$$

It follows that the trajectory of the pulse is the same as that of a particle moving in the gravitational potential $-c^2 n^2/2$. Thus, if n^2 decreases linearly with increasing height above the ground [which is the case if $N(z)$ increases linearly with z] then the trajectory of the pulse is a parabola.

6.10 Exercises

6.1 The electric polarization, **P**, in a linear dielectric medium is related to the electric field-strength, **E**, according to

$$\mathbf{P} = \epsilon_0\,(n^2 - 1)\,\mathbf{E},$$

where n is the refractive index. Any divergence of the polarization field is associated with a bound charge density

$$\rho = -\nabla\cdot\mathbf{P},$$

whereas any time variation generates a polarization current whose density is

$$\mathbf{j} = \frac{\partial\mathbf{P}}{\partial t}.$$

Consider an electromagnetic wave propagating through a quasi-neutral, linear, dielectric medium. Assuming a common $\exp(-\mathrm{i}\,\omega\,t)$ time variation of the wave fields, demonstrate from Maxwell's equations that

$$\nabla\times c\,\mathbf{B} = -\mathrm{i}\,k_0\,n^2\,\mathbf{E},$$

$$\nabla\times\mathbf{E} = \mathrm{i}\,k_0\,c\,\mathbf{B},$$

where $k_0 = \omega/c$.

6.2 Consider an electromagnetic wave, polarized in the y-direction, that propagates in the z-direction through a medium of refractive index $n(z)$. Assuming that

$$\mathbf{E} = E_y(z)\,\exp(-\mathrm{i}\,\omega\,t)\,\mathbf{e}_y,$$

$$\mathbf{B} = B_x(z)\,\exp(-\mathrm{i}\,\omega\,t)\,\mathbf{e}_x,$$

demonstrate that

$$\frac{d^2E_y}{dz^2} + k_0^2\,n^2\,E_y = 0,$$

$$\frac{d\,(c\,B_x)}{dz} + \mathrm{i}\,k_0\,n^2\,E_y = 0,$$

where $k_0 = \omega/c$.

6.3 Consider an electromagnetic wave, polarized in the y-direction, that propagates in the x-z plane through a medium of refractive index $n(z)$. Assuming that

$$\mathbf{E} = E_y(z)\,\mathrm{e}^{\,\mathrm{i}\,(k_x\,x-\omega\,t)}\,\mathbf{e}_y,$$

$$\mathbf{B} = B_x(z)\,\mathrm{e}^{\,\mathrm{i}\,(k_x\,x-\omega\,t)}\,\mathbf{e}_x + B_z(z)\,\mathrm{e}^{\,\mathrm{i}\,(k_x\,x-\omega\,t)}\,\mathbf{e}_z,$$

demonstrate that

$$\frac{d^2 E_y}{dz^2} + k_0^2\, q^2\, E_y = 0,$$

$$\frac{d\,(c\, B_x)}{dz} + \mathrm{i}\, k_0\, q^2\, E_y = 0,$$

where

$$q^2 = n^2 - S^2,$$

and $S = k_x/k_0$, and $k_0 = \omega/c$.

Show that the WKB solutions take the form

$$E_y(z) \simeq q^{-1/2} \exp\left(\pm \mathrm{i}\, k_0 \int_0^z q\, dz'\right),$$

$$c\, B_x(z) \simeq \mp q^{1/2} \exp\left(\pm \mathrm{i}\, k_0 \int_0^z q\, dz'\right),$$

and that the criterion for these solutions to be valid is

$$\frac{1}{k_0^2}\left|\frac{3}{4}\left(\frac{1}{q^2}\frac{dq}{dz}\right)^2 - \frac{1}{2\,q^3}\frac{d^2 q}{dz^2}\right| \ll 1.$$

6.4 Consider an electromagnetic wave, polarized in the x-z-plane, that propagates in the x-z plane through a medium of refractive index $n(z)$. Assuming that

$$\mathbf{E} = E_x(z)\,\mathrm{e}^{\,\mathrm{i}\,(k_x\, x - \omega\, t)}\,\mathbf{e}_x + E_z(z)\,\mathrm{e}^{\,\mathrm{i}\,(k_x\, x - \omega\, t)}\,\mathbf{e}_z,$$

$$\mathbf{B} = B_y(z)\,\mathrm{e}^{\,\mathrm{i}\,(k_x\, x - \omega\, t)}\,\mathbf{e}_y,$$

demonstrate that

$$\frac{dE_x}{dz} - \mathrm{i}\, k_0\, \frac{q^2}{n^2}\, c\, B_y = 0,$$

$$\frac{d^2 (c\, B_y)}{dz^2} - \frac{1}{n^2}\frac{dn^2}{dz}\frac{d\,(c\, B_y)}{dz} + k_0^2\, q^2\, c\, B_y = 0,$$

where

$$q^2 = n^2 - S^2,$$

and $S = k_x/k_0$, and $k_0 = \omega/c$.

Show that the WKB solutions take the form

$$c\, B_y(z) \simeq n\, q^{-1/2} \exp\left(\pm \mathrm{i}\, k_0 \int_0^z q\, dz'\right),$$

$$E_x(z) \simeq \pm n^{-1}\, q^{1/2} \exp\left(\pm \mathrm{i}\, k_0 \int_0^z q\, dz'\right),$$

and that the criterion for these solutions to be valid is

$$\frac{1}{k_0^2}\left|\frac{3}{4}\left(\frac{1}{q^2}\frac{dq}{dz}\right)^2-\frac{1}{2\,q^3}\frac{d^2q}{dz^2}+\frac{1}{q^2}\left[\frac{1}{n}\frac{d^2n}{dz^2}-2\left(\frac{1}{n}\frac{dn}{dz}\right)^2\right]\right|\ll 1.$$

6.5 An electromagnetic wave pulse of frequency ω is launched vertically from ground level, travels upward into the ionosphere, is reflected, and returns to ground level. If $\tau(\omega)$ is the net travel time of the pulse then the so-called *equivalent height of reflection* is defined $h(\omega) = c\,\tau(\omega)/2$. It follows that h is the altitude of the reflection layer calculated on the assumption that the pulse always travels at the velocity of light in vacuum. Let $\Pi_e(z)$ be the ionospheric plasma frequency, where z measures altitude above the ground. Neglect collisions and the Earth's magnetic field.

(a) Demonstrate that

$$h(\omega)=\int_0^{z_0(\omega)}\frac{\omega}{[\omega^2-\Pi_e^2(z)]^{1/2}}\,dz,$$

where $\Pi_e(z_0)=\omega$.

(b) Show that if $z_0(\omega)$ is a monotonically increasing function of ω then the previous integral can be inverted to give

$$z_0(\omega)=\frac{2}{\pi}\int_0^{\pi/2}h(\omega\,\sin\alpha)\,d\alpha,$$

or, equivalently,

$$z(\Pi_e)=\frac{2}{\pi}\int_0^{\pi/2}h(\Pi_e\,\sin\alpha)\,d\alpha.$$

(Hint: This is a form of Abel inversion. See Budden 1985.)

(c) Demonstrate that if

$$h(\omega)=h_0+\delta\left(\frac{\omega}{\Pi_0}\right)^p,$$

where h_0, δ, and Π_0 are positive constants, then $\Pi_e(z)=0$ for $z<h_0$, and

$$\Pi_e(z)=\left[\frac{\pi\,\Gamma(1+p)}{\Gamma(1/2+p/2)\,\Gamma(1/2+p/2)}\right]^{1/p}\frac{\Pi_0}{2}\left(\frac{z-h_0}{\delta}\right)^{1/p}$$

for $z\geq h_0$. Here, $\Gamma(z)$ is a Gamma function (Abramowitz and Stegun 1965a).

6.6 Suppose that the refractive index, $n(z)$, of the ionosphere is given by $n^2 = 1-\alpha\,(z-h_0)$ for $z\geq h_0$, and $n^2=1$ for $z<h_0$, where α and h_0 are positive constants, and the Earth's magnetic field and curvature are both neglected. Here, z measures altitude above the Earth's surface.

(a) A point transmitter sends up a wave packet at an angle θ to the vertical. Show that the packet returns to Earth a distance

$$x_0 = 2\,h_0\,\tan\theta + \frac{2}{\alpha}\,\sin 2\theta$$

from the transmitter. Demonstrate that if $\alpha\,h_0 < 1/4$ then for some values of x_0 the previous equation is satisfied by three different values of θ. In other words, wave packets can travel from the transmitter to the receiver via one of three different paths. Show that the critical case $\alpha\,h_0 = 1/4$ corresponds to $\theta = \pi/3$ and $x_0 = 6\sqrt{3}\,h_0$.

(b) A point radio transmitter emits a pulse of radio waves uniformly in all directions. Show that the pulse first returns to the Earth a distance $4\,h_0\,(2/\alpha\,h_0 - 1)^{1/2}$ from the transmitter, provided that $\alpha\,h_0 < 2$.

7

Magnetohydrodynamic Fluids

7.1 Introduction

As we saw in Section 4.13, the MHD equations are written

$$\frac{d\rho}{dt} + \rho \nabla \cdot \mathbf{V} = 0, \tag{7.1}$$

$$\rho \frac{d\mathbf{V}}{dt} + \nabla p - \mathbf{j} \times \mathbf{B} = \mathbf{0}, \tag{7.2}$$

$$\mathbf{E} + \mathbf{V} \times \mathbf{B} = \mathbf{0}, \tag{7.3}$$

$$\frac{d}{dt}\left(\frac{p}{\rho^{\Gamma}}\right) = 0, \tag{7.4}$$

where ρ is the plasma mass density, $\mathbf{V}$ the center of mass velocity, p the pressure, $\mathbf{E}$ the electric field-strength, $\mathbf{B}$ the magnetic field-strength, and $\Gamma = 5/3$ the ratio of specific heats.

It is often remarked that Equations (7.1)–(7.4) are identical to the equations governing the motion of an inviscid, adiabatic, perfectly conducting, electrically neutral, liquid. Indeed, this observation is sometimes used as the sole justification for adopting the MHD equations. After all, a hot, tenuous, quasi-neutral plasma is highly conducting, and if the motion is sufficiently rapid then viscosity and heat conduction can both plausibly be neglected (which implies that the motion is adiabatic). However, as should be clear from the discussion in Section 4.11, this is a highly oversimplified and misleading argument. The problem, of course, is that a weakly coupled plasma is a far more complicated dynamical system than a conducting liquid.

According to the analysis of Section 4.11, the MHD equations are only valid when

$$\delta^{-1} v_t \gg V \gg \delta\, v_t \tag{7.5}$$

for both species. Here, V is the typical fluid velocity associated with the plasma dynamics under investigation, v_t is the typical thermal velocity, and δ is the typical magnetization parameter (i.e., the ratio of a particle gyroradius to the scalelength of the motion). Clearly, the previous inequality is most likely to be satisfied in a highly magnetized (i.e., $\delta \to 0$) plasma.

If the plasma dynamics becomes too rapid (i.e., $V \sim \delta^{-1} v_t$) then resonances occur with the motions of individual particles (e.g., the cyclotron resonances), which invalidate the MHD equations. Furthermore, effects, such as electron inertia and the Hall current, that are not (usually) taken into account in the MHD equations, become important.

MHD is essentially a single-fluid plasma theory. A single-fluid approach is justified because the perpendicular motion is dominated by $\mathbf{E}\times\mathbf{B}$ drifts, which are the same for both plasma species. Furthermore, the relative streaming velocity, $U_{\parallel}$, of both species parallel to the magnetic field is strongly constrained by the fundamental MHD ordering (see Section 4.11)

$$U \sim \delta\, V. \tag{7.6}$$

However, if the plasma dynamics become too slow (i.e., $V \sim \delta\, v_t$) then the motions of the electron and ion fluids become sufficiently different that a single-fluid approach is no longer tenable. This occurs because the diamagnetic velocities, which are quite different for different plasma species, become comparable to the $\mathbf{E}\times\mathbf{B}$ velocity. (See Section 4.14.) Furthermore, effects such as plasma resistivity, viscosity, and thermal conductivity, which are not (usually) taken into account in the MHD equations, become important in this limit.

It follows, from the previous discussion, that the MHD equations describe relatively violent, large-scale motions of highly magnetized plasmas.

Strictly speaking, the MHD equations are only valid in collisional plasmas (i.e., plasmas in which the mean-free-path is much smaller than the typical variation scale-length). However, as is discussed in Section 4.15, the MHD equations also describe the perpendicular (but not the parallel) motions of collisionless plasmas fairly accurately.

Assuming that the MHD equations are valid, let us now investigate their properties.

7.2 Magnetic Pressure

The MHD equations can be combined with the Ampère- and Faraday-Maxwell equations,

$$\nabla\times\mathbf{B} = \mu_0\,\mathbf{j}, \tag{7.7}$$

$$\nabla\times\mathbf{E} = -\frac{\partial\mathbf{B}}{\partial t}, \tag{7.8}$$

respectively, to form a closed set. The displacement current is neglected in Equation (7.7) on the reasonable assumption that MHD motions are slow compared to the velocity of light. Equation (7.8) guarantees that $\nabla\cdot\mathbf{B} = 0$, provided that this relation is presumed to hold initially. Furthermore, the assumption of quasi-neutrality renders the Poisson-Maxwell equation, $\nabla\cdot\mathbf{E} = \rho/\epsilon_0$, redundant.

Equations (7.2) and (7.7) can be combined to give the MHD equation of motion:

$$\rho\,\frac{d\mathbf{V}}{dt} = -\nabla p + \nabla\cdot\mathbf{T}, \tag{7.9}$$

where

$$T_{ij} = \frac{B_i\,B_j - \delta_{ij}\,B^2/2}{\mu_0}. \tag{7.10}$$

Suppose that the magnetic field is approximately uniform, and directed along the z-axis. In this case, the previous equation of motion reduces to

$$\rho\,\frac{d\mathbf{V}}{dt} = -\nabla\cdot\mathbf{P}, \tag{7.11}$$

where

$$\mathbf{P} = \left(\begin{array}{ccc} p + B^2/2\,\mu_0, & 0, & 0 \\ 0, & p + B^2/2\,\mu_0, & 0 \\ 0, & 0, & p - B^2/2\,\mu_0 \end{array}\right). \tag{7.12}$$

It can be seen that the magnetic field increases the plasma pressure, by an amount $B^2/(2\,\mu_0)$, in directions perpendicular to the magnetic field, and decreases the plasma pressure, by the same amount, in the parallel direction. Thus, the magnetic field gives rise to a *magnetic pressure*, $B^2/(2\,\mu_0)$, acting perpendicular to field-lines, and a *magnetic tension*, $B^2/(2\,\mu_0)$, acting along field-lines. Because, as will become apparent in the next section, the plasma is tied to magnetic field-lines, it follows that magnetic field-lines embedded in an MHD plasma act rather like mutually repulsive elastic bands.

7.3 Flux Freezing

The MHD Ohm's law,

$$\mathbf{E} + \mathbf{V}\times\mathbf{B} = \mathbf{0}, \tag{7.13}$$

is sometimes referred to as the *perfect conductivity* equation (for obvious reasons), and sometimes as the *flux freezing* equation. The latter nomenclature comes about because Equation (7.13) implies that the magnetic flux through any loop in the plasma, each element of which moves with the local plasma velocity, is a conserved quantity.

In order to verify the previous assertion, let us consider the magnetic flux, Ψ, through a loop, C, that is co-moving with the plasma:

$$\Psi = \int_S \mathbf{B}\cdot d\mathbf{S}. \tag{7.14}$$

Here, $\mathbf{S}$ is some surface that spans C. The time rate of change of Ψ is made up of two parts. First, there is the part due to the time variation of $\mathbf{B}$ over the surface $\mathbf{S}$, which can be written

$$\left(\frac{\partial \Psi}{\partial t}\right)_1 = \int_{\mathbf{S}} \frac{\partial \mathbf{B}}{\partial t} \cdot d\mathbf{S}. \tag{7.15}$$

Using the Faraday-Maxwell equation, this reduces to

$$\left(\frac{\partial \Psi}{\partial t}\right)_1 = -\int_{\mathbf{S}} \nabla \times \mathbf{E} \cdot d\mathbf{S}. \tag{7.16}$$

Second, there is the part due to the motion of C. If $d\mathbf{r}$ is an element of C then $\mathbf{V}\times d\mathbf{r}$ is the area swept out by $d\mathbf{r}$ per unit time. Hence, the flux crossing this area is $\mathbf{B}\cdot\mathbf{V}\times d\mathbf{r}$. It follows that

$$\left(\frac{\partial \Psi}{\partial t}\right)_2 = \int_C \mathbf{B} \cdot \mathbf{V} \times d\mathbf{r} = \int_C \mathbf{B} \times \mathbf{V} \cdot d\mathbf{r}. \tag{7.17}$$

Using the curl theorem, we obtain

$$\left(\frac{\partial \Psi}{\partial t}\right)_2 = \int_{\mathbf{S}} \nabla \times (\mathbf{B} \times \mathbf{V}) \cdot d\mathbf{S}. \tag{7.18}$$

Hence, the total time rate of change of Ψ is given by

$$\frac{d\Psi}{dt} = -\int_{\mathbf{S}} \nabla \times (\mathbf{E} + \mathbf{V} \times \mathbf{B}) \cdot d\mathbf{S}. \tag{7.19}$$

The condition

$$\mathbf{E} + \mathbf{V} \times \mathbf{B} = \mathbf{0} \tag{7.20}$$

clearly implies that Ψ remains constant in time for any arbitrary co-moving loop, C. This, in turn, implies that magnetic field-lines must move with the plasma. In other words, the field-lines are frozen into the plasma.

A *flux-tube* is defined as a topologically cylindrical volume whose sides are defined by magnetic field-lines. Suppose that, at some initial time, a flux-tube is embedded in the plasma. According to the flux-freezing constraint,

$$\frac{d\Psi}{dt} = 0, \tag{7.21}$$

the subsequent motion of the plasma and the magnetic field is always such that it maintains the integrity of the flux-tube. Because magnetic field-lines can be regarded as infinitely thin flux-tubes, we conclude that MHD plasma motion also maintains the integrity of field-lines. In other words, magnetic field-lines embedded in an MHD plasma can never break and reconnect: that is, MHD forbids any change in topology of the field-lines. It turns out that this is an extremely restrictive constraint. Later on, we shall discuss situations in which this constraint is relaxed. (See Section 7.14.)

7.4 MHD Waves

Let us investigate the small amplitude waves that propagate through a spatially uniform MHD plasma. We start by combining Equations (7.1)–(7.4) and (7.7)–(7.8) to form a closed set of equations:

$$\frac{d\rho}{dt} + \rho\,\nabla\cdot\mathbf{V} = 0, \tag{7.22}$$

$$\rho\,\frac{d\mathbf{V}}{dt} + \nabla p - \frac{(\nabla\times\mathbf{B})\times\mathbf{B}}{\mu_0} = \mathbf{0}, \tag{7.23}$$

$$-\frac{\partial\mathbf{B}}{\partial t} + \nabla\times(\mathbf{V}\times\mathbf{B}) = \mathbf{0}, \tag{7.24}$$

$$\frac{d}{dt}\left(\frac{p}{\rho^{\Gamma}}\right) = 0. \tag{7.25}$$

Next, we linearize these equations (assuming, for the sake of simplicity, that the equilibrium flow velocity and equilibrium plasma current are both zero) to give

$$\frac{\partial\rho}{\partial t} + \rho_0\,\nabla\cdot\mathbf{V} = 0, \tag{7.26}$$

$$\rho_0\,\frac{\partial\mathbf{V}}{\partial t} + \nabla p - \frac{(\nabla\times\mathbf{B})\times\mathbf{B}_0}{\mu_0} = \mathbf{0}, \tag{7.27}$$

$$-\frac{\partial\mathbf{B}}{\partial t} + \nabla\times(\mathbf{V}\times\mathbf{B}_0) = \mathbf{0}, \tag{7.28}$$

$$\frac{\partial}{\partial t}\left(\frac{p}{p_0} - \frac{\Gamma\,\rho}{\rho_0}\right) = 0. \tag{7.29}$$

Here, the subscript 0 denotes an equilibrium quantity. Perturbed quantities are written without subscripts. Of course, ρ_0, p_0, and $\mathbf{B}_0$ are constants in a spatially uniform plasma.

Let us search for wave-like solutions to Equations (7.26)–(7.29) in which perturbed quantities vary like $\exp[\,\mathrm{i}\,(\mathbf{k}\cdot\mathbf{r} - \omega\,t)]$. It follows that

$$-\omega\,\rho + \rho_0\,\mathbf{k}\cdot\mathbf{V} = 0, \tag{7.30}$$

$$-\omega\,\rho_0\,\mathbf{V} + p\,\mathbf{k} - \frac{(\mathbf{k}\times\mathbf{B})\times\mathbf{B}_0}{\mu_0} = \mathbf{0}, \tag{7.31}$$

$$\omega\,\mathbf{B} + \mathbf{k}\times(\mathbf{V}\times\mathbf{B}_0) = \mathbf{0}, \tag{7.32}$$

$$-\omega\left(\frac{p}{p_0} - \frac{\Gamma\,\rho}{\rho_0}\right) = 0. \tag{7.33}$$

Assuming that $\omega \neq 0$, the previous equations yield

$$\rho = \rho_0 \frac{\mathbf{k}\cdot\mathbf{V}}{\omega}, \tag{7.34}$$

$$p = \Gamma\, p_0 \frac{\mathbf{k}\cdot\mathbf{V}}{\omega}, \tag{7.35}$$

$$\mathbf{B} = \frac{(\mathbf{k}\cdot\mathbf{V})\,\mathbf{B}_0 - (\mathbf{k}\cdot\mathbf{B}_0)\,\mathbf{V}}{\omega}. \tag{7.36}$$

Substitution of these expressions into the linearized equation of motion, Equation (7.31), gives

$$\left[\omega^2 - \frac{(\mathbf{k}\cdot\mathbf{B}_0)^2}{\mu_0\,\rho_0}\right]\mathbf{V} = \left[\left(\frac{\Gamma\,p_0}{\rho_0} + \frac{B_0^2}{\mu_0\,\rho_0}\right)\mathbf{k} - \frac{(\mathbf{k}\cdot\mathbf{B}_0)}{\mu_0\,\rho_0}\,\mathbf{B}_0\right](\mathbf{k}\cdot\mathbf{V})$$
$$- \frac{(\mathbf{k}\cdot\mathbf{B}_0)\,(\mathbf{V}\cdot\mathbf{B}_0)}{\mu_0\,\rho_0}\,\mathbf{k}. \tag{7.37}$$

We can assume, without loss of generality, that the equilibrium magnetic field, $\mathbf{B}_0$, is directed along the z-axis, and that the wavevector, $\mathbf{k}$, lies in the x-z plane. Let θ be the angle subtended between $\mathbf{B}_0$ and $\mathbf{k}$. Equation (7.37) reduces to the eigenvalue equation

$$\begin{pmatrix} \omega^2 - k^2\,V_A^2 - k^2\,V_S^2\,\sin^2\theta, & 0, & -k^2\,V_S^2\,\sin\theta\cos\theta \\ 0, & \omega^2 - k^2\,V_A^2\,\cos^2\theta, & 0 \\ -k^2\,V_S^2\,\sin\theta\,\cos\theta, & 0, & \omega^2 - k^2\,V_S^2\,\cos^2\theta \end{pmatrix}\begin{pmatrix} V_x \\ V_y \\ V_z \end{pmatrix} = \mathbf{0}. \tag{7.38}$$

Here,

$$V_A = \sqrt{\frac{B_0^2}{\mu_0\,\rho_0}} \tag{7.39}$$

is the *Alfvén speed*, and

$$V_S = \sqrt{\frac{\Gamma\,p_0}{\rho_0}} \tag{7.40}$$

is the *sound speed*. The solubility condition for Equation (7.38) is that the determinant of the square matrix is zero. This yields the dispersion relation

$$(\omega^2 - k^2\,V_A^2\,\cos^2\theta)\left[\omega^4 - \omega^2\,k^2\,(V_A^2 + V_S^2) + k^4\,V_A^2\,V_S^2\,\cos^2\theta\right] = 0. \tag{7.41}$$

There are three independent roots of the previous dispersion relation, corresponding to the three different types of wave that can propagate through an MHD plasma. The first, and most obvious, root is

$$\omega = k\,V_A\,\cos\theta, \tag{7.42}$$

which has the associated eigenvector $(0,\, V_y,\, 0)$. This root is characterized by both

$\mathbf{k} \cdot \mathbf{V} = 0$ and $\mathbf{V} \cdot \mathbf{B}_0 = 0$. It immediately follows from Equations (7.34) and (7.35) that there is zero perturbation of the plasma density or pressure associated with the root. In fact, this particular root can easily be identified as the shear-Alfvén wave introduced in Section 5.8. The properties of the shear-Alfvén wave in a warm (i.e., non-zero pressure) plasma are unchanged from those found earlier in a cold plasma. Finally, because the shear-Alfvén wave only involves plasma motion perpendicular to the magnetic field, we would expect the dispersion relation (7.42) to hold good in a collisionless, as well as a collisional, plasma.

The remaining two roots of the dispersion relation (7.41) are written

$$\omega = k\,V_+, \tag{7.43}$$

and

$$\omega = k\,V_-, \tag{7.44}$$

respectively. Here,

$$V_\pm = \left\{\frac{1}{2}\left[V_A^2 + V_S^2 \pm \sqrt{(V_A^2 + V_S^2)^2 - 4\,V_A^2\,V_S^2\,\cos^2\theta}\right]\right\}^{1/2}. \tag{7.45}$$

Note that $V_+ \geq V_-$. The first root is generally termed the *fast magnetosonic wave*, or fast wave, for short, whereas the second root is usually called the *slow magnetosonic wave*, or slow wave. The eigenvectors for these waves are $(V_x,\,0,\,V_z)$. It follows that $\mathbf{k} \cdot \mathbf{V} \neq 0$ and $\mathbf{V} \cdot \mathbf{B}_0 \neq 0$. Hence, these waves are associated with non-zero perturbations in the plasma density and pressure, and also involve plasma motion parallel, as well as perpendicular, to the magnetic field. The latter observation suggests that the dispersion relations (7.43) and (7.44) are likely to undergo significant modification in collisionless plasmas.

In order to better understand the nature of the fast and slow waves, let us consider the cold plasma limit, which is obtained by letting the sound speed, V_S, tend to zero. In this limit, the slow wave ceases to exist (in fact, its phase-velocity tends to zero), whereas the dispersion relation for the fast wave reduces to

$$\omega = k\,V_A. \tag{7.46}$$

This can be identified as the dispersion relation for the compressional-Alfvén wave introduced in Section 5.8. Thus, we can identify the fast wave as the compressional-Alfvén wave modified by a non-zero plasma pressure.

In the limit $V_A \gg V_S$, which is appropriate to low-β plasmas (see Section 4.15), the dispersion relation for the slow wave reduces to

$$\omega \simeq k\,V_S\,\cos\theta. \tag{7.47}$$

This is actually the dispersion relation of a sound wave propagating along magnetic field-lines. Thus, in low-β plasmas, the slow wave is a sound wave modified by the presence of the magnetic field.

The distinction between the fast and slow waves can be further understood by

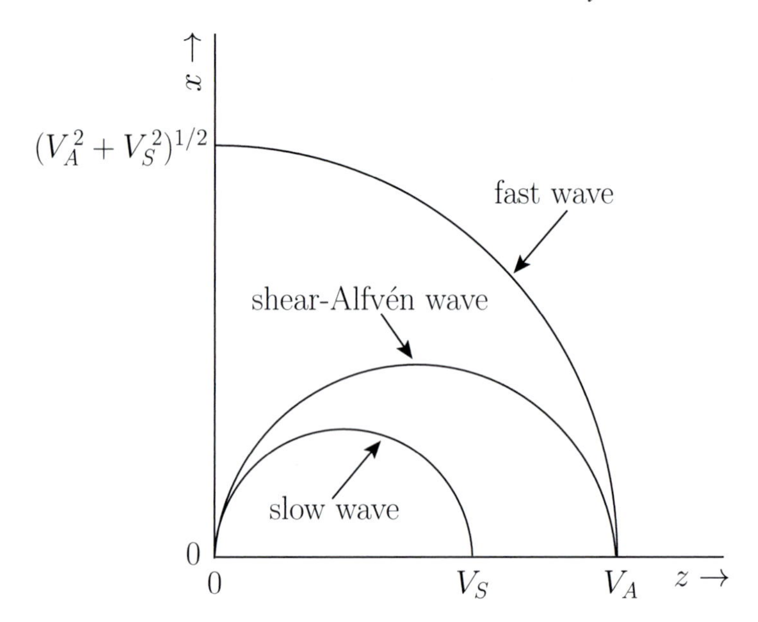

Figure 7.1
Schematic diagram showing the variation of the phase velocities of the three MHD waves with direction of propagation in the x-z plane.

comparing the signs of the wave-induced fluctuations in the plasma and magnetic pressures: p and $\mathbf{B}_0 \cdot \mathbf{B}/\mu_0$, respectively. It follows from Equation (7.36) that

$$\frac{\mathbf{B}_0 \cdot \mathbf{B}}{\mu_0} = \frac{(\mathbf{k} \cdot \mathbf{V})\, B_0^2 - (\mathbf{k} \cdot \mathbf{B}_0)\,(\mathbf{B}_0 \cdot \mathbf{V})}{\mu_0\, \omega}. \tag{7.48}$$

Now, the z-component of Equation (7.31) yields

$$\omega\, \rho_0\, V_z = k\, \cos\theta\, p. \tag{7.49}$$

Combining Equations (7.35), (7.39), (7.40), (7.48), and (7.49), we obtain

$$\frac{\mathbf{B}_0 \cdot \mathbf{B}}{\mu_0} = \frac{V_A^2}{V_S^2}\left(1 - \frac{k^2\, V_S^2\, \cos^2\theta}{\omega^2}\right) p. \tag{7.50}$$

Hence, p and $\mathbf{B}_0 \cdot \mathbf{B}/\mu_0$ have the same sign if $V > V_S\, \cos\theta$, and the opposite sign if $V < V_S\, \cos\theta$. Here, $V = \omega/k$ is the phase-velocity. It is straightforward to show that $V_+ > V_S\, \cos\theta$, and $V_- < V_S\, \cos\theta$. Thus, we conclude that the plasma pressure and magnetic pressure fluctuations reinforce one another in the fast magnetosonic wave, whereas the fluctuations oppose one another in the slow magnetosonic wave.

Figure 7.1 shows the variation of the phase velocities of the three MHD waves with direction of propagation in the x-z plane for a low-β plasma in which $V_S < V_A$.

It can be seen that the slow wave always has a smaller phase-velocity than the shear-Alfvén wave, which, in turn, always has a smaller phase-velocity than the fast wave.

The existence of MHD waves was first predicted theoretically by Alfvén (Alfvén 1942). These waves were subsequently observed in the laboratory—first in magnetized conducting fluids (e.g., mercury) (Lundquist 1949), and then in magnetized plasmas (Wilcox, Boley, and DeSilva 1960).

7.5 Solar Wind

The *solar wind* is a high-speed particle stream continuously blown out from the Sun into interplanetary space (Priest 1984). It extends far beyond the orbit of the Earth, and terminates in a shock front, called the *heliopause*, where it interfaces with the weakly ionized interstellar medium. The heliopause is predicted to lie between 110 and 160 AU (1 astronomical unit, which is the mean Earth-Sun distance, is 1.5×10^{11} m) from the center of the Sun (Suess 1990). The Voyager 1 spacecraft is inferred to have crossed the heliopause in August of 2012 (Webber and McDonald 2013).

In the vicinity of the Earth, (i.e., at about 1 AU from the Sun), the solar wind velocity typically ranges between 300 and 1400 km s^{-1} (Priest 1984). The average value is approximately 500 km s^{-1}, which corresponds to about a 4 day time-of-flight from the Sun. Note that the solar wind is both super-sonic and super-Alfvénic, and is predominately composed of protons and electrons.

The solar wind was predicted theoretically by Eugine Parker (Parker 1958) a number of years before its existence was confirmed by means of satellite data (Neugebauer and Snyder 1966). Parker's prediction of a super-sonic outflow of gas from the Sun is a fascinating application of plasma physics.

The solar wind originates from the *solar corona*, which is a hot, tenuous plasma, surrounding the Sun, with characteristic temperatures and particle densities of about 10^6 K and 10^{14} m^{-3}, respectively (Priest 1984). The corona is actually far hotter than the solar atmosphere, or *photosphere*. In fact, the temperature of the photosphere is only about 6000 K. It is thought that the corona is heated by Alfvén waves emanating from the photosphere (Priest 1984). The solar corona is most easily observed during a total solar eclipse, when it is visible as a white filamentary region immediately surrounding the Sun.

Let us start, following Chapman (Chapman 1957), by attempting to construct a model for a static solar corona. The equation of hydrostatic equilibrium for the corona takes the form

$$\frac{dp}{dr} = -\rho \, \frac{G \, M_\odot}{r^2}, \tag{7.51}$$

where $G = 6.67 \times 10^{-11}$ m^3 s^{-2} kg^{-1} is the gravitational constant, $M_\odot = 2 \times 10^{30}$ kg the solar mass (Yoder 1995), and r the radial distance from the center of the Sun. The plasma density is written

$$\rho \simeq n \, m_p, \tag{7.52}$$

where n is the number density of protons. If both protons and electrons are assumed to possess a common temperature, $T(r)$, then the coronal pressure is given by

$$p = 2\,n\,T. \tag{7.53}$$

The thermal conductivity of the corona is dominated by the electron thermal conductivity, and takes the form [see Equations (4.89) and (4.108)]

$$\kappa = \kappa_0\,T^{\,5/2}, \tag{7.54}$$

where κ_0 is a relatively weak function of density and temperature. For typical coronal conditions, this conductivity is extremely high. In fact, it is about twenty times the thermal conductivity of copper at room temperature. The coronal heat flux density is written

$$\mathbf{q} = -\kappa\,\nabla T. \tag{7.55}$$

For a static corona, in the absence of energy sources or sinks, we require

$$\nabla\cdot\mathbf{q} = 0. \tag{7.56}$$

Assuming spherical symmetry, this expression reduces to (Huba 2000b)

$$\frac{1}{r^2}\frac{d}{dr}\left(r^2\,\kappa_0\,T^{\,5/2}\,\frac{dT}{dr}\right) = 0. \tag{7.57}$$

Adopting the sensible boundary condition that the coronal temperature must tend to zero at large distances from the Sun, we obtain

$$T(r) = T(a)\left(\frac{a}{r}\right)^{2/7}. \tag{7.58}$$

The reference level $r = a$ is conveniently taken to be the base of the corona, where $a \sim 7\times 10^5$ km, $n \sim 2\times 10^{14}\,\mathrm{m}^{-3}$, and $T \sim 2\times 10^6$ K (Priest 1984).

Equations (7.51), (7.52), (7.53), and (7.58) can be combined and integrated to give

$$p(r) = p(a)\exp\left\{\frac{7}{5}\,\frac{G\,M_\odot\,m_p}{2\,T(a)\,a}\left[\left(\frac{a}{r}\right)^{5/7} - 1\right]\right\}. \tag{7.59}$$

Observe that, as $r\rightarrow\infty$, the coronal pressure tends towards a finite constant value:

$$p(\infty) = p(a)\,\exp\left[-\frac{7}{5}\,\frac{G\,M_\odot\,m_p}{2\,T(a)\,a}\right] = p(a)\,\exp\left[-\frac{14}{5}\,\frac{T_0}{T(a)}\right], \tag{7.60}$$

where T_0 is defined in Equation (7.66). There is, of course, nothing at large distances from the Sun that could contain such a pressure (the pressure of the interstellar medium is negligibly small). Thus, we conclude, following Parker, that the static coronal model is unphysical.

We have just demonstrated that a static model of the solar corona is unsatisfactory. Let us, instead, attempt to construct a dynamic model in which material flows outward from the Sun.

7.6 Parker Model of Solar Wind

By symmetry, we expect a purely radial, steady-state, coronal outflow. The radial equation of motion of the corona [which is a modified version of Equation (7.2)] takes the form (Huba 2000b)

$$\rho\, u\, \frac{du}{dr} = -\frac{dp}{dr} - \rho\, \frac{G\, M_\odot}{r^2}, \tag{7.61}$$

where u is the radial expansion speed. The continuity equation [which is equivalent to Equation (7.1)] reduces to (Huba 2000b)

$$\frac{1}{r^2}\frac{d(r^2\, \rho\, u)}{dr} = 0. \tag{7.62}$$

In order to obtain a closed set of equations, we now need to adopt an equation of state for the corona, relating the pressure, p, and the density, ρ. For the sake of simplicity, we adopt the simplest conceivable equation of state, which corresponds to an isothermal corona. Thus, we have

$$p = \frac{2\, \rho\, T}{m_p}, \tag{7.63}$$

where T is a constant. More realistic equations of state complicate the analysis, but do not significantly modify any of the physics results (Priest 1984).

Equation (7.62) can be integrated to give

$$r^2\, \rho\, u = I, \tag{7.64}$$

where I is a constant. The previous expression simply states that the mass flux per unit solid angle, which takes the value I, is independent of the radius, r. Equations (7.61), (7.63), and (7.64) can be combined to give

$$\frac{1}{u}\frac{du}{dr}\left(u^2 - \frac{2\, T}{m_p}\right) = \frac{4\, T}{m_p\, r} - \frac{G\, M_\odot}{r^2}. \tag{7.65}$$

Let us restrict our attention to coronal temperatures that satisfy

$$T < T_0 \equiv \frac{G\, M_\odot\, m_p}{4\, a}, \tag{7.66}$$

where a is the radius of the base of the corona. For typical coronal parameters, $T_0 \simeq 5.8 \times 10^6$ K, which is certainly greater than the temperature of the corona at $r = a$. For $T < T_0$, the right-hand side of Equation (7.65) is negative for $a < r < r_c$, where

$$\frac{r_c}{a} = \frac{T_0}{T}, \tag{7.67}$$

and positive for $r_c < r < \infty$. The right-hand side of (7.65) is zero at $r = r_c$, implying that the left-hand side is also zero at this radius, which is usually termed the "critical radius." There are two ways in which the left-hand side of (7.65) can be zero at the critical radius. Either

$$u^2(r_c) = u_c^2 \equiv \frac{2\,T}{m_p}, \tag{7.68}$$

or

$$\frac{du(r_c)}{dr} = 0. \tag{7.69}$$

Note that u_c is the coronal sound speed.

As is easily demonstrated, if Equation (7.68) is satisfied then du/dr has the same sign for all r, and $u(r)$ is either a monotonically increasing, or a monotonically decreasing, function of r. On the other hand, if Equation (7.69) is satisfied then $u^2 - u_c^2$ has the same sign for all r, and $u(r)$ has an extremum close to $r = r_c$. The flow is either super-sonic for all r, or sub-sonic for all r. These possibilities lead to the existence of four classes of solutions to Equation (7.65), with the following properties:

1. $u(r)$ is sub-sonic throughout the domain $a < r < \infty$. $u(r)$ increases with r, attains a maximum value around $r = r_c$, and then decreases with r.

2. a unique solution for which $u(r)$ increases monotonically with r, and $u(r_c) = u_c$.

3. a unique solution for which $u(r)$ decreases monotonically with r, and $u(r_c) = u_c$.

4. $u(r)$ is super-sonic throughout the domain $a < r < \infty$. $u(r)$ decreases with r, attains a minimum value around $r = r_c$, and then increases with r.

These four classes of solutions are illustrated in Figure 7.2.

Each of the classes of solutions described previously fits a different set of boundary conditions at $r = a$ and $r \to \infty$. The physical acceptability of these solutions depends on these boundary conditions. For example, both Class 3 and Class 4 solutions can be ruled out as plausible models for the solar corona because they predict super-sonic flow at the base of the corona, which is not observed, and is also not consistent with a static solar photosphere. Class 1 and Class 2 solutions remain acceptable models for the solar corona on the basis of their properties around $r = a$, because they both predict sub-sonic flow in this region. However, the Class 1 and Class 2 solutions behave quite differently as $r \to \infty$, and the physical acceptability of these two classes hinges on this difference.

Equation (7.65) can be rearranged to give

$$\frac{du^2}{dr}\left(1 - \frac{u_c^2}{u^2}\right) = \frac{4\,u_c^2}{r}\left(1 - \frac{r_c}{r}\right), \tag{7.70}$$

where use has been made of Equations (7.66) and (7.67). The previous expression can be integrated to give

$$\left(\frac{u}{u_c}\right)^2 - \ln\left(\frac{u}{u_c}\right)^2 = 4\,\ln\left(\frac{r}{r_c}\right) + 4\,\frac{r_c}{r} + C, \tag{7.71}$$

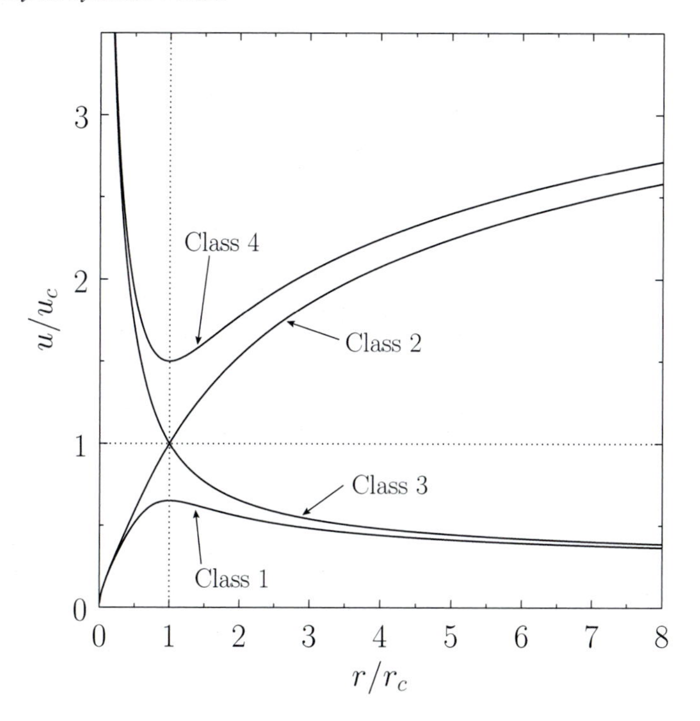

Figure 7.2
The four classes of Parker outflow solutions for the solar wind.

where C is a constant of integration.

Let us consider the behavior of Class 1 solutions in the limit $r \to \infty$. It is clear from Figure 7.2 that, for Class 1 solutions, u/u_c is less than unity and monotonically decreasing as $r \to \infty$. In the large-r limit, Equation (7.71) reduces to

$$\ln\left(\frac{u}{u_c}\right) \simeq -2 \ln\left(\frac{r}{r_c}\right), \tag{7.72}$$

so that

$$u \propto \frac{1}{r^2}. \tag{7.73}$$

It follows from Equation (7.64) that the coronal density, ρ, approaches a finite, constant value, ρ_∞, as $r \to \infty$. Thus, the Class 1 solutions yield a finite pressure,

$$p_\infty = \frac{2\,\rho_\infty\,T}{m_p}, \tag{7.74}$$

at large r, which cannot be matched to the much smaller pressure of the interstellar medium. Obviously, Class 1 solutions are unphysical.

Let us consider the behavior of the Class 2 solution in the limit $r \to \infty$. It is clear from Figure 7.2 that, for the Class 2 solution, u/u_c is greater than unity and monotonically increasing as $r \to \infty$. In the large-r limit, Equation (7.71) reduces to

$$\left(\frac{u}{u_c}\right)^2 \simeq 4 \ln\left(\frac{r}{r_c}\right), \tag{7.75}$$

so that

$$u \simeq 2\, u_c \left[\ln\left(\frac{r}{r_c}\right)\right]^{1/2}. \tag{7.76}$$

It follows from Equation (7.64) that $\rho \to 0$ as $r \to \infty$. Thus, the Class 2 solution yields $p \to 0$ at large r, and can, therefore, be matched to the low pressure interstellar medium.

We conclude that the only solution to Equation (7.65) that is consistent with the physical boundary conditions at $r = a$ and $r \to \infty$ is the Class 2 solution. This solution predicts that the solar corona expands radially outward at relatively modest, sub-sonic velocities close to the Sun, and gradually accelerates to super-sonic velocities as it moves further away from the Sun. Parker termed this continuous, super-sonic expansion of the corona the "solar wind."

Equation (7.71) can be rewritten

$$\left(\frac{u^2}{u_c^2} - 1\right) - \ln\left(\frac{u}{u_c}\right)^2 = 4 \ln\left(\frac{r}{r_c}\right) + 4\left(\frac{r_c}{r} - 1\right), \tag{7.77}$$

where the constant C is determined by demanding that $u = u_c$ when $r = r_c$. Note that both u_c and r_c can be evaluated in terms of the coronal temperature, T, via Equations (7.67) and (7.68). Figure 7.3 shows $u(r)$ calculated from Equation (7.77) for various values of the coronal temperature. It can be seen that plausible values of T (i.e., $T \sim 1\text{–}2 \times 10^6$ K) yield expansion speeds of several hundreds of kilometers per second at 1 AU, which accords well with satellite observations. The critical surface where the solar wind makes the transition from sub-sonic to super-sonic flow is predicted to lie a few solar radii away from the Sun (i.e., $r_c \sim 5\, R_\odot$, where $R_\odot$ is the solar radius). Unfortunately, the Parker model's prediction for the density of the solar wind at the Earth is significantly too high compared to satellite observations. Consequently, there have been many further developments of this model. In particular, the unrealistic assumption that the solar wind plasma is isothermal has been relaxed, and two-fluid effects have been incorporated into the analysis (Priest 1984).

7.7 Interplanetary Magnetic Field

Let us now investigate how the solar wind and the interplanetary magnetic field affect one another.

The hot coronal plasma making up the solar wind possesses an extremely high

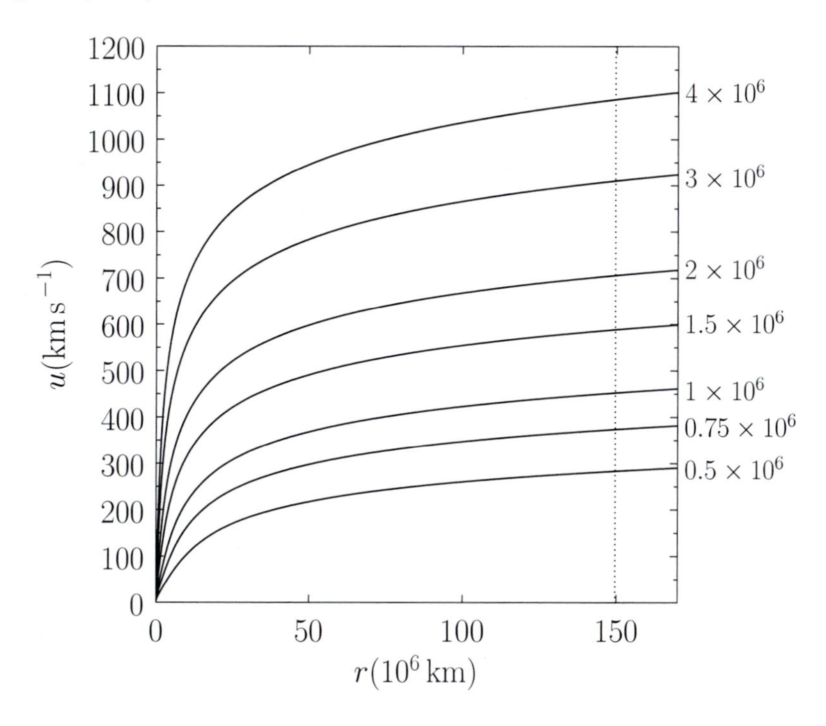

Figure 7.3
Parker outflow solutions for the solar wind. Each curve is labelled by the corresponding coronal temperature in degrees kelvin. The vertical dashed line indicates the mean radius of the Earth's orbit.

electrical conductivity. In such a plasma, we expect the concept of "frozen-in" magnetic field-lines, discussed in Section 7.3, to be applicable. The continuous flow of coronal material into interplanetary space must, therefore, result in the transport of the solar magnetic field into the interplanetary region. If the Sun did not rotate then the resulting magnetic configuration would be very simple. The radial coronal expansion considered previously (with the neglect of any magnetic forces) would produce magnetic field-lines extending radially outward from the Sun.

Of course, the Sun does rotate, with a (latitude dependent) period of about 25 days.[1] Because the solar photosphere is an excellent electrical conductor, the magnetic field at the base of the corona is frozen into the rotating frame of reference of the Sun. A magnetic field-line starting from a given location on the surface of the Sun is drawn out along the path followed by the element of the solar wind emanating from that location. As before, let us suppose that the coronal expansion is purely radial in a stationary frame of reference. Consider a spherical coordinate system (r, θ, ϕ) that co-rotates with the Sun. Of course, the symmetry axis of the coordinate system is assumed to coincide with the axis of the Sun's rotation. In the rotating coordinate

[1] To an observer orbiting with the Earth, the rotation period appears to be about 27 days.

system, the velocity components of the solar wind are written

$$u_r = u, \tag{7.78}$$

$$u_\theta = 0, \tag{7.79}$$

$$u_\phi = -\Omega\, r\, \sin\theta, \tag{7.80}$$

where $\Omega = 2.7\times 10^{-6}$ rad sec^{-1} is the angular velocity of solar rotation (Yoder 1995). The azimuthal velocity u_ϕ is entirely due to the transformation to the rotating frame of reference. The stream-lines of the flow satisfy the differential equation (Huba 2000b)

$$\frac{1}{r\,\sin\theta}\frac{dr}{d\phi} \simeq \frac{u_r}{u_\phi} = -\frac{u}{\Omega\, r\,\sin\theta} \tag{7.81}$$

at constant θ. The stream-lines are also magnetic field-lines, so Equation (7.81) can be regarded as the differential equation of a magnetic field-line. For radii, r, greater than several times the critical radius, r_c, the solar wind solution (7.77) predicts that $u(r)$ is almost constant. (See Figure 7.3.) Thus, for $r \gg r_c$, it is reasonable to write $u(r) = u_s$, where u_s is a constant. Equation (7.81) can then be integrated to give the equation of a magnetic field-line,

$$r - r_0 = -\frac{u_s}{\Omega}\,(\phi - \phi_0), \tag{7.82}$$

where the field-line is assumed to pass through the point $(r_0,\ \theta,\ \phi_0)$. Maxwell's equation $\nabla\cdot\mathbf{B} = 0$, plus the assumption of a spherically symmetric magnetic field, easily yield the following expressions for the components of the interplanetary magnetic field:

$$B_r(r,\theta,\phi) = B(r_0,\theta,\phi_0)\left(\frac{r_0}{r}\right)^2, \tag{7.83}$$

$$B_\theta(r,\theta,\phi) = 0, \tag{7.84}$$

$$B_\phi(r,\theta,\phi) = -B(r_0,\theta,\phi_0)\,\frac{\Omega\, r_0}{u_s}\,\frac{r_0}{r}\,\sin\theta. \tag{7.85}$$

Figure 7.4 illustrates the interplanetary magnetic field close to the ecliptic plane (i.e., $\theta = \pi/2$). The magnetic field-lines of the Sun are drawn into Archimedean spirals by the solar rotation. Transformation to a stationary frame of reference gives the same magnetic field configuration, with the addition of an electric field

$$\mathbf{E} = -\mathbf{u}\times\mathbf{B} = u_s\, B_\phi\, \mathbf{e}_\theta \tag{7.86}$$

The latter field arises because the radial plasma flow is no longer parallel to magnetic field-lines in the stationary frame.

The interplanetary magnetic field at 1 AU is observed to lie in the ecliptic plane, and is directed at an angle of approximately 45° from the radial direction to the Sun (Priest 1984). This is in basic agreement with the spiral configuration predicted previously.

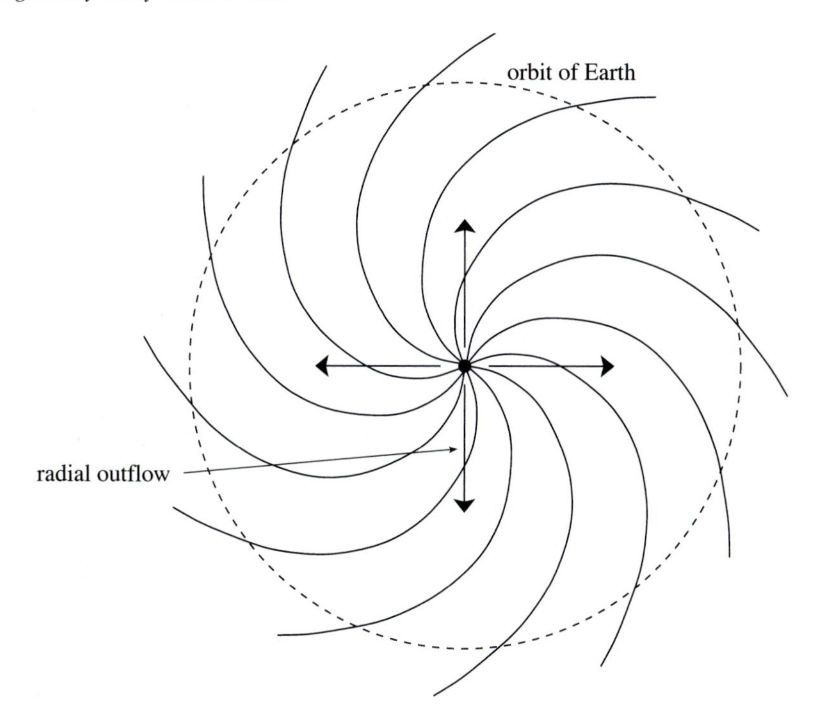

Figure 7.4
The interplanetary magnetic field in the ecliptic plane.

The previous analysis is premised on the assumption that the interplanetary magnetic field is too weak to affect the coronal outflow, and is, therefore, passively convected by the solar wind. In fact, this is only the case if the interplanetary magnetic energy density, $B^2/(2\,\mu_0)$, is much less that the kinetic energy density, $\rho\,u^2/2$, of the solar wind. Rearrangement yields the condition

$$u > V_A, \tag{7.87}$$

where V_A is the Alfvén speed. It turns out that $u \sim 10\,V_A$ at 1 AU. On the other hand, $u \ll V_A$ close to the base of the corona. In fact, the solar wind becomes super-Alfvénic at a radius, denoted r_A, which is typically $50\,R_\odot$, or $1/4$ of an astronomical unit (Priest 1984). We conclude that the previous analysis is only valid well outside the Alfvén radius: that is, in the region $r \gg r_A$.

Well inside the Alfvén radius (i.e., in the region $r \ll r_A$), the solar wind is too weak to modify the structure of the solar magnetic field. In fact, in this region, we expect the solar magnetic field to force the solar wind to co-rotate with the Sun. Observe that flux-freezing is a two-way-street: if the energy density of the flow greatly exceeds that of the magnetic field then the magnetic field is passively convected by the flow, but if the energy density of the magnetic field greatly exceeds that of the flow then the flow is forced to conform to the magnetic field.

The previous discussion leads us to the following, rather crude, picture of the interaction of the solar wind and the interplanetary magnetic field. We expect the interplanetary magnetic field to be the undistorted continuation of the Sun's magnetic field for $r < r_A$. On the other hand, we expect the interplanetary field to be dragged out into a spiral pattern for $r > r_A$. Furthermore, we expect the Sun's magnetic field to impart a non-zero azimuthal velocity $u_\phi(r)$ to the solar wind. In the ecliptic plane ($\theta = \pi/2$), we infer that

$$u_\phi = \Omega\, r \tag{7.88}$$

for $r < r_A$, and

$$u_\phi = \Omega\, r_A \left(\frac{r_A}{r}\right) \tag{7.89}$$

for $r > r_A$. This corresponds to co-rotation with the Sun inside the Alfvén radius, and outflow at constant angular velocity outside the Alfvén radius. We, therefore, expect the solar wind at 1 AU to possess a small azimuthal velocity component. This is indeed the case. In fact, the direction of the solar wind at 1 AU deviates from purely radial outflow by about 1.5° (Priest 1984).

7.8 Mass and Angular Momentum Loss

Let us consider what impact the solar wind has on solar evolution. The most obvious question is whether the mass loss due to the wind is actually significant. Using typical measured values [i.e., a typical solar wind velocity and particle density at 1 AU of $500\,\mathrm{km\,s^{-1}}$ and $7\times10^6\ \mathrm{m^{-3}}$, respectively (Priest 1984)], the Sun is apparently losing mass at a rate of $3\times10^{-14}\,M_\odot$ per year, where $M_\odot = 2\times10^{30}\,\mathrm{kg}$ is the solar mass (Yoder 1995), implying a timescale for significant mass loss of 3×10^{13} years, or some 6,000 times longer than the estimated 5×10^9 year age of the Sun (Hansen, Kawaler, and Trimble 2004). Clearly, the mass carried off by the solar wind has a negligible effect on the Sun's evolution. Note, however, that many stars in the universe exhibit significant mass loss via stellar winds. This is particularly the case for late-type stars (Mestel 2012).

Let us now consider the angular momentum carried off by the solar wind. Angular momentum loss is a crucially important topic in astrophysics, because only by losing angular momentum can large, diffuse objects, such as interstellar gas clouds, collapse under the influence of gravity to produce small, compact objects, such as stars and proto-stars (Mestel 2012). Magnetic fields generally play a crucial role in angular momentum loss. This is certainly the case for the solar wind, where the solar magnetic field enforces co-rotation with the Sun out to the Alfvén radius, r_A. Thus, the angular momentum carried away by a particle of mass m is $\Omega\, r_A^2\, m$, rather than $\Omega\, R_\odot^2\, m$. The angular momentum loss timescale is, therefore, shorter than the mass loss timescale by a factor $(R_\odot/r_A)^2 \simeq 1/2500$, making the angular momentum loss timescale comparable to the solar lifetime. It is clear that magnetized stellar winds represent a very important vehicle for angular momentum loss in the universe

(Mestel 2012). Let us investigate angular momentum loss via stellar winds in more detail.

Under the assumption of spherical symmetry and steady flow, the azimuthal momentum evolution equation for the solar wind, taking into account the influence of the interplanetary magnetic field, is written (Huba 2000a)

$$\rho\,[(\mathbf{V}\cdot\nabla)\mathbf{V}]_\phi \equiv \rho\,\frac{u_r}{r}\frac{d(r\,u_\phi)}{dr} = (\mathbf{j}\times\mathbf{B})_\phi \equiv \frac{B_r}{\mu_0\,r}\frac{d(r\,B_\phi)}{dr}. \tag{7.90}$$

The constancy of the mass flux [see Equation (7.64)] and the $1/r^2$ dependence of B_r [see Equation (7.83)] permit the immediate integration of the previous equation to give

$$r\,u_\phi - \frac{r\,B_r\,B_\phi}{\mu_0\,\rho\,u_r} = L, \tag{7.91}$$

where L is the angular momentum per unit mass carried off by the solar wind. In the presence of an azimuthal wind velocity, the magnetic field and velocity components are related by an expression similar to Equation (7.81):

$$\frac{B_r}{B_\phi} = \frac{u_r}{u_\phi - \Omega\,r\,\sin\theta}. \tag{7.92}$$

The fundamental physics assumption underlying the previous expression is the absence of an electric field in the frame of reference co-rotating with the Sun. Using Equation (7.92) to eliminate B_ϕ from Equation (7.91), we obtain (in the ecliptic plane, where $\sin\theta = 1$)

$$r\,u_\phi = \frac{L\,M_A^2 - \Omega\,r^2}{M_A^2 - 1}, \tag{7.93}$$

where

$$M_A = \sqrt{\frac{u_r^2}{B_r^2/\mu_0\,\rho}} \tag{7.94}$$

is the *radial Alfvén Mach number*. The radial Alfvén Mach number is small near the base of the corona, and about 10 at 1 AU: it passes through unity at the Alfvén radius, r_A, which is about 0.25 AU from the Sun. The zero denominator on the right-hand side of Equation (7.93) at $r = r_A$ implies that u_ϕ is finite and continuous only if the numerator is also zero at the Alfvén radius. This condition then determines the angular momentum content of the outflow via

$$L = \Omega\,r_A^2. \tag{7.95}$$

Note that the angular momentum carried off by the solar wind is indeed equivalent to that which would be carried off were coronal plasma to co-rotate with the Sun out to the Alfvén radius, and subsequently outflow at constant angular velocity. Of course, the solar wind does not actually rotate rigidly with the Sun in the region $r < r_A$: much of the angular momentum in this region is carried in the form of electromagnetic stresses.

It is easily demonstrated that the quantity $M_A^2/(u_r\,r^2)$ is a constant (because $B_r \propto r^{-2}$, and $r^2\,\rho\,u_r$ is constant), and can, therefore, be evaluated at $r = r_A$ to give

$$M_A^2 = \frac{u_r\,r^2}{u_{rA}\,r_A^2}, \tag{7.96}$$

where $u_{rA} \equiv u_r(r_A)$. Equations (7.93), (7.95), and (7.96) can be combined to produce

$$u_\phi = \frac{\Omega\,r}{u_{rA}}\,\frac{u_{rA} - u_r}{1 - M_A^2}. \tag{7.97}$$

In the limit $r \to \infty$, we have $M_A \gg 1$, so the previous expression yields

$$u_\phi \to \Omega\,r_A\left(\frac{r_A}{r}\right)\left(1 - \frac{u_{rA}}{u_r}\right) \tag{7.98}$$

at large distances from the Sun. Recall, from Section 7.7, that if the coronal plasma were to simply co-rotate with the Sun out to $r = r_A$, and experience no torque beyond this radius, then we would expect

$$u_\phi \to \Omega\,r_A\left(\frac{r_A}{r}\right) \tag{7.99}$$

at large distances from the Sun. The difference between the previous two expressions is the factor $1 - u_{rA}/u_r$, which is a correction for the angular momentum retained by the magnetic field at large r.

The previous analysis presented was first incorporated into a quantitative coronal expansion model by Weber and Davis (Weber and Davis 1967). The model of Weber and Davis is very complicated. For instance, the solar wind is required to flow smoothly through no less than three critical points. These are associated with the sound speed (as in Parker's original model), the radial Alfvén speed, $B_r/\sqrt{\mu_0\,\rho}$, (as previously described), and the total Alfvén speed, $B/\sqrt{\mu_0\,\rho}$. Nevertheless, the simplified analysis outlined in this section captures most of the essential features of the outflow.

7.9 MHD Dynamo Theory

Many stars, planets, and galaxies possess magnetic fields whose origins are not easily explained. Even the "solid" planets could not possibly be sufficiently ferromagnetic to account for their magnetism, because the temperatures of their interiors lie above the Curie temperature at which permanent magnetism disappears (Reif 1965). It goes without saying that stars and galaxies are not ferromagnetic. Magnetic fields cannot be dismissed as transient phenomena that just happen to be present today. For instance, *paleomagnetism*, the study of magnetic fields "fossilized" in rocks at the time of their formation in the remote geological past, shows that the Earth's magnetic field has existed at much its present strength for at least the past 3×10^9 years (Dunlop and Özdemir 2001; Ogg 2012). The problem is that, in the absence of an internal source of electric currents, magnetic fields contained in a conducting body decay ohmically on a timescale

$$\tau_{\rm ohm} = \mu_0 \, \sigma \, L^2, \tag{7.100}$$

where σ is the typical electrical conductivity, and L is the typical lengthscale of the body, and this decay timescale is generally very small compared to the inferred lifetimes of astrophyiscal magnetic fields. For instance, the Earth contains a highly conducting region: namely, its molten core, of radius $L \sim 3.5 \times 10^6$ m, and conductivity $\sigma \sim 4 \times 10^5 \, {\rm S\,m^{-1}}$ (Yoder 1995). This yields an ohmic decay time for the terrestrial magnetic field of only $\tau_{\rm ohm} \sim 2 \times 10^5$ years, which is obviously far shorter than the inferred lifetime of this field. Clearly, some process inside the Earth must be actively maintaining the terrestrial magnetic field (Roberts and King 2013). Such a process is conventionally termed a *dynamo*. Similar considerations lead us to postulate the existence of dynamos acting inside stars and galaxies, in order to account for the persistence of stellar and galactic magnetic fields over cosmological timescales (Mestel 2012).

The basic premise of dynamo theory is that all astrophysical bodies that contain anomalously long-lived magnetic fields also contain convecting, highly conducting, fluids (e.g., the Earth's molten core, the ionized gas that makes up the Sun), and it is the electric currents associated with the motions of these fluids that maintain the observed magnetic fields. At first sight, this proposal, first made by Larmor in 1919 (Larmor 1919), sounds suspiciously like pulling yourself up by your own shoelaces. However, there is really no conflict with the demands of energy conservation. The magnetic energy irreversibly lost via ohmic heating is replenished at the rate (per unit volume) $\mathbf{V} \cdot (\mathbf{j} \times \mathbf{B})$: in other words, by the rate of work done against the Lorentz force. The flow field, $\mathbf{V}$, is assumed to be driven via thermal convention. If the flow is sufficiently vigorous then it is, at least, plausible that the energy input to the magnetic field can overcome the losses due to ohmic heating, thus permitting the field to persist over timescales far longer than the characteristic ohmic decay time.

Paleomagnetic data from marine sediment cores shows that the Earth's magnetic field is quite variable, and actually reversed polarity about 700,000 years ago (Dunlop and Özdemir 2001; Valet, Meynadier, and Guyodo 2005). In fact, more extensive data shows that the Earth's magnetic field reverses polarity about once every ohmic decay timescale (i.e., a few times every million years) (Ogg 2012). The Sun's magnetic field exhibits similar behavior, reversing polarity about once every 11 years (Jones, Thompson, and Tobais 2010; Mestel 2012). An examination of this type of data reveals that dynamo magnetic fields (and velocity fields) are essentially chaotic in nature, exhibiting strong random variability superimposed on more regular quasi-periodic oscillations.

A thorough investigation of dynamo theory would be a far too difficult and time consuming task. Instead, we shall examine a far simpler version of this theory, known as *kinematic dynamo theory*, in which the velocity field, $\mathbf{V}$, is prescribed (Moffatt 1978; Krause and Rädler 1980). In order for this approach to be self-consistent, it must be assumed that the magnetic field is sufficiently weak that it does not affect the velocity field. Let us start from the MHD Ohm's law, modified by resistivity:

$$\mathbf{E} + \mathbf{V} \times \mathbf{B} = \eta\, \mathbf{j}. \tag{7.101}$$

Here, the resistivity η is assumed to be a constant, for the sake of simplicity. Taking the curl of the previous equation, and making use of Maxwell's equations, we obtain

$$\frac{\partial \mathbf{B}}{\partial t} - \nabla \times (\mathbf{V} \times \mathbf{B}) = \frac{\eta}{\mu_0} \nabla^2 \mathbf{B}. \tag{7.102}$$

If the velocity field, $\mathbf{V}$, is prescribed, and unaffected by the presence of the magnetic field, then the previous equation is essentially a linear eigenvalue equation for the magnetic field, $\mathbf{B}$. The question we wish to address is as follows. For what sort of velocity fields, if any, does the previous equation possess solutions in which the magnetic field grows exponentially in time? In trying to formulate an answer to this question, we hope to learn what type of motion of an MHD fluid is capable of self-generating a magnetic field.

7.10 Homopolar Disk Dynamo

Some of the peculiarities of dynamo theory are well illustrated by the prototype example of self-excited dynamo action, which is the *homopolar disk dynamo*. As illustrated in Figure 7.5, this device consists of a conducting disk that rotates at some angular frequency Ω about its axis under the action of an applied torque. A wire, twisted about the axis in the manner shown, makes sliding contact with the disc at A, and with the axis at B, and carries a current $I(t)$. The magnetic field, $\mathbf{B}$, associated with this current has a flux $\Phi = M\, I$ across the disc, where M is the mutual inductance between the wire and the rim of the disc. The rotation of the disc in the presence of this flux generates a radial electromotive force

$$\frac{\Omega}{2\pi}\, \Phi = \frac{\Omega}{2\pi}\, M\, I, \tag{7.103}$$

because a radius of the disc cuts the magnetic flux Φ once every $2\pi/\Omega$ seconds. According to this simplistic description, the equation for I is written

$$L\, \frac{dI}{dt} + R\, I = \frac{M}{2\pi}\, \Omega\, I, \tag{7.104}$$

where R is the total resistance of the circuit, and L is its self-inductance.

Suppose that the angular velocity Ω is maintained by suitable adjustment of the driving torque. It follows that Equation (7.104) possesses an exponential solution $I(t) = I(0)\, \exp(\gamma\, t)$, where

$$\gamma = L^{-1} \left(\frac{M}{2\pi}\, \Omega - R \right). \tag{7.105}$$

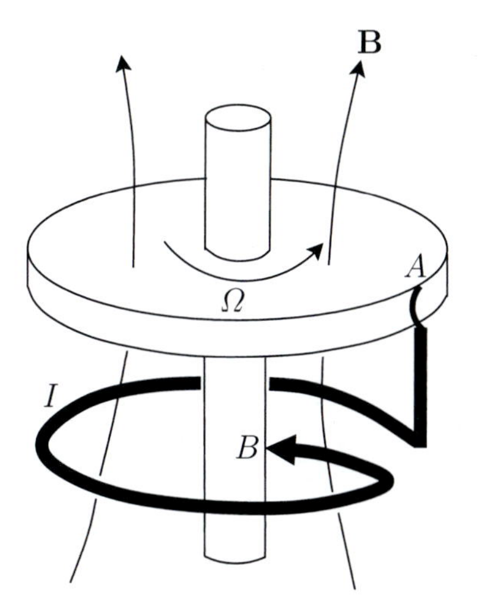

Figure 7.5
The homopolar disk dynamo.

Clearly, we have exponential growth of $I(t)$—and, hence, of the magnetic field to which it gives rise (i.e., we have dynamo action)—provided that

$$\Omega > \frac{2\pi R}{M}: \tag{7.106}$$

that is, provided that the disk rotates sufficiently rapidly. Note that the homopolar disk dynamo depends for its success on its built-in axial asymmetry. If the disk rotates in the opposite direction to that shown in Figure 7.5, then $\Omega < 0$, and the electromotive force generated by the rotation of the disk always acts to reduce I. In this case, dynamo action is impossible (i.e., γ is always negative). This is a troubling observation, because most astrophysical objects, such as stars and planets, possess very good axial symmetry. We conclude that if such bodies are to act as dynamos then the asymmetry of their internal motions must somehow compensate for their lack of built-in asymmetry. It is far from obvious how this is going to happen.

Incidentally, although the previous treatment of a homopolar disk dynamo (which is the standard analysis found in most textbooks) is very appealing in its simplicity, it cannot be entirely correct. Consider the limiting situation of a perfectly conducting disk and wire, in which $R = 0$. On the one hand, Equation (7.105) yields $\gamma = M\,\Omega/(2\pi\,L)$, so that we still have dynamo action. But, on the other hand, the rim of the disk is a closed circuit embedded in a perfectly conducting medium, so the flux freezing constraint requires that the flux, Φ, through this circuit must remain a constant. There is an obvious contradiction. The problem is that we have neglected the currents that flow azimuthally in the disc: that is, the currents that control the diffusion of magnetic flux across the rim of the disk. These currents become particularly important in the limit $R \to \infty$.

The previous paradox can be resolved by supposing that the azimuthal current,

$J(t)$, is constrained to flow around the rim of the disk (e.g., by a suitable distribution of radial insulating strips). In this case, the fluxes through the I and J circuits are

$$\Phi_1 = L\,I + M\,J, \tag{7.107}$$

$$\Phi_2 = M\,I + L'\,J, \tag{7.108}$$

and the equations governing the current flow become

$$\frac{d\Phi_1}{dt} = \frac{\Omega}{2\pi}\,\Phi_2 - R\,I, \tag{7.109}$$

$$\frac{d\Phi_2}{dt} = -R'\,J, \tag{7.110}$$

where R', and L' refer to the J circuit. Let us search for exponential solutions, $(I, J) \propto \exp(\gamma\,t)$, of the previous system of equations. It is easily demonstrated that

$$\gamma = \frac{-(L R' + L' R) \pm \sqrt{(L R' + L' R)^2 + 4\,R'\,(L\,L' - M^2)\,(M\,\Omega/2\pi - R)}}{2\,(L\,L' - M^2)}. \tag{7.111}$$

Recall the standard result in electromagnetic theory that $L\,L' > M^2$ for two non-coincident circuits (Jackson 1998). It is clear, from the previous expression, that the condition for dynamo action (i.e., $\gamma > 0$) is

$$\Omega > \frac{2\pi\,R}{M}, \tag{7.112}$$

as before. Note, however, that $\gamma \rightarrow 0$ as $R' \rightarrow 0$. In other words, if the rotating disk is a perfect conductor then dynamo action is impossible. The previous system of equations can be transformed into the well-known Lorenz system, which exhibits chaotic behavior in certain parameter regimes (Knobloch 1981). It is noteworthy that this simplest prototype dynamo system already contains the seeds of chaos (provided that the formulation is self-consistent).

The previous discussion implies that, while dynamo action requires the resistance, R, of the circuit to be low, we lose dynamo action altogether if we go to the perfectly conducting limit, $R \rightarrow 0$, because magnetic fields are unable to diffuse into the region in which magnetic induction is operating. Thus, an efficient dynamo requires a conductivity that is large, but not too large.

7.11 Slow and Fast Dynamos

Let us search for solutions of the MHD kinematic dynamo equation,

$$\frac{\partial \mathbf{B}}{\partial t} = \nabla\times(\mathbf{V}\times\mathbf{B}) + \frac{\eta}{\mu_0}\nabla^2\mathbf{B}, \tag{7.113}$$

for a prescribed steady-state velocity field, $\mathbf{V}(\mathbf{r})$, subject to certain practical constraints. First, we require a self-contained solution; in other words, a solution in which the magnetic field is maintained by the motion of the MHD fluid, rather than by currents at infinity. This suggests that $V, B \to 0$ as $r \to \infty$. Second, we require an exponentially growing solution: that is, a solution for which $\mathbf{B} \propto \exp(\gamma\, t)$, where $\gamma > 0$.

In most MHD fluids that occur in astrophysical contexts, the resistivity, η, is extremely small. Let us consider the perfectly conducting limit, $\eta \to 0$. In this limit, Vainshtein and Zel'dovich introduced an important distinction between two fundamentally different classes of dynamo solution (Vainshtein and Zel'dovich 1972). Suppose that we solve the eigenvalue equation (7.113) to obtain the growth-rate, γ, of the magnetic field in the limit $\eta \to 0$. We expect that

$$\lim_{\eta\to 0} \gamma \propto \eta^{\alpha}, \tag{7.114}$$

where $0 \leq \alpha \leq 1$. There are two possibilities. Either $\alpha > 0$, in which case the growth-rate depends on the resistivity; or $\alpha = 0$, in which case the growth-rate is independent of the resistivity. The former case is termed a *slow dynamo*, whereas the latter case is termed a *fast dynamo*. By definition, slow dynamos are unable to operate in the perfectly conducting limit, because $\gamma \to 0$ as $\eta \to 0$. On the other hand, fast dynamos can, in principle, operate when $\eta = 0$.

It is clear, from the discussion in the previous section, that a homopolar disk dynamo is an example of a slow dynamo. In fact, it is easily seen that any dynamo that depends on the motion of a rigid conductor for its operation is bound to be a slow dynamo—in the perfectly conducting limit, the magnetic flux linking the conductor could never change, so there would be no magnetic induction. So, why do we believe that fast dynamo action is even a possibility for an MHD fluid? The answer, of course, is that an MHD fluid is a non-rigid body, and, thus, its motion possesses degrees of freedom not accessible to rigid conductors.

We know that in the perfectly conducting limit ($\eta \to 0$) magnetic field-lines are frozen into an MHD fluid. If the motion is incompressible (i.e., $\nabla \cdot \mathbf{V} = 0$) then the stretching of field-lines implies a proportionate intensification of the field-strength. The simplest heuristic fast dynamo, first described by Vainshtein and Zel'dovich, is based on this effect (Vainshtein and Zel'dovich 1972). As illustrated in Figure 7.6, a magnetic flux-tube can be doubled in intensity by taking it around a stretch-twist-fold cycle. The doubling time for this process clearly does not depend on the resistivity—in this sense, the dynamo is a fast dynamo. However, under repeated application of the cycle, the magnetic field develops increasingly fine-scale structure. In fact, in the limit $\eta \to 0$, both the $\mathbf{V}$ and $\mathbf{B}$ fields eventually become chaotic and non-differentiable. A little resistivity is always required to smooth out the fields on small lengthscales. Even in this case, the fields remain chaotic.

At present, the physical existence of fast dynamos has not been conclusively established, because most of the literature on this subject is based on mathematical paradigms rather than actual solutions of the dynamo equation (Childress and Gilbert 1995). It should be noted, however, that the need for fast dynamo solutions is fairly

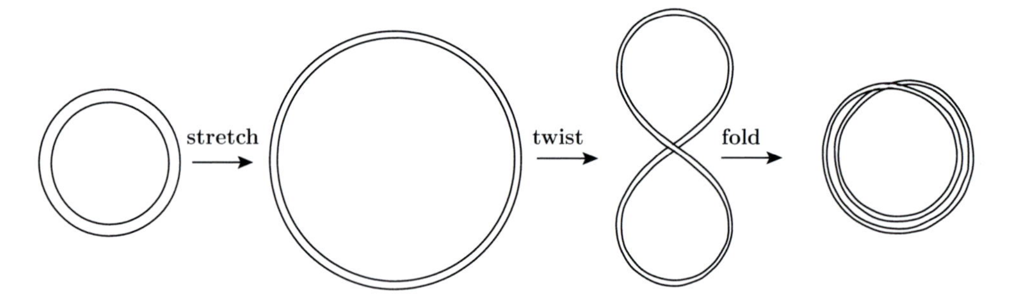

Figure 7.6
The stretch-twist-fold cycle of a fast dynamo.

acute, especially in stellar dynamo theory. For instance, consider the Sun. The ohmic decay time for the Sun is about 10^{12} years, whereas the reversal time for the solar magnetic field is only 11 years (Mestel 2012). It is obviously a little difficult to believe that resistivity is playing any significant role in the solar dynamo.

In the following, we shall restrict our analysis to slow dynamos, which undoubtably exist in nature, and which are characterized by non-chaotic $\mathbf{V}$ and $\mathbf{B}$ fields.

7.12 Cowling Anti-Dynamo Theorem

One of the most important results in slow, kinematic dynamo theory is credited to Cowling (Cowling 1934; Cowling 1957b). The so-called *Cowling anti-dynamo theorem* states that:

An axisymmetric magnetic field cannot be maintained via dynamo action.

Let us attempt to prove this proposition.

We adopt standard cylindrical coordinates: (ϖ, θ, z). The system is assumed to possess axial symmetry, so that $\partial/\partial\theta \equiv 0$. For the sake of simplicity, the plasma flow is assumed to be incompressible, which implies that $\nabla\cdot\mathbf{V} = 0$.

It is convenient to split the magnetic and velocity fields into poloidal and toroidal components:

$$\mathbf{B} = \mathbf{B}_p + \mathbf{B}_t, \tag{7.115}$$

$$\mathbf{V} = \mathbf{V}_p + \mathbf{V}_t. \tag{7.116}$$

Here, a "poloidal" vector only possesses non-zero ϖ- and z-components, whereas a "toroidal" vector only possesses a non-zero θ-component.

The poloidal components of the magnetic and velocity fields are written (Huba 2000a),

$$\mathbf{B}_p = \nabla\times\left(\frac{\psi}{\varpi}\,\mathbf{e}_\theta\right) \equiv \frac{\nabla\psi\times\mathbf{e}_\theta}{\varpi}, \tag{7.117}$$

$$\mathbf{V}_p = \nabla\times\left(\frac{\phi}{\varpi}\,\mathbf{e}_\theta\right) \equiv \frac{\nabla\phi\times\mathbf{e}_\theta}{\varpi}, \tag{7.118}$$

where $\psi = \psi(\varpi, z, t)$ and $\phi = \phi(\varpi, z, t)$. The toroidal components are given by

$$\mathbf{B}_t = B_t(\varpi, z, t)\,\mathbf{e}_\theta, \tag{7.119}$$

$$\mathbf{V}_t = V_t(\varpi, z, t)\,\mathbf{e}_\theta. \tag{7.120}$$

Note that by writing the $\mathbf{B}$ and $\mathbf{V}$ fields in the previous form we ensure that the constraints $\nabla\cdot\mathbf{B} = 0$ and $\nabla\cdot\mathbf{V} = 0$ are automatically satisfied. Note, further, that because $\mathbf{B}\cdot\nabla\psi = 0$ and $\mathbf{V}\cdot\nabla\phi = 0$, we can regard ψ and ϕ as stream-functions for the magnetic and velocity fields, respectively.

The condition for the magnetic field to be maintained by dynamo currents, rather than by currents at infinity, is

$$\psi \to \frac{1}{r} \qquad \text{as } r\to\infty, \tag{7.121}$$

where $r = (\varpi^2 + z^2)^{1/2}$. We also require the flow stream-function, ϕ, to remain bounded as $r\to\infty$.

Consider the MHD Ohm's law for a resistive plasma:

$$\mathbf{E} + \mathbf{V}\times\mathbf{B} = \eta\,\mathbf{j}. \tag{7.122}$$

Taking the toroidal component of this equation, we obtain

$$E_t + (\mathbf{V}_p\times\mathbf{B}_p)\cdot\mathbf{e}_\theta = \eta\, j_t. \tag{7.123}$$

It is easily demonstrated from the Faraday-Maxwell equation that

$$E_t = -\frac{1}{\varpi}\frac{\partial\psi}{\partial t}. \tag{7.124}$$

Furthermore,

$$(\mathbf{V}_p\times\mathbf{B}_p)\cdot\mathbf{e}_\theta = \frac{(\nabla\phi\times\nabla\psi)\cdot\mathbf{e}_\theta}{\varpi^2} = \frac{1}{\varpi^2}\left(\frac{\partial\psi}{\partial\varpi}\frac{\partial\phi}{\partial z} - \frac{\partial\phi}{\partial\varpi}\frac{\partial\psi}{\partial z}\right), \tag{7.125}$$

and (Huba 2000a)

$$\mu_0\, j_t = \nabla\times\mathbf{B}_p\cdot\mathbf{e}_\theta = -\left[\nabla^2\left(\frac{\psi}{\varpi}\right) - \frac{\psi}{\varpi^3}\right] = -\frac{1}{\varpi}\left(\frac{\partial^2\psi}{\partial\varpi^2} - \frac{1}{\varpi}\frac{\partial\psi}{\partial\varpi} + \frac{\partial^2\psi}{\partial z^2}\right). \tag{7.126}$$

Thus, Equation (7.123) reduces to

$$\frac{\partial\psi}{\partial t} - \frac{1}{\varpi}\left(\frac{\partial\psi}{\partial\varpi}\frac{\partial\phi}{\partial z} - \frac{\partial\phi}{\partial\varpi}\frac{\partial\psi}{\partial z}\right) = \frac{\eta}{\mu_0}\left(\frac{\partial^2\psi}{\partial\varpi^2} - \frac{1}{\varpi}\frac{\partial\psi}{\partial\varpi} + \frac{\partial^2\psi}{\partial z^2}\right). \tag{7.127}$$

Multiplying the previous equation by ψ and integrating over all space, we obtain

$$\frac{1}{2}\frac{d}{dt}\int \psi^2\,dV - \int\!\!\int 2\pi\,\psi\left(\frac{\partial\psi}{\partial\varpi}\frac{\partial\phi}{\partial z} - \frac{\partial\phi}{\partial\varpi}\frac{\partial\psi}{\partial z}\right)d\varpi\,dz$$
$$= \frac{\eta}{\mu_0}\int\!\!\int 2\pi\,\varpi\,\psi\left(\frac{\partial^2\psi}{\partial\varpi^2} - \frac{1}{\varpi}\frac{\partial\psi}{\partial\varpi} + \frac{\partial^2\psi}{\partial z^2}\right)d\varpi\,dz. \tag{7.128}$$

The second term on the left-hand side of the previous expression can be integrated by parts to give

$$-\int\!\!\int 2\pi\left[-\phi\,\frac{\partial}{\partial z}\left(\psi\,\frac{\partial\psi}{\partial\varpi}\right) + \phi\,\frac{\partial}{\partial\varpi}\left(\psi\,\frac{\partial\psi}{\partial z}\right)\right]d\varpi\,dz = 0, \tag{7.129}$$

where surface terms have been neglected, in accordance with Equation (7.121). Likewise, the term on the right-hand side of Equation (7.128) can be integrated by parts to give

$$\frac{\eta}{\mu_0}\int\!\!\int 2\pi\left[-\frac{\partial(\varpi\,\psi)}{\partial\varpi}\,\frac{\partial\psi}{\partial\varpi} - \varpi\left(\frac{\partial\psi}{\partial z}\right)^2\right]d\varpi\,dz =$$
$$-\frac{\eta}{\mu_0}\int\!\!\int 2\pi\,\varpi\left[\left(\frac{\partial\psi}{\partial\varpi}\right)^2 + \left(\frac{\partial\psi}{\partial z}\right)^2\right]d\varpi\,dz. \tag{7.130}$$

Thus, Equation (7.128) reduces to

$$\frac{d}{dt}\int \psi^2\,dV = -2\,\frac{\eta}{\mu_0}\int |\nabla\psi|^2\,dV \leq 0. \tag{7.131}$$

It is clear, from the previous expression, that the poloidal stream-function, ψ—and, hence, the poloidal magnetic field, $\mathbf{B}_p$—decays to zero under the influence of resistivity. We conclude that the poloidal magnetic field cannot be maintained via dynamo action.

Of course, we have not ruled out the possibility that the toroidal magnetic field can be maintained via dynamo action. In the absence of a poloidal field, the curl of the poloidal component of Equation (7.122) yields

$$-\frac{\partial\mathbf{B}_t}{\partial t} + \nabla\times(\mathbf{V}_p\times\mathbf{B}_t) = \eta\,\nabla\times\mathbf{j}_p, \tag{7.132}$$

which reduces to

$$-\frac{\partial B_t}{\partial t} + \nabla\times(\mathbf{V}_p\times\mathbf{B}_t)\cdot\mathbf{e}_\theta = -\frac{\eta}{\mu_0}\,\nabla^2(B_t\,\mathbf{e}_\theta)\cdot\mathbf{e}_\theta. \tag{7.133}$$

Now (Huba 2000a),

$$\nabla^2(B_t\,\mathbf{e}_\theta)\cdot\mathbf{e}_\theta = \frac{\partial^2 B_t}{\partial\varpi^2} + \frac{1}{\varpi}\frac{\partial B_t}{\partial\varpi} + \frac{\partial^2 B_t}{\partial z^2} - \frac{B_t}{\varpi^2}, \tag{7.134}$$

and (Huba 2000a)

$$\nabla\times(\mathbf{V}_p\times\mathbf{B}_t)\cdot\mathbf{e}_\theta = \frac{\partial}{\partial\varpi}\left(\frac{B_t}{\varpi}\right)\frac{\partial\phi}{\partial z} - \frac{\partial}{\partial z}\left(\frac{B_t}{\varpi}\right)\frac{\partial\phi}{\partial\varpi}. \tag{7.135}$$

Thus, Equation (7.133) yields

$$\frac{\partial\chi}{\partial t} - \frac{1}{\varpi}\left(\frac{\partial\chi}{\partial\varpi}\frac{\partial\phi}{\partial z} - \frac{\partial\phi}{\partial\varpi}\frac{\partial\chi}{\partial z}\right) = \frac{\eta}{\mu_0}\left(\frac{\partial^2\chi}{\partial\varpi^2} + \frac{3}{\varpi}\frac{\partial\chi}{\partial\varpi} + \frac{\partial^2\chi}{\partial z^2}\right), \tag{7.136}$$

where

$$B_t = \varpi\,\chi. \tag{7.137}$$

Multiply Equation (7.136) by χ, integrating over all space, and then integrating by parts, we obtain

$$\frac{d}{dt}\int\chi^2\,dV = -2\,\frac{\eta}{\mu_0}\int|\nabla\chi|^2\,dV \leq 0. \tag{7.138}$$

It is clear, from this equation, that χ—and, hence, the toroidal magnetic field, $\mathbf{B}_t$—decays to zero under the influence of resistivity. We conclude that no axisymmetric magnetic field—either poloidal or toroidal—can be maintained by dynamo action, which proves Cowling's theorem.

Cowling's theorem is the earliest, and most significant, of a number of anti-dynamo theorems that severely restrict the types of magnetic fields that can be maintained via dynamo action. For instance, it is possible to prove that a two-dimensional magnetic field cannot be maintained by dynamo action (Moffatt 1978). Here, "two-dimensional" implies that in some Cartesian coordinate system, $(x,\,y,\,z)$, the magnetic field is independent of z. The suite of anti-dynamo theorems can be summed up by saying that successful dynamos possess a rather low degree of symmetry.

7.13 Ponomarenko Dynamo

The simplest known kinematic dynamo is that of Ponomarenko (Ponomarenko 1973). Consider a conducting fluid of resistivity η that fills all space. The motion of the fluid is confined to a cylinder of radius a. Adopting standard cylindrical coordinates $(r,\,\theta,\,z)$ aligned with this cylinder, the flow field is written

$$\mathbf{V} = \begin{cases} (0,\,r\,\Omega,\,U) & \text{for } r \leq a \\ \mathbf{0} & \text{for } r > a \end{cases}, \tag{7.139}$$

where Ω and U are constants. Note that the flow is incompressible. In other words, $\nabla\cdot\mathbf{V} = 0$.

The MHD kinematic dynamo equation, (7.113), can be written

$$\frac{\partial\mathbf{B}}{\partial t} = (\mathbf{B}\cdot\nabla)\,\mathbf{V} - (\mathbf{V}\cdot\nabla)\,\mathbf{B} + \frac{\eta}{\mu_0}\,\nabla^2\mathbf{B}, \tag{7.140}$$

where use has been made of $\nabla \cdot \mathbf{B} = \nabla \cdot \mathbf{V} = 0$. Let us search for solutions to this equation of the form

$$\mathbf{B}(r, \theta, z, t) = \mathbf{B}(r)\, \exp[\,\mathrm{i}\,(m\,\theta - k\,z) + \gamma\,t]. \tag{7.141}$$

The r- and θ-components of Equation (7.140) are written (Huba 2000a)

$$\gamma\,B_r = -\mathrm{i}\,(m\,\Omega - k\,U)\,B_r + \frac{\eta}{\mu_0}\left[\frac{d^2B_r}{dr^2} + \frac{1}{r}\frac{dB_r}{dr} - \frac{(m^2 + k^2r^2 + 1)\,B_r}{r^2} - \frac{\mathrm{i}\,2\,m\,B_\theta}{r^2}\right], \tag{7.142}$$

and

$$\gamma\,B_\theta = r\,\frac{d\Omega}{dr}\,B_r - \mathrm{i}\,(m\,\Omega - k\,U)\,B_\theta + \frac{\eta}{\mu_0}\left[\frac{d^2B_\theta}{dr^2} + \frac{1}{r}\frac{dB_\theta}{dr} - \frac{(m^2 + k^2r^2 + 1)\,B_\theta}{r^2} + \frac{\mathrm{i}\,2\,m\,B_r}{r^2}\right], \tag{7.143}$$

respectively. In general, the term involving $d\Omega/dr$ is zero. In fact, this term is only included in the analysis to enable us to evaluate the correct matching conditions at $r = a$. We do not need to write the z-component of Equation (7.140), because B_z can be obtained more directly from B_r and B_θ via the constraint $\nabla \cdot \mathbf{B} = 0$.

Let

$$B_\pm = B_r \pm \mathrm{i}\,B_\theta, \tag{7.144}$$

$$y = \frac{r}{a}, \tag{7.145}$$

$$\tau_R = \frac{\mu_0\,a^2}{\eta}, \tag{7.146}$$

$$q^2 = k^2\,a^2 + \gamma\,\tau_R + \mathrm{i}\,(m\,\Omega - k\,U)\,\tau_R, \tag{7.147}$$

$$s^2 = k^2\,a^2 + \gamma\,\tau_R. \tag{7.148}$$

Here, τ_R is the typical time required for magnetic flux to diffuse a distance a under the action of resistivity. Equations (7.142)–(7.148) can be combined to give

$$y^2\,\frac{d^2B_\pm}{dy^2} + y\,\frac{dB_\pm}{dy} - \left[(m \pm 1)^2 + q^2\,y^2\right]B_\pm = 0 \tag{7.149}$$

for $y \leq 1$, and

$$y^2\,\frac{d^2B_\pm}{dy^2} + y\,\frac{dB_\pm}{dy} - \left[(m \pm 1)^2 + s^2\,y^2\right]B_\pm = 0 \tag{7.150}$$

for $y > 1$. The previous equations are modified Bessel's equations of order $m \pm 1$ (Abramowitz and Stegun 1965b). Thus, the physical solutions of Equations (7.149) and (7.150) that are well behaved as $y \to 0$ and $y \to \infty$ can be written

$$B_\pm(y) = C_\pm\,\frac{I_{m\pm1}(q\,y)}{I_{m\pm1}(q)} \tag{7.151}$$

for $y \leq 1$, and

$$B_\pm(y) = D_\pm \, \frac{K_{m\pm 1}(s\, y)}{K_{m\pm 1}(s)} \tag{7.152}$$

for $y > 1$. Here, $C_\pm$ and $D_\pm$ are arbitrary constants. Note that the arguments of q and s are both constrained to lie in the range $-\pi/2$ to $+\pi/2$.

The first matching condition at $y = 1$ is the continuity of $B_\pm$, which yields

$$C_\pm = D_\pm. \tag{7.153}$$

The second matching condition is obtained by integrating Equation (7.143) from $r = a - \delta$ to $r = a - \delta$, where δ is an infinitesimal quantity, and making use of the fact that the angular velocity Ω jumps discontinuously to zero at $r = a$. It follows that

$$a\,\Omega\,B_r = \frac{\eta}{\mu_0}\left[\frac{dB_\theta}{dr}\right]_{r=a_-}^{r=a_+}. \tag{7.154}$$

Furthermore, integration of Equation (7.142) tells us that dB_r/dr is continuous at $r = a$. We can combine this information to give the matching condition

$$\left[\frac{dB_\pm}{dy}\right]_{y=1_-}^{y=1_+} = \pm \mathrm{i}\,\Omega\,\tau_R\left(\frac{B_+ + B_-}{2}\right). \tag{7.155}$$

Equations (7.151)–(7.155) yield the dispersion relation

$$G_+\,G_- = \frac{\mathrm{i}}{2}\,\Omega\,\tau_R\,(G_+ - G_-), \tag{7.156}$$

where

$$G_\pm = q\,\frac{I'_{m\pm 1}(q)}{I_{m\pm 1}(q)} - s\,\frac{K'_{m\pm 1}(s)}{K_{m\pm 1}(s)}. \tag{7.157}$$

Here, $'$ denotes a derivative with respect to argument.

Unfortunately, despite the fact that we are investigating the simplest known kinematic dynamo, the dispersion relation (7.156) is sufficiently complicated that it can only be solved numerically. We can simplify matters considerably taking the limit $|q|, |s| \gg 1$, which corresponds to that of small wavelength (i.e., $k\,a \gg 1$). The large argument asymptotic behavior of the Bessel functions is specified by (Abramowitz and Stegun 1965b)

$$\sqrt{\frac{2\,z}{\pi}}\,K_m(z) = \mathrm{e}^{-z}\left[1 + \frac{4\,m^2 - 1}{8\,z} + O\!\left(\frac{1}{z^2}\right)\right], \tag{7.158}$$

$$\sqrt{2\pi\,z}\,I_m(z) = \mathrm{e}^{+z}\left[1 - \frac{4\,m^2 - 1}{8\,z} + O\!\left(\frac{1}{z^2}\right)\right], \tag{7.159}$$

where $|\arg(z)| < \pi/2$. It follows that

$$G_\pm = q + s + \left(\frac{m^2}{2} \pm m + \frac{3}{8}\right)\left(\frac{1}{q} + \frac{1}{s}\right) + O\!\left(\frac{1}{q^2} + \frac{1}{s^2}\right). \tag{7.160}$$

Thus, the dispersion relation (7.156) reduces to

$$(q+s)\,q\,s = \mathrm{i}\,m\,\Omega\,\tau_R, \tag{7.161}$$

where $|\arg(q)|, |\arg(s)| < \pi/2$.

In the limit $\mu \to 0$, where

$$\mu = (m\,\Omega - k\,U)\,\tau_R, \tag{7.162}$$

which corresponds to $(\mathbf{V}\cdot\nabla)\,\mathbf{B} \to 0$, the simplified dispersion relation (7.161) can be solved to give

$$\gamma\,\tau_R \simeq \mathrm{e}^{\,\mathrm{i}\,\pi/3}\left(\frac{m\,\Omega\,\tau_R}{2}\right)^{2/3} - k^2\,a^2 - \mathrm{i}\,\frac{\mu}{2}. \tag{7.163}$$

Dynamo behavior [i.e., $\mathrm{Re}(\gamma) > 0$] takes place when

$$\Omega\,\tau_R > \frac{2^{\,5/2}\,(k\,a)^{\,3}}{m}. \tag{7.164}$$

Observe that $\mathrm{Im}(\gamma) \neq 0$, implying that the dynamo mode oscillates, or rotates, as well as growing exponentially in time. The dynamo generated magnetic field is both non-axisymmetric [note that dynamo activity is impossible, according to Equation (7.163), if $m = 0$] and three-dimensional, and is, thus, not subject to either of the anti-dynamo theorems mentioned in the preceding section.

It is clear, from Equation (7.164), that dynamo action occurs whenever the flow is made sufficiently rapid. But, what is the minimum amount of flow needed to give rise to dynamo action? In order to answer this question, we have to solve the full dispersion relation, (7.156), for various values of m and k, in order to find the dynamo mode that grows exponentially in time for the smallest values of Ω and U. It is conventional to parameterize the flow in terms of the *magnetic Reynolds number*,

$$S = \frac{\tau_R}{\tau_H}, \tag{7.165}$$

where

$$\tau_H = \frac{L}{V} \tag{7.166}$$

is the typical timescale for convective motion across the system. Here, V is a typical flow velocity, and L is the characteristic lengthscale of the system. Taking $V = |\mathbf{V}(a)| = (\Omega^2\,a^2 + U^2)^{1/2}$, and $L = a$, we have

$$S = \frac{\tau_R\,(\Omega^2\,a^2 + U^2)^{1/2}}{a} \tag{7.167}$$

for the Ponomarenko dynamo. The critical value of the Reynolds number above which dynamo action occurs is found to be (Ponomarenko 1973)

$$S_c = 17.7. \tag{7.168}$$

The most unstable dynamo mode is characterized by $m = 1$, $U/(\Omega\,a) = 1.3$, $k\,a =$

0.39, and $\mathrm{Im}(\gamma)\,\tau_R = 0.41$. As the magnetic Reynolds number, S, is increased above the critical value, S_c, other dynamo modes are eventually destabilized.

In 2000, the Ponomarenko dynamo was realized experimentally by means of a tall cylinder filled with liquid sodium in which helical flow was excited by a propeller (Gailitis et al. 2000). More information on laboratory dynamo experiments can be found in Verhille et alia 2009.

7.14 Magnetic Reconnection

Magnetic reconnection is a phenomenon that is of particular importance in solar system plasmas. In the solar corona, it results in the rapid release to the plasma of energy stored in the large-scale structure of the coronal magnetic field, an effect that is thought to give rise to *solar flares* (Priest 1984). Small-scale reconnection may play a role in heating the corona, and, thereby, driving the outflow of the solar wind (Priest 1984). In the Earth's magnetosphere, magnetic reconnection in the magnetotail is thought to be the precursor for *auroral sub-storms* (Ratcliffe 1972).

The evolution of the magnetic field in a resistive-MHD plasma is governed by the following well-known equation [see Equation (7.102)]:

$$\frac{\partial \mathbf{B}}{\partial t} = \nabla\times(\mathbf{V}\times\mathbf{B}) + \frac{\eta}{\mu_0}\,\nabla^2\mathbf{B}. \tag{7.169}$$

The first term on the right-hand side of this equation describes the convection of the magnetic field by the plasma flow. The second term describes the resistive diffusion of the field through the plasma. If the first term dominates, then magnetic flux is frozen into the plasma, and the topology of the magnetic field cannot change. On the other hand, if the second term dominates, then there is little coupling between the field and the plasma flow, and the topology of the magnetic field is free to change.

The relative magnitude of the two terms on the right-hand side of Equation (7.169) is conventionally measured in terms of *magnetic Reynolds number*, or *Lundquist number*:

$$S = \frac{\mu_0\,V\,L}{\eta} \simeq \frac{|\nabla\times(\mathbf{V}\times\mathbf{B})|}{|(\eta/\mu_0)\,\nabla^2\mathbf{B}|}, \tag{7.170}$$

where V is the characteristic flow speed, and L the characteristic lengthscale of the plasma. If S is much larger than unity then convection dominates, and the frozen flux constraint prevails, whereas if S is much less than unity then diffusion dominates, and the coupling between the plasma flow and the magnetic field is relatively weak.

It turns out that very large S-values are virtually guaranteed to occur in the solar system because of the extremely large lengthscales of solar system plasmas. For instance, $S \sim 10^8$ for solar flares, whereas $S \sim 10^{11}$ is appropriate for the solar wind and the Earth's magnetosphere (Priest 1984). Of course, in calculating these values, we have identified the lengthscale L with the characteristic size of the plasma under investigation.

On the basis of the previous discussion, it seems reasonable to neglect diffusive processes altogether in solar system plasmas. Of course, this leads to very strong constraints on the behavior of such plasmas, because, in this limit, all cross-field mixing of plasma elements is suppressed. Particles may freely mix along field-lines (within limitations imposed by magnetic mirroring, etc.), but are completely ordered perpendicular to the field, because they always remain tied to the same field-lines as they convect in the plasma flow.

Let us consider what happens when two initially separate plasma regions come into contact with one another, as occurs, for example, in the interaction between the solar wind and the Earth's magnetic field. Assuming that each plasma is frozen to its own magnetic field, and that cross-field diffusion is absent, we conclude that the two plasmas will not mix, but, instead, that a thin boundary layer will form between them, separating the two plasmas and their respective magnetic fields. In equilibrium, the location of the boundary layer will be determined by pressure balance. Because, in general, the frozen fields on either side of the boundary will have differing strengths, and differing orientations tangential to the boundary, the layer must also constitute a current sheet. Thus, flux freezing leads inevitably to the prediction that in plasma systems space becomes divided into separate cells, wholly containing the plasma and magnetic field from individual sources, and separated from each other by thin current sheets.

The "separate cell" picture constitutes an excellent zeroth-order approximation to the interaction of solar system plasmas, as witnessed, for example, by the well-defined planetary magnetospheres (Russell 1991). It must be noted, however, that the large S-values upon which the applicability of the frozen flux constraint was justified were derived using the large overall spatial scales of the systems involved. However, strict application of this constraint to the problem of the interaction of separate plasma systems leads to the inevitable conclusion that structures will form having small spatial scales, at least in one dimension: that is, the thin current sheets constituting the cell boundaries. It is certainly not guaranteed, from the previous discussion, that the effects of diffusion can be neglected in these boundary layers. In fact, we shall demonstrate that the localized breakdown of the flux freezing constraint in the boundary regions, due to diffusion, not only has an impact on the properties of the boundary regions themselves, but can also have a decisive impact on the large lengthscale plasma regions in which the flux freezing constraint remains valid. This observation illustrates both the subtlety and the significance of the magnetic reconnection process.

7.15 Linear Tearing Mode Theory

Consider the interface between two plasmas containing magnetic fields of different orientations. The simplest imaginable field configuration is that illustrated in Figure 7.7. Here, the field varies in the x-direction, and is parallel to the y-axis. The

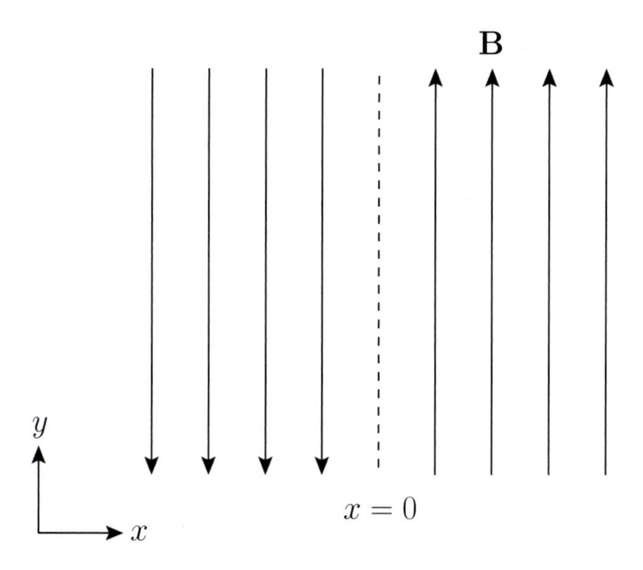

Figure 7.7
A reconnecting magnetic field configuration.

field is directed in the $-y$-direction for $x < 0$, and in the $+y$-direction for $x > 0$. The interface is situated at $x = 0$. The sudden reversal of the field direction across the interface gives rise to a z-directed current sheet at $x = 0$.

With the neglect of plasma resistivity, the field configuration shown in Figure 7.7 represents a stable equilibrium state, assuming, of course, that we have normal pressure balance across the interface. But, does the field configuration remain stable when we take resistivity into account? If not, then we expect an instability to develop that relaxes the configuration to one possessing lower magnetic energy. As we shall see, this type of relaxation process inevitably entails the breaking and reconnection of magnetic field-lines, and is, therefore, termed *magnetic reconnection*. The magnetic energy released during the reconnection process eventually appears as plasma thermal energy. Thus, magnetic reconnection also involves plasma heating.

In the following, we shall outline the standard method for determining the linear stability of the type of magnetic field configuration shown in Figure 7.7, taking into account the effect of plasma resistivity. We are particularly interested in plasma instabilities that are stable in the absence of resistivity, and only grow when the resistivity is non-zero. Such instabilities are conventionally termed *tearing modes*. Because magnetic reconnection is, in fact, a nonlinear process, we shall then proceed to investigate the nonlinear development of tearing modes.

The equilibrium magnetic field is written

$$\mathbf{B}_0 = B_{0\,y}(x)\,\mathbf{e}_y, \tag{7.171}$$

where $B_{0\,y}(-x) = -B_{0\,y}(x)$. There is assumed to be no equilibrium plasma flow.

The linearized equations of resistive-MHD [i.e., Equations (7.1)–(7.4), with Equation (7.3) replaced by Equation (7.102)], assuming incompressible flow, take the form

$$\frac{\partial \mathbf{B}}{\partial t} = \nabla\times(\mathbf{V}\times\mathbf{B}_0) + \frac{\eta}{\mu_0}\,\nabla^2\mathbf{B}, \tag{7.172}$$

$$\rho_0\,\frac{\partial \mathbf{V}}{\partial t} = -\nabla p + \frac{(\nabla\times\mathbf{B})\times\mathbf{B}_0}{\mu_0} + \frac{(\nabla\times\mathbf{B}_0)\times\mathbf{B}}{\mu_0}, \tag{7.173}$$

$$\nabla\cdot\mathbf{B} = 0, \tag{7.174}$$

$$\nabla\cdot\mathbf{V} = 0, \tag{7.175}$$

Here, ρ_0 is the equilibrium plasma density, $\mathbf{B}$ the perturbed magnetic field, $\mathbf{V}$ the perturbed plasma velocity, p the perturbed plasma pressure, and use has been made of Maxwell's equation. The assumption of incompressible plasma flow is valid provided that the plasma velocity associated with the instability remains significantly smaller than both the Alfvén velocity and the sonic velocity.

Suppose that all perturbed quantities vary like

$$A(x,y,z,t) = A(x)\,\exp(\mathrm{i}\,k\,y + \gamma\,t)\,, \tag{7.176}$$

where γ is the instability growth-rate. The x-component of Equation (7.172), and the z-component of the curl of Equation (7.173), reduce to

$$\gamma\,B_x = \mathrm{i}\,k\,B_{0\,y}\,V_x + \frac{\eta}{\mu_0}\left(\frac{d^2}{dx^2} - k^2\right)B_x, \tag{7.177}$$

$$\gamma\,\rho_0\left(\frac{d^2}{dx^2} - k^2\right)V_x = \frac{\mathrm{i}\,k\,B_{0\,y}}{\mu_0}\left(\frac{d^2}{dx^2} - k^2 - \frac{B_{0\,y}''}{B_{0\,y}}\right)B_x, \tag{7.178}$$

respectively, where use has been made of Equations (7.174) and (7.175). Here, $'$ denotes d/dx.

It is convenient to normalize Equations (7.177) and (7.178) using a typical magnetic field-strength, B_0, and a typical lengthscale, a. Let us define the *Alfvén timescale*

$$\tau_A = \frac{a}{V_A}, \tag{7.179}$$

where $V_A = B_0/\sqrt{\mu_0\,\rho_0}$ is the Alfvén velocity, and the *resistive diffusion timescale*

$$\tau_R = \frac{\mu_0\,a^2}{\eta}. \tag{7.180}$$

The ratio of these two timescales is the Lundquist number:

$$S = \frac{\tau_R}{\tau_A}. \tag{7.181}$$

Let $\bar{x} = x/a$, $\psi(\bar{x}) = B_x/B_0$, $\phi(\bar{x}) = \mathrm{i}\,k\,V_x/\gamma$, $F(\bar{x}) = B_{0\,y}/B_0$, $F' \equiv dF/d\bar{x}$, $\bar{\gamma} = \gamma\,\tau_A$, and $\bar{k} = k\,a$. It follows that

$$\bar{\gamma}\,(\psi - F\,\phi) = S^{-1}\left(\frac{d^2}{d\bar{x}^2} - \bar{k}^2\right)\psi, \tag{7.182}$$

$$\bar{\gamma}^2\left(\frac{d^2}{d\bar{x}^2} - \bar{k}^2\right)\phi = -\bar{k}^2\,F\left(\frac{d^2}{d\bar{x}^2} - \bar{k}^2 - \frac{F''}{F}\right)\psi. \tag{7.183}$$

The term on the right-hand side of Equation (7.182) represents plasma resistivity, whereas the term on the left-hand side of Equation (7.183) represents plasma inertia.

It is assumed that the tearing instability grows on a hybrid timescale that is much less than τ_R, but much greater than τ_A. It follows that

$$\bar{\gamma} \ll 1 \ll S\,\bar{\gamma}. \tag{7.184}$$

Thus, throughout most of the plasma, we can neglect the right-hand side of Equation (7.182), and the left-hand side of Equation (7.183), which is equivalent to the neglect of plasma resistivity and inertia. In this case, Equations (7.182) and (7.183) reduce to

$$\phi = \frac{\psi}{F}, \tag{7.185}$$

$$\frac{d^2\psi}{d\bar{x}^2} - \bar{k}^2\,\psi - \frac{F''}{F}\,\psi = 0. \tag{7.186}$$

Equation (7.185) is simply the flux-freezing constraint, which requires the plasma to move with the magnetic field. Equation (7.186) is the linearized, static force balance criterion: $\nabla \times (\mathbf{j} \times \mathbf{B}) = \mathbf{0}$. Equations (7.185) and (7.186) are known collectively as the equations of *ideal-MHD*, and are valid throughout virtually the whole plasma. However, it is clear that these equations break down in the immediate vicinity of the interface, where $F = 0$ (i.e., where the magnetic field reverses direction). Observe, for instance, that the normalized "radial" velocity, ϕ, becomes infinite as $F \rightarrow 0$, according to Equation (7.185).

The ideal-MHD equations break down close to the interface because the neglect of plasma resistivity and inertia becomes untenable as $F \rightarrow 0$. Thus, there is a thin layer, in the immediate vicinity of the interface, $\bar{x} = 0$, where the behavior of the plasma is governed by the resistive-MHD equations, (7.182) and (7.183). We can simplify these equations, making use of the fact that $\bar{x} \ll 1$ and $d/d\bar{x} \gg 1$ in a thin layer, to obtain the following layer equations:

$$\bar{\gamma}\,(\psi - \bar{x}\,\phi) = S^{-1}\,\frac{d^2\psi}{d\bar{x}^2}, \tag{7.187}$$

$$\bar{\gamma}^2\,\frac{d^2\phi}{d\bar{x}^2} = -\bar{x}\,\frac{d^2\psi}{d\bar{x}^2}. \tag{7.188}$$

Note that we have redefined the variables ϕ, $\bar{\gamma}$, and S, such that $\phi \rightarrow F'(0)\,\phi$, $\bar{\gamma} \rightarrow \gamma\,\tau_H$, and $S \rightarrow \tau_R/\tau_H$. Here,

$$\tau_H = \frac{\tau_A}{k\,a\,F'(0)} \tag{7.189}$$

is the so-called *hydromagnetic timescale*.

The tearing mode stability problem reduces to solving the resistive-MHD layer equations, (7.187) and (7.188), in the immediate vicinity of the interface, $\bar{x} = 0$, solving the ideal-MHD equations, (7.185) and (7.186), everywhere else in the plasma, matching the two solutions at the edge of the layer, and applying physical boundary conditions as $|\bar{x}| \to \infty$. This method of solution was first described in a classic paper by Furth, Killeen, and Rosenbluth (Furth, Killeen, and Rosenbluth 1963).

Let us consider the solution of the ideal-MHD equation (7.186) throughout the bulk of the plasma. We could imagine launching a solution $\psi(\bar{x})$ at large positive $\bar{x}$, which satisfies physical boundary conditions as $\bar{x} \to \infty$, and integrating this solution to the right-hand boundary of the resistive-MHD layer at $\bar{x} = 0_+$. Likewise, we could also launch a solution at large negative $\bar{x}$, which satisfies physical boundary conditions as $\bar{x} \to -\infty$, and integrate this solution to the left-hand boundary of the resistive-MHD layer at $\bar{x} = 0_-$. Maxwell's equations demand that ψ must be continuous on either side of the layer. Hence, we can multiply our two solutions by appropriate factors, so as to ensure that ψ matches to the left and right of the layer. This leaves the function $\psi(\bar{x})$ undetermined to an overall arbitrary multiplicative constant, just as we would expect in a linear problem. In general, $d\psi/d\bar{x}$ is not continuous to the left and right of the layer. Thus, the ideal solution can be characterized by the real number

$$\Delta' = \left[\frac{1}{\psi}\frac{d\psi}{d\bar{x}}\right]_{\bar{x}=0_-}^{\bar{x}=0_+}: \tag{7.190}$$

that is, by the jump in the logarithmic derivative of ψ to the left and right of the layer. This parameter is known as the *tearing stability index*, and is solely a property of the plasma equilibrium, the wavenumber, k, and the boundary conditions imposed at infinity.

The layer equations, (7.187) and (7.188), possess a trivial solution ($\phi = \phi_0$, $\psi = \bar{x}\,\phi_0$, where ϕ_0 is independent of $\bar{x}$), and a nontrivial solution for which $\psi(-\bar{x}) = \psi(\bar{x})$ and $\phi(-\bar{x}) = -\phi(\bar{x})$. The asymptotic behavior of the nontrivial solution at the edge of the layer is

$$\psi(x) \to \left(\frac{\Delta}{2}\,|\bar{x}| + 1\right)\Psi, \tag{7.191}$$

$$\phi(x) \to \frac{\psi}{\bar{x}}, \tag{7.192}$$

where the parameter $\Delta(\bar{\gamma}, S)$ is determined by solving the layer equations, subject to the previous boundary conditions. Finally, the growth-rate, γ, of the tearing instability is determined by the matching criterion

$$\Delta(\bar{\gamma}, S) = \Delta'. \tag{7.193}$$

The layer equations, (7.187) and (7.188), can be solved in a fairly straightforward manner in Fourier transform space. Let

$$\phi(\bar{x}) = \int_{-\infty}^{\infty} \hat{\phi}(t)\, \mathrm{e}^{\,\mathrm{i}\,S^{1/3}\,\bar{x}\,t}\, dt, \tag{7.194}$$

$$\psi(\bar{x}) = \int_{-\infty}^{\infty} \hat{\psi}(t)\, \mathrm{e}^{\,\mathrm{i}\,S^{1/3}\,\bar{x}\,t}\, dt, \tag{7.195}$$

where $\hat{\phi}(-t) = -\hat{\phi}(t)$. Equations (7.187) and (7.188) can be Fourier transformed, and the results combined, to give

$$\frac{d}{dt}\left(\frac{t^2}{Q+t^2}\frac{d\hat{\phi}}{dt}\right) - Q\,t^2\,\hat{\phi} = 0, \tag{7.196}$$

where

$$Q = \gamma\,\tau_H^{2/3}\,\tau_R^{1/3}. \tag{7.197}$$

The most general small-t asymptotic solution of Equation (7.196) is written

$$\hat{\phi}(t) \rightarrow \frac{a_{-1}}{t} + a_0 + O(t), \tag{7.198}$$

where a_{-1} and a_0 are independent of t, and it is assumed that $t > 0$. When inverse Fourier transformed, the previous expression leads to the following expression for the asymptotic behavior of ϕ at the edge of the resistive-MHD layer (Erdéyli 1954):

$$\phi(\bar{x}) \rightarrow a_{-1}\,\frac{\pi}{2}\,S^{1/3}\,\mathrm{sgn}(x) + \frac{a_0}{\bar{x}} + O\left(\frac{1}{\bar{x}^2}\right). \tag{7.199}$$

It follows from a comparison with Equations (7.191) and (7.192) that

$$\Delta = \pi\,\frac{a_{-1}}{a_0}\,S^{1/3}. \tag{7.200}$$

Thus, the matching parameter Δ is determined from the small-t asymptotic behavior of the Fourier transformed layer solution.

Let us search for an unstable tearing mode, characterized by $Q > 0$. It is convenient to assume that

$$Q \ll 1. \tag{7.201}$$

This ordering, which is known as the *constant-ψ approximation* [because it implies that $\psi(\bar{x})$ is approximately constant across the layer] will be justified later on.

In the limit $t \gg Q^{1/2}$, Equation (7.196) reduces to

$$\frac{d^2\hat{\phi}}{dt^2} - Q\,t^2\,\hat{\phi} = 0. \tag{7.202}$$

The solution to this equation that is well behaved in the limit $t \rightarrow \infty$ is written $U(0, \sqrt{2}\,Q^{1/4}\,t)$, where $U(a, x)$ is a standard parabolic cylinder function (Abramowitz and Stegun 1965e). In the limit

$$Q^{1/2} \ll t \ll Q^{-1/4} \tag{7.203}$$

we can make use of the standard small argument asymptotic expansion of $U(a, x)$ to write the most general solution to Equation (7.196) in the form (Abramowitz and Stegun 1965e)

$$\hat{\phi}(t) = A\left[1 - 2\,\frac{\Gamma(3/4)}{\Gamma(1/4)}\,Q^{1/4}\,t + O(t^2)\right]. \tag{7.204}$$

Here, A is an arbitrary constant.

In the limit

$$t \ll Q^{-1/4}, \tag{7.205}$$

Equation (7.196) reduces to

$$\frac{d}{dt}\left(\frac{t^2}{Q+t^2}\,\frac{d\hat{\phi}}{dt}\right) = 0. \tag{7.206}$$

The most general solution to this equation is written

$$\hat{\phi}(t) = B\left(-\frac{Q}{t} + t\right) + C + O(t^2), \tag{7.207}$$

where B and C are arbitrary constants. Matching coefficients between Equations (7.204) and (7.207) in the range of t satisfying the inequality (7.203) yields the following expression for the most general solution to Equation (7.196) in the limit $t \ll Q^{1/2}$:

$$\hat{\phi} = A\left[2\,\frac{\Gamma(3/4)}{\Gamma(1/4)}\,\frac{Q^{5/4}}{t} + 1 + O(t)\right]. \tag{7.208}$$

Finally, a comparison of Equations (7.198), (7.200), and (7.208) gives the result

$$\Delta = 2\pi\,\frac{\Gamma(3/4)}{\Gamma(1/4)}\,S^{1/3}\,Q^{5/4}. \tag{7.209}$$

The asymptotic matching condition (7.193) can be combined with the previous expression for Δ to give the tearing mode dispersion relation

$$\gamma = \left[\frac{\Gamma(1/4)}{2\pi\,\Gamma(3/4)}\right]^{4/5}\frac{(\Delta')^{4/5}}{\tau_H^{2/5}\,\tau_R^{3/5}}. \tag{7.210}$$

Here, use has been made of the definitions of S and Q. According to the dispersion relation, the tearing mode is unstable whenever $\Delta' > 0$, and grows on the hybrid timescale $\tau_H^{2/5}\,\tau_R^{3/5}$. It is easily demonstrated that the tearing mode is stable whenever $\Delta' < 0$. According to Equations (7.193), (7.201), and (7.209), the constant-ψ approximation holds provided that

$$\Delta' \ll S^{1/3}: \tag{7.211}$$

that is, provided that the tearing mode does not become too unstable.

According to Equation (7.202), the thickness of the resistive-MHD layer in t-space is

$$\delta_t \sim \frac{1}{Q^{1/4}}. \tag{7.212}$$

It follows from Equations (7.194) and (7.195) that the thickness of the layer in $\bar{x}$-space is

$$\bar{\delta} \sim \frac{1}{S^{1/3}\,\delta_t} \sim \left(\frac{\bar{\gamma}}{S}\right)^{1/4}. \tag{7.213}$$

When $\Delta' \sim O(1)$ then $\bar{\gamma} \sim S^{-3/5}$, according to Equation (7.210), giving $\bar{\delta} \sim S^{-2/5}$. It is clear, therefore, that if the Lundquist number, S, is very large then the resistive-MHD layer centered on the interface, $\bar{x} = 0$, is extremely narrow.

The timescale for magnetic flux to diffuse across a layer of thickness $\bar{\delta}$ (in $\bar{x}$-space) is [see Equation (7.180)]

$$\tau \sim \tau_R\,\bar{\delta}^2. \tag{7.214}$$

If

$$\gamma\,\tau \ll 1 \tag{7.215}$$

then the tearing mode grows on a timescale that is far longer than the timescale on which magnetic flux diffuses across the resistive layer. In this case, we would expect the normalized "radial" magnetic field, ψ, to be approximately constant across the layer, because any non-uniformities in ψ would be smoothed out via resistive diffusion. It follows from Equations (7.213) and (7.214) that the constant-ψ approximation holds provided that

$$\bar{\gamma} \ll S^{-1/3} \tag{7.216}$$

(i.e., $Q \ll 1$), which is in agreement with Equation (7.201).

7.16 Nonlinear Tearing Mode Theory

We have seen that if $\Delta' > 0$ then a magnetic field configuration of the type shown in Figure 7.7 is unstable to a tearing mode. Let us now investigate how a tearing instability affects the field configuration as it develops.

It is convenient to write the magnetic field in terms of a flux-function:

$$\mathbf{B} = B_0\,a\,\nabla\psi \times \mathbf{e}_z. \tag{7.217}$$

Note that $\mathbf{B}\cdot\nabla\psi = 0$. It follows that magnetic field-lines run along contours of $\psi(x, y)$.

We can write

$$\psi(\bar{x}, \bar{y}) \simeq \psi_0(\bar{x}) + \psi_1(\bar{x}, \bar{y}), \tag{7.218}$$

where ψ_0 generates the equilibrium magnetic field, and ψ_1 generates the perturbed magnetic field associated with the tearing mode. Here, $\bar{y} = y/a$. In the vicinity of the interface, we have

$$\psi(\bar{x}, \bar{y}) \simeq -\frac{F'(0)}{2}\,\bar{x}^2 + \Psi\,\cos(\bar{k}\,\bar{y}), \tag{7.219}$$

where Ψ is a constant. Here, we have made use of the fact that $\psi_1(\bar{x}, \bar{y}) \simeq \psi_1(\bar{y})$ if the constant-ψ approximation holds good (which is assumed to be the case).

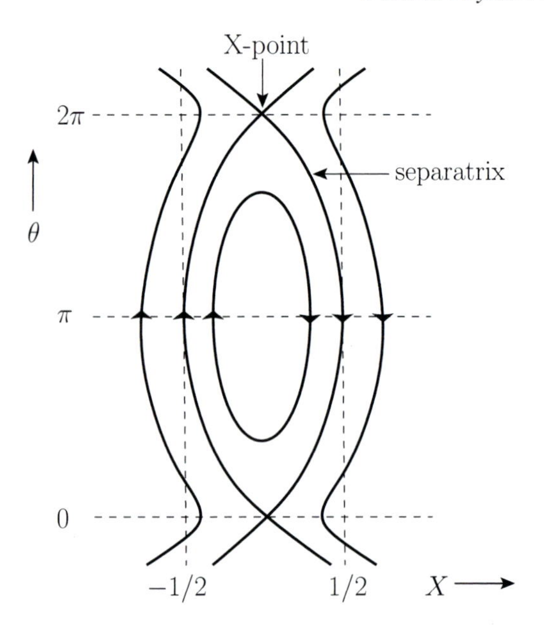

Figure 7.8
Magnetic field-lines in the vicinity of a magnetic island.

Let $\chi = -\psi/\Psi$ and $\theta = \bar{k}\,\bar{y}$. It follows that the normalized perturbed magnetic flux function, χ, in the vicinity of the interface takes the form

$$\chi = 8\,X^2 - \cos\theta, \tag{7.220}$$

where $X = \bar{x}/\overline{W}$, and

$$\overline{W} = 4\sqrt{\frac{\Psi}{F'(0)}}. \tag{7.221}$$

Figure 7.8 shows the contours of χ plotted in X-θ space. It can be seen that the tearing mode gives rise to the formation of a *magnetic island* centered on the interface, $X = 0$. Magnetic field-lines situated outside the "separatrix" are displaced by the tearing mode, but still retain their original topology. By contrast, field-lines inside the separatrix have been broken and reconnected, and now possess quite different topology. The reconnection obviously takes place at the "X-points," which are located at $X = 0$ and $\theta = j\,2\pi$, where j is an integer. The maximum width of the reconnected region (in x-space) is given by the *island width*, $a\,\overline{W}$. Note that the island width is proportional to the square root of the perturbed "radial" magnetic field at the interface (i.e., $\overline{W} \propto \sqrt{\Psi}$).

According to a result first established in a very elegant paper by Rutherford (Rutherford 1973), the nonlinear evolution of the island width is governed by

$$0.823\,\tau_R\,\frac{d\overline{W}}{dt} = \Delta'(\overline{W}), \tag{7.222}$$

where

$$\Delta'(\overline{W}) = \left[\frac{1}{\psi}\frac{d\psi}{d\bar{x}}\right]_{-\overline{W}/2}^{+\overline{W}/2} \tag{7.223}$$

is the jump in the logarithmic derivative of ψ taken across the island (White, Monticello, Rosenbluth, and Waddell 1977). It is clear that once the tearing mode enters the nonlinear regime (i.e., once the normalized island width, $\overline{W}$, exceeds the normalized linear layer width, $S^{-2/5}$), the growth-rate of the instability slows down considerably, until the mode eventually ends up growing on the extremely slow resistive timescale, τ_R. The tearing mode stops growing when it has attained a saturated island width $\overline{W}_0$, satisfying

$$\Delta'(\overline{W}_0) = 0. \tag{7.224}$$

The saturated width is a function of the original plasma equilibrium, but is independent of the resistivity. There is no particular reason why $\overline{W}_0$ should be small. In general, the saturated island width is comparable with the characteristic length-scale of the magnetic field configuration. We conclude that, although ideal-MHD only breaks down in a narrow region of relative width $S^{-2/5}$, centered on the interface, $x = 0$, the reconnection of magnetic field-lines that takes place in this region is capable of significantly modifying the whole magnetic field configuration.

7.17 Fast Magnetic Reconnection

Up to now, we have only considered magnetic reconnection that develops spontaneously from a plasma instability. As we have seen, such reconnection takes place at a fairly leisurely pace. Let us now consider so-called *forced magnetic reconnection*, in which the reconnection takes place as a consequence of an externally imposed flow, or magnetic perturbation, rather than occurring spontaneously. The principal difference between forced and spontaneous reconnection is the development of extremely large, positive Δ' values in the former case. Generally speaking, we expect Δ' to be $O(1)$ for spontaneous reconnection. By analogy with the previous analysis, we would expect forced reconnection to proceed faster than spontaneous reconnection (because the reconnection rate increases with increasing Δ'). The question is—how much faster? To be more exact, if we take the limit $\Delta' \to \infty$, which corresponds to the limit of extreme forced reconnection, how fast can we make the magnetic field reconnect? At present, this is a controversial question that is far from being completely resolved. In the following, we shall content ourselves with a discussion of the two "classical" fast reconnection models. These models form the starting point of virtually all recent research on this subject (Yamada, Kulsrud, and Ji 2010).

Let us start off by considering the *Sweet-Parker model*, which was first proposed by Sweet (Sweet 1958) and Parker (Parker 1957). The main features of the envisioned magnetic and plasma flow fields are illustrated in Figure 7.9. The system is two-dimensional and steady-state (i.e., $\partial/\partial z \equiv 0$ and $\partial/\partial t \equiv 0$). The reconnecting

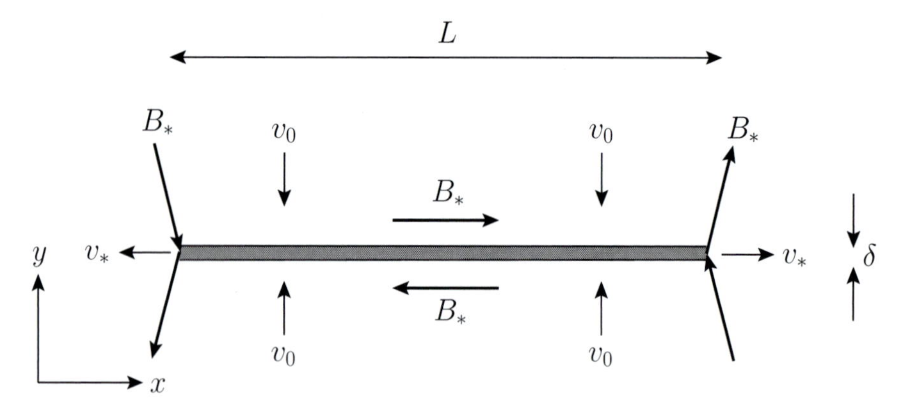

Figure 7.9
The Sweet-Parker magnetic reconnection scenario.

magnetic fields are anti-parallel, and of equal strength, B_*. We imagine that these fields are being forcibly pushed together via the action of some external agency. We expect a strong current sheet to form at the boundary between the two fields, where the direction of **B** suddenly changes. This current sheet is assumed to be of thickness δ and length L.

Plasma is assumed to diffuse into the current layer, along its whole length, at some relatively small inflow velocity, v_0. The plasma is accelerated along the layer, and eventually expelled from its two ends at some relatively large exit velocity, v_*. The inflow velocity is simply an $\mathbf{E}\times\mathbf{B}$ velocity, so

$$v_0 \sim \frac{E_z}{B_*}. \tag{7.225}$$

The z-component of Ohm's law yields

$$E_z \sim \frac{\eta\, B_*}{\mu_0\,\delta}. \tag{7.226}$$

Continuity of plasma flow inside the layer gives

$$L\, v_0 \sim \delta\, v_*, \tag{7.227}$$

assuming incompressible flow. Finally, pressure balance along the length of the layer yields

$$\frac{B_*^2}{\mu_0} \sim \rho\, v_*^2. \tag{7.228}$$

Here, we have balanced the magnetic pressure at the center of the layer against the dynamic pressure of the outflowing plasma at the ends of the layer. Note that η and ρ are the plasma resistivity and density, respectively.

We can measure the rate of reconnection via the inflow velocity, v_0, because all of the magnetic field-lines that are convected into the layer, with the plasma, are eventually reconnected. The Alfvén velocity is written

$$V_A = \frac{B_*}{\sqrt{\mu_0\,\rho}}. \tag{7.229}$$

Likewise, we can write the Lundquist number of the plasma as

$$S = \frac{\mu_0\,L\,V_A}{\eta}, \tag{7.230}$$

where we have assumed that the length of the reconnecting layer, L, is commensurate with the macroscopic lengthscale of the system. The reconnection rate is parameterized via the *Alfvénic Mach number* of the inflowing plasma, which is defined

$$M_0 = \frac{v_0}{V_A}. \tag{7.231}$$

The previous equations can be rearranged to give

$$v_* \sim V_A. \tag{7.232}$$

In other words, the plasma is squirted out of the ends of the reconnecting layer at the Alfvén velocity. Furthermore,

$$\delta \sim M_0\,L, \tag{7.233}$$

and

$$M_0 \sim S^{-1/2}. \tag{7.234}$$

We conclude that the reconnecting layer is extremely narrow, assuming that the Lundquist number of the plasma is very large. The magnetic reconnection takes place on the hybrid timescale $\tau_A^{1/2}\,\tau_R^{1/2}$, where τ_A is the Alfvén transit timescale across the plasma, and τ_R is the resistive diffusion timescale across the plasma.

The Sweet-Parker reconnection ansatz is undoubtedly correct. It has been simulated numerically many times, and was confirmed experimentally in the Magnetic Reconnection Experiment (MRX) operated by Princeton Plasma Physics Laboratory (PPPL) (Ji, Yamada, Hsu, and Kulsrud 1998). The problem is that Sweet-Parker reconnection takes place far too slowly to account for many reconnection processes that are thought to take place in the solar system. For instance, in solar flares $S \sim 10^8$, $V_A \sim 100\,\mathrm{km\,s^{-1}}$, and $L \sim 10^4\,\mathrm{km}$ (Priest 1984). According to the Sweet-Parker model, magnetic energy is released to the plasma via reconnection on a typical timescale of a few tens of days. In reality, the energy is released in a few minutes to an hour (Priest 1984). Clearly, we can only hope to account for solar flares using a reconnection mechanism that operates far more rapidly than the Sweet-Parker mechanism.

One possible resolution of this problem was suggested by Petschek (Petschek 1964), who pointed out that magnetic energy can be converted into plasma thermal energy as a result of shock waves being set up in the plasma, in addition to the conversion due to the action of resistive diffusion. The configuration envisaged by

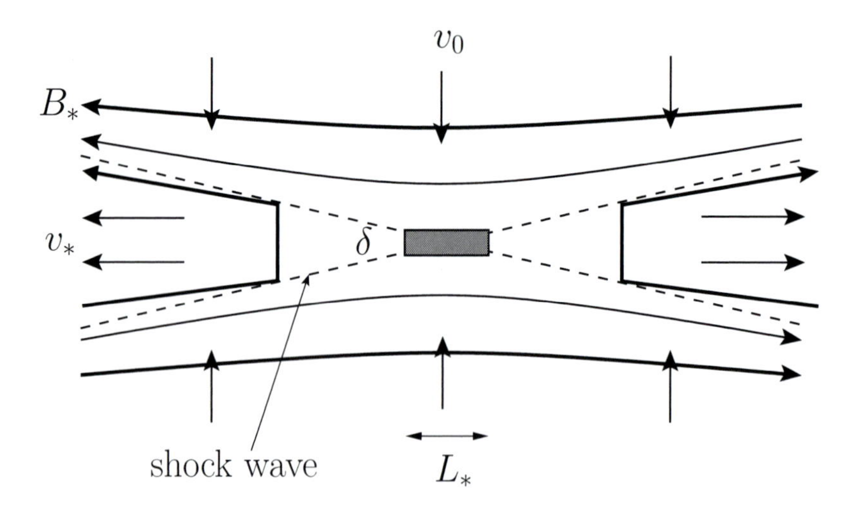

Figure 7.10
The Petschek magnetic reconnection scenario.

Petschek is sketched in Figure 7.10. Two waves (slow mode shocks) stand in the flow on either side of the interface, where the direction of **B** reverses, marking the boundaries of the plasma outflow regions. A small diffusion region still exists on the interface, but now constitutes a miniature (in length) Sweet-Parker system. The width of the reconnecting layer is given by

$$\delta = \frac{L}{M_0\, S}, \tag{7.235}$$

just as in the Sweet-Parker model. However, we do not now assume that the length, L_*, of the layer is comparable to the characteristic lengthscale, L, of the system. Rather, the length may be considerably smaller than L, and is determined self-consistently from the continuity condition

$$L_* = \frac{\delta}{M_0}, \tag{7.236}$$

where we have assumed incompressible flow, and an outflow speed of order the Alfvén speed, as before. Thus, if the inflow speed, v_0, is much less than V_A then the length of the reconnecting layer is much larger than its width, as assumed by Sweet and Parker. On the other hand, if we allow the inflow velocity to approach the Alfvén velocity then the layer shrinks in length, so that L_* becomes comparable with δ.

It follows that for reasonably large reconnection rates (i.e., $M_0 \to 1$) the length of the diffusion region becomes much smaller than the characteristic lengthscale of the system, L, so that most of the plasma flowing into the boundary region does so across the standing waves, rather than through the central diffusion region. The angle

θ that the shock waves make with the interface is given approximately by

$$\tan\theta \sim M_0. \tag{7.237}$$

Thus, for small inflow speeds, the outflow is confined to a narrow wedge along the interface, but as the inflow speed increases, the angle of the outflow wedges increases to accommodate the increased flow.

It turns out that there is a maximum inflow speed beyond which Petschek-type solutions cease to exist. The corresponding maximum Alfvénic Mach number,

$$(M_0)_{\rm max} = \frac{\pi}{8\ln S}, \tag{7.238}$$

can be regarded as specifying the maximum allowable rate of magnetic reconnection according to the Petschek model. Clearly, because the maximum reconnection rate depends inversely on the logarithm of the Lundquist number, rather than its square root, it is much larger than that predicted by the Sweet-Parker model.

It must be pointed out that the Petschek model is controversial. Many researchers think that it is incorrect, and that the maximum rate of magnetic reconnection allowed by resistive-MHD is that predicted by the Sweet-Parker model. In particular, Biskamp wrote an influential, and widely quoted, paper reporting the results of a numerical experiment that appeared to disprove the Petschek model (Biskamp 1986). When the plasma inflow exceeded that allowed by the Sweet-Parker model, there was no acceleration of the reconnection rate. Instead, magnetic flux "piled up" in front of the reconnecting layer, and the rate of reconnection never deviated significantly from that predicted by the Sweet-Parker model. Priest and Forbes later argued that Biskamp imposed boundary conditions in his numerical experiment, which precluded Petschek reconnection (Priest and Forbes 1992). Probably the most powerful argument against the validity of the Petschek model is the fact that, more than 50 years after it was first proposed, nobody has ever managed to simulate Petschek reconnection numerically (except by artificially increasing the resistivity in the reconnecting region—which is not a legitimate approach) (Yamada, Kulsrud, and Ji 2010).

7.18 MHD Shocks

Consider a subsonic disturbance moving through a conventional neutral fluid. As is well known, sound waves propagating ahead of the disturbance give advance warning of its arrival, and, thereby, allow the response of the fluid to be both smooth and adiabatic. Now, consider a supersonic distrurbance. In this case, sound waves are unable to propagate ahead of the disturbance, and so there is no advance warning of its arrival, and, consequently, the fluid response is sharp and non-adiabatic. This type of response is generally known as a *shock*.

Let us investigate shocks in MHD fluids. Because information in such fluids is carried via three different waves—namely, fast, or compressional-Alfvén, waves; intermediate, or shear-Alfvén, waves; and slow, or magnetosonic, waves (see Section 7.4)—we might expect MHD fluids to support three different types of shock, corresponding to disturbances traveling faster than each of the aforementioned waves. This is indeed the case.

In general, a shock propagating through an MHD fluid produces a significant difference in plasma properties on either side of the shock front. The thickness of the front is determined by a balance between convective and dissipative effects. However, dissipative effects in high temperature plasmas are only comparable to convective effects when the spatial gradients in plasma variables become extremely large. Hence, MHD shocks in such plasmas tend to be extremely narrow, and are well approximated as discontinuous changes in plasma parameters. The MHD equations, combined with Maxwell's equations, can be integrated across a shock to give a set of jump conditions that relate plasma properties on each side of the shock front. If the shock is sufficiently narrow then these relations become independent of its detailed structure. Let us derive the jump conditions for a narrow, planar, steady-state, MHD shock.

Maxwell's equations, and the MHD equations, (7.1)–(7.4), can be combined and written in the following convenient form (Boyd and Sanderson 2003):

$$\nabla\cdot\mathbf{B} = 0, \tag{7.239}$$

$$\frac{\partial\mathbf{B}}{\partial t} - \nabla\times(\mathbf{V}\times\mathbf{B}) = \mathbf{0}, \tag{7.240}$$

$$\frac{\partial\rho}{\partial t} + \nabla\cdot(\rho\,\mathbf{V}) = 0, \tag{7.241}$$

$$\frac{\partial(\rho\,\mathbf{V})}{\partial t} + \nabla\cdot\mathbf{T} = \mathbf{0}, \tag{7.242}$$

$$\frac{\partial U}{\partial t} + \nabla\cdot\mathbf{u} = 0, \tag{7.243}$$

where

$$\mathbf{T} = \rho\,\mathbf{V}\,\mathbf{V} + \left(p + \frac{B^2}{2\,\mu_0}\right)\mathbf{I} - \frac{\mathbf{B}\,\mathbf{B}}{\mu_0} \tag{7.244}$$

is the total (i.e., including electromagnetic, as well as plasma, contributions) stress tensor, $\mathbf{I}$ the identity tensor,

$$U = \frac{1}{2}\,\rho\,V^2 + \frac{p}{\Gamma-1} + \frac{B^2}{2\,\mu_0} \tag{7.245}$$

the total energy density, and

$$\mathbf{u} = \left(\frac{1}{2}\,\rho\,V^2 + \frac{\Gamma}{\Gamma-1}\,p\right)\mathbf{V} + \frac{\mathbf{B}\times(\mathbf{V}\times\mathbf{B})}{\mu_0} \tag{7.246}$$

the total energy flux density.

Let us transform into the rest frame of the shock. Suppose that the shock front

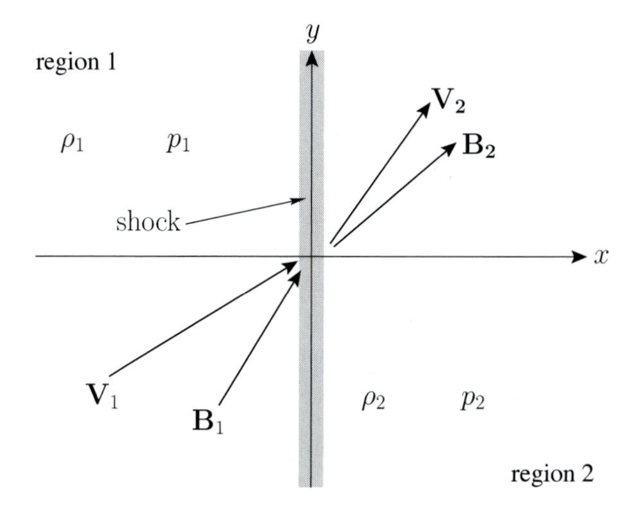

Figure 7.11
A planar MHD shock.

coincides with the y-z plane. Furthermore, let the regions of the plasma upstream and downstream of the shock, which are termed regions 1 and 2, respectively, be spatially uniform and non-time-varying. It follows that $\partial/\partial t = \partial/\partial y = \partial/\partial z = 0$. Moreover, $\partial/\partial x = 0$, except in the immediate vicinity of the shock. Finally, let the velocity and magnetic fields upstream and downstream of the shock all lie in the x-y plane. The situation under discussion is illustrated in Figure 7.11. Here, ρ_1, p_1, $\mathbf{V}_1$, and $\mathbf{B}_1$ are the upstream mass density, pressure, velocity, and magnetic field, respectively, whereas ρ_2, p_2, $\mathbf{V}_2$, and $\mathbf{B}_2$ are the corresponding downstream quantities. In the immediate vicinity of the shock, Equations (7.239)–(7.243) reduce to

$$\frac{dB_x}{dx} = 0, \qquad \frac{d}{dx}(V_x\,B_y - V_y\,B_x) = 0, \tag{7.247}$$

$$\frac{d(\rho\,V_x)}{dx} = 0, \qquad \frac{dT_{xx}}{dx} = 0, \tag{7.248}$$

$$\frac{dT_{xy}}{dx} = 0, \qquad \frac{du_x}{dx} = 0. \tag{7.249}$$

Integration across the shock yields the desired jump conditions:

$$[B_x]_1^2 = 0, \tag{7.250}$$

$$[V_x\,B_y - V_y\,B_x]_1^2 = 0, \tag{7.251}$$

$$[\rho\,V_x]_1^2 = 0, \tag{7.252}$$

$$[\rho\,V_x^2 + p + B_y^2/2\,\mu_0]_1^2 = 0, \tag{7.253}$$

$$[\rho\,V_x\,V_y - B_x\,B_y/\mu_0]_1^2 = 0, \tag{7.254}$$

$$\left[\frac{1}{2}\rho\,V^2\,V_x + \frac{\Gamma}{\Gamma-1}\,p\,V_x + \frac{B_y\,(V_x\,B_y - V_y\,B_x)}{\mu_0}\right]_1^2 = 0, \tag{7.255}$$

where $[A]_1^2 \equiv A_2 - A_1$. These relations are known as the *Rankine-Hugoniot relations* for MHD (Boyd and Sanderson 2003). Assuming that all of the upstream plasma parameters are known, there are six unknown parameters in the problem—namely, $B_{x\,2}$, $B_{y\,2}$, $V_{x\,2}$, $V_{y\,2}$, ρ_2, and p_2. These six unknowns are fully determined by the six jump conditions. Unfortunately, the general case is very complicated. So, before tackling it, let us examine a couple of relatively simple special cases.

7.19 Parallel MHD Shocks

The first special case is the so-called *parallel MHD shock*, in which both the upstream and downstream plasma flows are parallel to the magnetic field, as well as perpendicular to the shock front. In other words,

$$\mathbf{V}_1 = (V_1,\,0,\,0), \qquad \mathbf{V}_2 = (V_2,\,0,\,0), \tag{7.256}$$

$$\mathbf{B}_1 = (B_1,\,0,\,0), \qquad \mathbf{B}_2 = (B_2,\,0,\,0). \tag{7.257}$$

Substitution into the general jump conditions (7.250)–(7.255) yields

$$\frac{B_2}{B_1} = 1, \qquad \frac{\rho_2}{\rho_1} = r, \tag{7.258}$$

$$\frac{V_2}{V_1} = r^{-1}, \qquad \frac{p_2}{p_1} = R, \tag{7.259}$$

with

$$r = \frac{(\Gamma+1)\,M_1^2}{2+(\Gamma-1)\,M_1^2}, \tag{7.260}$$

$$R = 1 + \Gamma\,M_1^2\,(1-r^{-1}) = \frac{(\Gamma+1)\,r-(\Gamma-1)}{(\Gamma+1)-(\Gamma-1)\,r}. \tag{7.261}$$

Here, $M_1 = V_1/V_{S\,1}$, where $V_{S\,1} = (\Gamma\,p_1/\rho_1)^{1/2}$ is the upstream sound speed. Thus, the upstream flow is supersonic if $M_1 > 1$, and subsonic if $M_1 < 1$. Incidentally, as is clear from the previous expressions, a parallel shock is unaffected by the presence of a magnetic field. In fact, this type of shock is identical to that which occurs in neutral fluids, and is, therefore, usually called a *hydrodynamic shock*.

It is easily seen from Equations (7.258)–(7.261) that there is no shock (i.e., no jump in plasma parameters across the shock front) when the upstream flow is exactly sonic: that is, when $M_1 = 1$. In other words, $r = R = 1$ when $M_1 = 1$. However, if $M_1 \neq 1$ then the upstream and downstream plasma parameters become different (i.e., $r \neq 1, R \neq 1$), and a true shock develops. In fact, it can be demonstrated that

$$\frac{\Gamma - 1}{\Gamma + 1} \leq r \leq \frac{\Gamma + 1}{\Gamma - 1}, \tag{7.262}$$

$$0 \leq R \leq \infty, \tag{7.263}$$

$$\frac{\Gamma - 1}{2\Gamma} \leq M_1^2 \leq \infty. \tag{7.264}$$

Note that the upper and lower limits in the previous inequalities are all attained simultaneously.

The previous discussion seems to imply that a parallel shock can be either compressive (i.e., $r > 1$) or expansive (i.e., $r < 1$). However, there is one additional physics principle that needs to be factored into our analysis—namely, the second law of thermodynamics. This law states that the entropy of a closed system can spontaneously increase, but can never spontaneously decrease (Reif 1965). Now, in general, the entropy per particle is different on either side of a hydrodynamic shock front. Accordingly, the second law of thermodynamics mandates that the downstream entropy must exceed the upstream entropy, so as to ensure that the shock generates a net increase, rather than a net decrease, in the overall entropy of the system, as the plasma flows through it.

The (suitably normalized) entropy per particle of an ideal plasma takes the form [see Equation (4.51)]

$$S = \ln\left(\frac{p}{\rho^{\Gamma}}\right). \tag{7.265}$$

Hence, the difference between the upstream and downstream entropies is

$$[S\,]_1^2 = \ln R - \Gamma \ln r. \tag{7.266}$$

Now, using (7.261),

$$r\,\frac{d[S\,]_1^2}{dr} = \frac{r}{R}\,\frac{dR}{dr} - \Gamma = \frac{\Gamma\,(\Gamma^2 - 1)\,(r-1)^2}{[(\Gamma+1)\,r - (\Gamma-1)]\,[(\Gamma+1) - (\Gamma-1)\,r]}. \tag{7.267}$$

Furthermore, it is easily seen from Equations (7.262)–(7.264) that $d[S\,]_1^2/dr \geq 0$ in all situations of physical interest. However, $[S\,]_1^2 = 0$ when $r = 1$, because, in this case, there is no discontinuity in plasma parameters across the shock front. We conclude that $[S\,]_1^2 < 0$ for $r < 1$, and $[S\,]_1^2 > 0$ for $r > 1$. It follows that the second law of thermodynamics requires hydrodynamic shocks to be compressive: that is, $r > 1$. In other words, the plasma density must always increase when a shock front is crossed in the direction of the relative plasma flow. It turns out that this is a general rule that applies to all three types of MHD shock (Boyd and Sanderson 2003).

The *upstream Mach number*, M_1, is a good measure of shock strength: that is, if $M_1 = 1$ then there is no shock, if $M_1 - 1 \ll 1$ then the shock is weak, and if

$M_1 \gg 1$ then the shock is strong. We can define an analogous *downstream Mach number*, $M_2 = V_2/(\Gamma\, p_2/\rho_2)^{1/2}$. It is easily demonstrated from the jump conditions that if $M_1 > 1$ then $M_2 < 1$. In other words, in the shock rest frame, the shock is associated with an irreversible (because the entropy suddenly increases) transition from supersonic to subsonic flow. Note that $r \equiv \rho_2/\rho_1 \to (\Gamma+1)/(\Gamma-1)$, whereas $R \equiv p_2/p_1 \to \infty$, in the limit $M_1 \to \infty$. In other words, as the shock strength increases, the *compression ratio*, r, asymptotes to a finite value, whereas the *pressure ratio*, R, increases without limit. For a conventional plasma with $\Gamma = 5/3$, the limiting value of the compression ratio is 4: in other words, the downstream density can never be more than four times the upstream density. We conclude that, in the strong shock limit, $M_1 \gg 1$, the large jump in the plasma pressure across the shock front must be predominately a consequence of a large jump in the plasma temperature, rather than the plasma density. In fact, Equations (7.260) and (7.261) imply that

$$\frac{T_2}{T_1} \equiv \frac{R}{r} \to \frac{2\,\Gamma\,(\Gamma-1)\,M_1^2}{(\Gamma+1)^2} \gg 1 \tag{7.268}$$

as $M_1 \to \infty$. Thus, a strong parallel, or hydrodynamic, shock is associated with intense plasma heating.

As we have seen, the condition for the existence of a hydrodynamic shock is $M_1 > 1$, or $V_1 > V_{S\,1}$. In other words, in the shock frame, the upstream plasma velocity, V_1, must be supersonic. However, by Galilean invariance, V_1 can also be interpreted as the propagation velocity of the shock through an initially stationary plasma. It follows that, in a stationary plasma, a parallel, or hydrodynamic, shock propagates along the magnetic field with a supersonic velocity.

7.20 Perpendicular MHD Shocks

The second special case is the so-called *perpendicular MHD shock* in which both the upstream and downstream plasma flows are perpendicular to the magnetic field, as well as the shock front. In other words,

$$\mathbf{V}_1 = (V_1,\,0,\,0), \qquad \mathbf{V}_2 = (V_2,\,0,\,0), \tag{7.269}$$

$$\mathbf{B}_1 = (0,\,B_1,\,0), \qquad \mathbf{B}_2 = (0,\,B_2,\,0). \tag{7.270}$$

Substitution into the general jump conditions (7.250)–(7.255) yields

$$\frac{B_2}{B_1} = r, \qquad \frac{\rho_2}{\rho_1} = r, \tag{7.271}$$

$$\frac{V_2}{V_1} = r^{-1}, \qquad \frac{p_2}{p_1} = R, \tag{7.272}$$

where

$$R = 1 + \Gamma\,M_1^2\,(1 - r^{-1}) + \beta_1^{-1}\,(1 - r^2), \tag{7.273}$$

and r is a real positive root of the quadratic

$$F(r) = 2\,(2-\Gamma)\,r^2+\Gamma\left[2\,(1+\beta_1)+(\Gamma-1)\,\beta_1\,M_1^2\right]r-\Gamma\,(\Gamma+1)\,\beta_1\,M_1^2 = 0. \quad (7.274)$$

Here, $\beta_1 = 2\,\mu_0\,p_1/B_1^2$.

If r_1 and r_2 are the two roots of Equation (7.274) then

$$r_1\,r_2 = -\frac{\Gamma\,(\Gamma+1)\,\beta_1\,M_1^2}{2\,(2-\Gamma)}. \quad (7.275)$$

Assuming that $\Gamma < 2$, we conclude that one of the roots is negative, and, hence, that Equation (7.274) only possesses one physical solution: that is, there is only one type of MHD shock that is consistent with Equations (7.269) and (7.270). Now, it is easily demonstrated that $F(0) < 0$ and $F[(\Gamma+1)/(\Gamma-1)] > 0$. Hence, the physical root lies between $r = 0$ and $r = (\Gamma+1)/(\Gamma-1)$.

Using similar analysis to that employed in the previous section, it can be demonstrated that the second law of thermodynamics requires a perpendicular shock to be compressive: that is, $r > 1$ (Boyd and Sanderson 2003). It follows that a physical solution is only obtained when $F(1) < 0$, which reduces to

$$M_1^2 > 1 + \frac{2}{\Gamma\,\beta_1}. \quad (7.276)$$

This condition can also be written

$$V_1^2 > V_{S\,1}^2 + V_{A\,1}^2, \quad (7.277)$$

where $V_{A\,1} = B_1/(\mu_0\,\rho_1)^{1/2}$ is the upstream Alfvén velocity. Now, $V_{+\,1} = (V_{S\,1}^2 + V_{A\,1}^2)^{1/2}$ can be recognized as the velocity of a fast wave propagating perpendicular to the magnetic field. (See Section 7.4.) Thus, the condition for the existence of a perpendicular shock is that the relative upstream plasma velocity must be greater than the upstream fast wave velocity. Incidentally, it is easily demonstrated that if this is the case then the downstream plasma velocity is less than the downstream fast wave velocity. We can also deduce that, in a stationary plasma, a perpendicular shock propagates across the magnetic field with a velocity that exceeds the fast wave velocity.

In the strong shock limit, $M_1 \gg 1$, Equations (7.273) and (7.274) become identical to Equations (7.260) and (7.261). Hence, a strong perpendicular shock is very similar to a strong hydrodynamic shock (except that the former shock propagates perpendicular, whereas the latter shock propagates parallel, to the magnetic field). In particular, just like a hydrodynamic shock, a perpendicular shock cannot compress the density by more than a factor $(\Gamma+1)/(\Gamma-1)$. However, according to Equation (7.271), a perpendicular shock compresses the magnetic field by the same factor that it compresses the plasma density. It follows that there is also an upper limit to the factor by which a perpendicular shock can compress the magnetic field.

7.21 Oblique MHD Shocks

Let us now consider the general case in which the plasma velocities and the magnetic fields on each side of the shock are neither parallel nor perpendicular to the shock front. It is convenient to transform into the so-called *de Hoffmann-Teller frame* in which $|\mathbf{V}_1 \times \mathbf{B}_1| = 0$, or

$$V_{x\,1}\,B_{y\,1} - V_{y\,1}\,B_{x\,1} = 0. \tag{7.278}$$

In other words, it is convenient to transform to a frame that moves at the local $\mathbf{E}\times\mathbf{B}$ velocity of the plasma. It immediately follows from the jump condition (7.251) that

$$V_{x\,2}\,B_{y\,2} - V_{y\,2}\,B_{x\,2} = 0, \tag{7.279}$$

or $|\mathbf{V}_2 \times \mathbf{B}_2| = 0$. Thus, in the de Hoffmann-Teller frame, the upstream plasma flow is parallel to the upstream magnetic field, and the downstream plasma flow is also parallel to the downstream magnetic field. Furthermore, the magnetic contribution to the jump condition (7.255) becomes identically zero, which is a considerable simplification.

Equations (7.278) and (7.279) can be combined with the general jump conditions (7.250)–(7.255) to give

$$\frac{\rho_2}{\rho_1} = r, \tag{7.280}$$

$$\frac{B_{x\,2}}{B_{x\,1}} = 1, \tag{7.281}$$

$$\frac{B_{y\,2}}{B_{y\,1}} = r\left(\frac{v_1^2 - \cos^2\theta_1\,V_{A\,1}^2}{v_1^2 - r\,\cos^2\theta_1\,V_{A\,1}^2}\right), \tag{7.282}$$

$$\frac{V_{x\,2}}{V_{x\,1}} = \frac{1}{r}, \tag{7.283}$$

$$\frac{V_{y\,2}}{V_{y\,1}} = \frac{v_1^2 - \cos^2\theta_1\,V_{A\,1}^2}{v_1^2 - r\,\cos^2\theta_1\,V_{A\,1}^2}, \tag{7.284}$$

$$\frac{p_2}{p_1} = 1 + \frac{\Gamma\,v_1^2\,(r-1)}{V_{S\,1}^2\,r}\left[1 - \frac{r\,V_{A\,1}^2\,\sin^2\theta_1\,([r+1]\,v_1^2 - 2\,r\,V_{A\,1}^2\,\cos^2\theta_1)}{2\,(v_1^2 - r\,V_{A\,1}^2\,\cos^2\theta_1)^2}\right]. \tag{7.285}$$

where $v_1 = V_{x\,1} = V_1\,\cos\theta_1$ is the component of the upstream velocity normal to the shock front, and θ_1 is the angle subtended between the upstream plasma flow and the shock front normal. Finally, given the compression ratio, r, the square of the normal upstream velocity, v_1^2, is a real root of a cubic equation known as the *shock adiabatic*:

$$\begin{aligned} 0 = {} & (v_1^2 - r\,\cos^2\theta_1\,V_{A\,1}^2)^2\left([(\Gamma+1) - (\Gamma-1)\,r]\,v_1^2 - 2\,r\,V_{S\,1}^2\right) \\ & - r\,\sin^2\theta_1\,v_1^2\,V_{A\,1}^2\left([\Gamma + (2-\Gamma)\,r]\,v_1^2 - [(\Gamma+1) - (\Gamma-1)\,r]\,r\,\cos^2\theta_1\,V_{A\,1}^2\right). \end{aligned} \tag{7.286}$$

As before, the second law of thermodynamics mandates that $r > 1$.

Let us first consider the weak shock limit $r \to 1$. In this case, it is easily seen that the three roots of the shock adiabatic reduce to

$$v_1^2 = V_{-1}^2 \equiv \frac{V_{A\,1}^2 + V_{S\,1}^2 - [(V_{A\,1} + V_{S\,1})^2 - 4\,\cos^2\theta_1\,V_{S\,1}^2\,V_{A\,1}^2]^{1/2}}{2}, \tag{7.287}$$

$$v_1^2 = \cos^2\theta_1\,V_{A\,1}^2, \tag{7.288}$$

$$v_1^2 = V_{+1}^2 \equiv \frac{V_{A\,1}^2 + V_{S\,1}^2 + [(V_{A\,1} + V_{S\,1})^2 - 4\,\cos^2\theta_1\,V_{S\,1}^2\,V_{A\,1}^2]^{1/2}}{2}. \tag{7.289}$$

However, from Section 7.4, we recognize these velocities as belonging to slow, intermediate (or shear-Alfvén), and fast waves, respectively, propagating in the normal direction to the shock front. We conclude that slow, intermediate, and fast MHD shocks degenerate into the associated MHD waves in the limit of small shock amplitude. Conversely, we can think of the various MHD shocks as nonlinear versions of the associated MHD waves. Now, it can be demonstrated that

$$V_{+\,1} > \cos\theta_1\,V_{A\,1} > V_{-\,1}. \tag{7.290}$$

In other words, a fast wave travels faster than an intermediate wave, which travels faster than a slow wave. It is reasonable to suppose that the same is true of the associated MHD shocks, at least at relatively low shock strength. It follows from Equation (7.282) that $B_{y\,2} > B_{y\,1}$ for a fast shock, whereas $B_{y\,2} < B_{y\,1}$ for a slow shock. For the case of an intermediate shock, we can show, after a little algebra, that $B_{y\,2} \to -B_{y\,1}$ in the limit $r \to 1$. We conclude that (in the de Hoffmann-Teller frame) fast shocks refract the magnetic field and plasma flow (recall that they are parallel in our adopted frame of the reference) away from the normal to the shock front, whereas slow shocks refract these quantities toward the normal. Moreover, the tangential magnetic field and plasma flow generally reverse across an intermediate shock front. This is illustrated in Figure 7.12.

When r is slightly larger than unity, it is easily demonstrated that the conditions for the existence of a slow, intermediate, and fast shock are $v_1 > V_{-\,1}$, $v_1 > \cos\theta_1\,V_{A\,1}$, and $v_1 > V_{+\,1}$, respectively.

Let us now consider the strong shock limit, $v_1^2 \to \infty$. In this case, the shock adiabatic yields $r \to r_m = (\Gamma + 1)/(\Gamma - 1)$, and

$$v_1^2 \simeq \frac{r_m}{\Gamma - 1}\,\frac{2\,V_{S\,1}^2 + \sin^2\theta_1\,[\Gamma + (2 - \Gamma)\,r_m]\,V_{A\,1}^2}{r_m - r}. \tag{7.291}$$

There are no other real roots. The previous root is clearly a type of fast shock. The fact that there is only one real root suggests that there exists a critical shock strength above which the slow and intermediate shock solutions cease to exist. In fact, they merge and annihilate one another (Gurnett and Bhattacharjee 2005). In other words, there is a limit to the strength of a slow or an intermediate shock. On the other hand, there is no limit to the strength of a fast shock. Note, however, that the plasma density and tangential magnetic field cannot be compressed by more than a factor $(\Gamma + 1)/(\Gamma - 1)$ by any type of MHD shock.

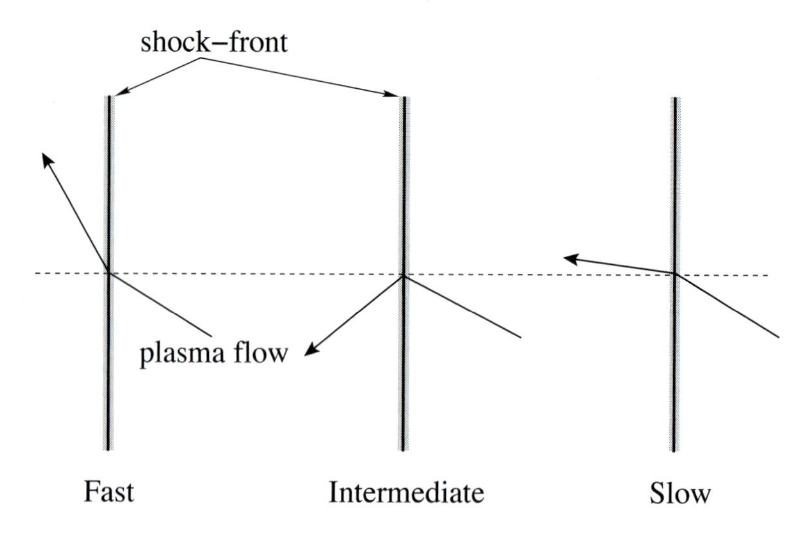

Figure 7.12
Characteristic plasma flow patterns across the three different types of MHD shock in the de Hoffmann-Teller frame.

Consider the special case $\theta_1 = 0$ in which both the plasma flow and the magnetic field are normal to the shock front. In this case, the three roots of the shock adiabatic are

$$v_1^2 = \frac{2\,r\,V_{S\,1}^2}{(\Gamma+1)-(\Gamma-1)\,r}, \tag{7.292}$$

$$v_1^2 = r\,V_{A\,1}^2, \tag{7.293}$$

$$v_1^2 = r\,V_{A\,1}^2. \tag{7.294}$$

We recognize the first of these roots as the hydrodynamic shock discussed in Section 7.19 [see Equation (7.260)]. This shock is classified as a slow shock when $V_{S\,1} < V_{A\,1}$, and as a fast shock when $V_{S\,1} > V_{A\,1}$. The other two roots are identical, and correspond to shocks that propagate at the velocity $v_1 = \sqrt{r}\,V_{A\,1}$ and "switch-on" the tangential components of the plasma flow and the magnetic field: that is, it can be seen from Equations (7.282) and (7.284) that $V_{y\,1} = B_{y\,1} = 0$ while $V_{y\,2} \neq 0$ and $B_{y\,2} \neq 0$ for these types of shock. Incidentally, it is also possible to have a "switch-off" shock that eliminates the tangential components of the plasma flow and the magnetic field. According to Equations (7.282) and (7.284), such a shock propagates at the velocity $v_1 = \cos\theta_1\,V_{A\,1}$. Switch-on and switch-off shocks are illustrated in Figure 7.13.

Let us, finally, consider the special case $\theta = \pi/2$. As is easily demonstrated, the

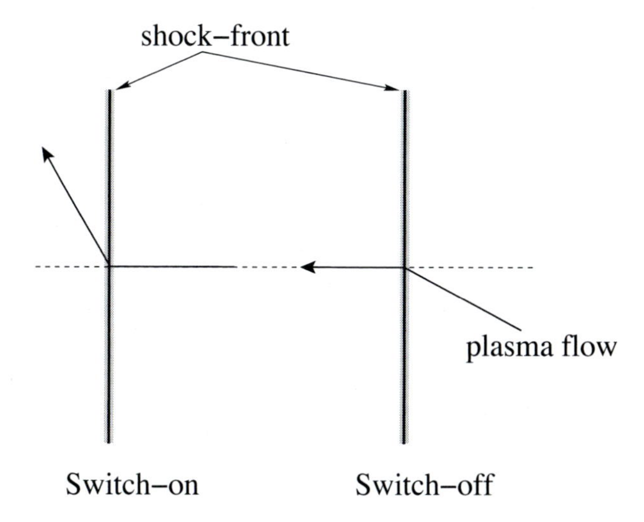

Figure 7.13
Characteristic plasma flow patterns across switch-on and switch-off shocks in the de Hoffmann-Teller frame.

three roots of the shock adiabatic are

$$v_1^2 = r\left(\frac{2\,V_{S\,1}^2 + [\Gamma + (2-\Gamma)\,r]\,V_{A\,1}^2}{[\Gamma+1] - [\Gamma-1]\,r}\right), \tag{7.295}$$

$$v_1^2 = 0, \tag{7.296}$$

$$v_1^2 = 0. \tag{7.297}$$

The first of these roots is clearly a fast shock, and is identical to the perpendicular shock discussed in Section 7.20, except that there is no plasma flow across the shock front in this case. The fact that the two other roots are zero indicates that, like the corresponding MHD waves, slow and intermediate MHD shocks do not propagate perpendicular to the magnetic field.

MHD shocks have been observed in a large variety of situations. For instance, shocks are known to be formed by supernova explosions, by strong stellar winds, by solar flares, and by the solar wind upstream of planetary magnetospheres (Gurnett and Bhattacharjee 2005).

7.22 Exercises

7.1 We can add viscous effects to the MHD momentum equation by including a term $\mu\,\nabla^2\mathbf{V}$, where μ is the dynamic viscosity, so that

$$\rho\,\frac{d\mathbf{V}}{dt} = \mathbf{j}\times\mathbf{b} - \nabla p + \mu\,\nabla^2\mathbf{V}.$$

Likewise, we can add finite conductivity effects to the Ohm's law by including the term $(1/\mu_0\,\sigma)\,\nabla^2\mathbf{B}$, to give

$$\frac{\partial\mathbf{B}}{\partial t} = \nabla\times(\mathbf{V}\times\mathbf{B}) + \frac{1}{\mu_0\,\sigma}\,\nabla^2\mathbf{B},$$

Show that the modified dispersion relation for Alfvén waves can be obtained from the standard one by multiplying both ω^2 and V_S^2 by a factor

$$[1 + {\rm i}\,k^2/(\mu_0\,\sigma\,\omega)],$$

and ω^2 by an additional factor

$$[1 + {\rm i}\,\mu\,k^2/(\rho_0\,\omega)].$$

If the finite conductivity and viscous corrections are small (i.e., $\sigma \to \infty$ and $\mu \to 0$), show that, for parallel ($\theta = 0$) propagation, the dispersion relation for the shear-Alfvén wave reduces to

$$k \simeq \frac{\omega}{V_A} + {\rm i}\,\frac{\omega^2}{2\,V_A^3}\left(\frac{1}{\mu_0\,\sigma} + \frac{\mu}{\rho_0}\right).$$

7.2 Demonstrate that $V_+ > V_S\,\cos\theta$, and $V_- < V_S\,\cos\theta$, where V_+ and V_- are defined in Equation (7.45).

7.3 Demonstrate that Equation (7.65) can be rearranged to give

$$\frac{du^2}{dr}\left(1 - \frac{u_c^2}{u^2}\right) = \frac{4\,u_c^2}{r}\left(1 - \frac{r_c}{r}\right),$$

Show that this expression can be integrated to give

$$\left(\frac{u}{u_c}\right)^2 - \ln\left(\frac{u}{u_c}\right)^2 = 4\,\ln\left(\frac{r}{r_c}\right) + 4\,\frac{r_c}{r} + C,$$

where C is a constant.

Let $r/r_c = 1 + x$. Demonstrate that, in the limit $|x| \ll 1$, the previous expression yields either

$$u^2 = u_c^2\left[1 \pm 2\,x + O(x^2)\right]$$

or

$$u^2 = u_0^2 \left[1 + \frac{2\,u_c^2\,x^2}{u_0^2 - u_c^2} + O(x^3) \right],$$

where $u_0 \neq u_c$ is an arbitrary constant. Deduce that the former solution with the plus sign is such that u is a monotonically increasing function of r with $u \lesseqgtr u_c$ as $r \lesseqgtr r_c$ (this is a Class 2 solution); that the former solution with the minus sign is such that u is a monotonically decreasing function of r with $u \gtreqless u_c$ as $r \lesseqgtr r_c$ (this is a Class 3 solution); that the latter solution with $u_0 < u_c$ is such that $u < u_c$ for all r (this is a Class 1 solution); and that the latter solution with $u_0 > u_c$ is such that $u > u_c$ for all r (this is a Class 4 solution).

7.4 Derive expression (7.111) from Equations (7.107)–(7.110).

7.5 Consider a "two-dimensional" MHD fluid whose magnetic and velocity fields take the divergence-free forms

$$\mathbf{B} = \nabla\psi \times \mathbf{e}_z + B_z\,\mathbf{e}_z,$$

$$\mathbf{V} = \nabla\phi \times \mathbf{e}_z + V_z\,\mathbf{e}_z,$$

respectively, where $\psi = \psi(x, y)$ and $\phi = \phi(x, y)$. Here, (x, y, z) are standard Cartesian coordinates. Demonstrate from the MHD Ohm's law and Maxwell's equations that

$$\frac{d}{dt}\int \psi^2\,dx\,dy = -\frac{2\,\eta}{\mu_0}\int\!\!\int |\nabla\psi|^2\,dx\,dy,$$

where η is the (spatially uniform) plasma resistivity. Hence, deduce that a two-dimensional "poloidal" magnetic field, $\mathbf{B}_p = \nabla\psi \times \mathbf{e}_z$, cannot be maintained against ohmic dissipation by dynamo action.

Given that $\mathbf{B}_p = \mathbf{0}$, show that

$$\frac{d}{dt}\int B_z^2\,dx\,dy = -\frac{2\,\eta}{\mu_0}\int\!\!\int |\nabla B_z|^2\,dx\,dy.$$

Hence, deduce that a two-dimensional "axial" magnetic field, $\mathbf{B}_t = B_z\,\mathbf{e}_z$, cannot be maintained against ohmic dissipation by dynamo action.

7.6 Derive Equations (7.142) and (7.143) from Equations (7.139)–(7.141).

7.7 Derive Equations (7.149) and (7.150) from Equations (7.142)–(7.148).

7.8 Derive Equation (7.156) from Equations (7.151)–(7.155).

7.9 Derive Equation (7.161) from Equations (7.156)–(7.159).

7.10 Derive Equation (7.163) from Equation (7.161).

7.11 Derive Equations (7.177) and (7.178) from Equations (7.171)–(7.176).

7.12 Consider the linear tearing stability of the following field configuration,

$$F(\bar{x}) = \begin{cases} F'(0)\,\bar{x} & |\bar{x}| < 1 \\ F'(0)\,\mathrm{sgn}(\bar{x}) & |\bar{x}| \geq 1 \end{cases}.$$

This configuration is generated by a uniform, z-directed current sheet of thickness a, centered at $x = 0$. Solve the ideal-MHD equation, (7.186), subject to the constraints $\psi(-\bar{x}) = \psi(\bar{x})$, and $\psi(\bar{x}) \to 0$ as $|\bar{x}| \to \infty$. Here, $\bar{x} = x/a$. Hence, deduce that the tearing stability index for this configuration is

$$\Delta' = \frac{2\,\bar{k}}{\tanh(\bar{k})}\left[\frac{\bar{k} + \bar{k}\,\tanh(\bar{k}) - 1}{1 - \bar{k}/\tanh(\bar{k}) - \bar{k}}\right].$$

Show that

$$\Delta' \to \frac{2}{\bar{k}} - \frac{8}{3} + O(\bar{k})$$

as $\bar{k} \to 0$, and

$$\Delta' \to -2\,\bar{k} + 2\left[1 + \frac{1}{2\,\bar{k}} + O\left(\frac{1}{\bar{k}^2}\right)\right]\exp(-2\,\bar{k})$$

as $\bar{k} \to \infty$. Demonstrate that the field configuration is tearing unstable (i.e., $\Delta' > 0$) provided that $\bar{k} < \bar{k}_c$, where

$$\bar{k}_c\,[1 + \tanh(\bar{k}_c)] = 1.$$

Show that $\bar{k}_c = 0.639$.

7.13 We can incorporate plasma viscosity into the linearized resistive-MHD equations, (7.172)–(7.175), by modifying Equation (7.173) to read

$$\rho_0\,\frac{\partial \mathbf{V}}{\partial t} = -\nabla p + \frac{(\nabla \times \mathbf{B}) \times \mathbf{B}_0}{\mu_0} + \frac{(\nabla \times \mathbf{B}_0) \times \mathbf{B}}{\mu_0} + \mu\,\nabla^2 \mathbf{V},$$

where μ is the dynamic viscosity.

(a) Show that, in this case, Equations (7.177) and (7.178) generalize to give

$$\gamma\,B_x = \mathrm{i}\,k\,B_{0y}\,V_x + \frac{\eta}{\mu_0}\left(\frac{d^2}{dx^2} - k^2\right)B_x,$$

$$\gamma\,\rho_0\left(\frac{d^2}{dx^2} - k^2\right)V_x = \frac{\mathrm{i}\,k\,B_{0y}}{\mu_0}\left(\frac{d^2}{dx^2} - k^2 - \frac{B_{0y}''}{B_{0y}}\right)B_x + \mu\left(\frac{d^2}{dx^2} - k^2\right)^2 V_x,$$

respectively.

(b) Show that Equations (7.182) and (7.183) generalize to give

$$\bar{\gamma}\,(\psi - F\,\phi) = S^{-1}\left(\frac{d^2}{d\bar{x}^2} - \bar{k}^2\right)\psi,$$

$$\bar{\gamma}^2\left(\frac{d^2}{d\bar{x}^2} - \bar{k}^2\right)\phi = -\bar{k}^2\,F\left(\frac{d^2}{d\bar{x}^2} - \bar{k}^2 - \frac{F''}{F}\right)\psi + \bar{\gamma}\,S^{-1}\,P\left(\frac{d^2}{d\bar{x}^2} - \bar{k}^2\right)^2\phi,$$

where

$$P = \frac{\tau_R}{\tau_M}$$

is the *magnetic Prandtl number*, and

$$\tau_M = \frac{\rho_0\,a^2}{\mu}$$

is the viscous diffusion time.

(c) Show that the resistive layer equations, (7.187) and (7.188), generalize to give

$$\bar{\gamma}\,(\psi - \bar{x}\,\phi) = S^{-1}\,\frac{d^2\psi}{d\bar{x}^2},$$

$$\bar{\gamma}^2\,\frac{d^2\phi}{d\bar{x}^2} = -\bar{x}\,\frac{d^2\psi}{d\bar{x}^2} + \bar{\gamma}\,S^{-1}\,P\,\frac{d^4\phi}{d\bar{x}^4}.$$

(d) Show that the Fourier transformed resistive layer equation, (7.196), generalizes to give

$$\frac{d}{dt}\left(\frac{t^2}{Q+t^2}\,\frac{d\hat{\phi}}{dt}\right) - (Q\,t^2 + P\,t^4)\,\hat{\phi} = 0.$$

(e) Finally, solve the Fourier transformed resistive layer equation to determine the layer matching parameter, Δ. Demonstrate that if $1 \gg Q \gg P^{2/3}$ then

$$\Delta = 2\pi\,\frac{\Gamma(3/4)}{\Gamma(1/4)}\,S^{1/3}\,Q^{5/4},$$

whereas if $Q \ll P^{-1/3}, P^{2/3}$ then

$$\Delta = 6^{2/3}\,\pi\,\frac{\Gamma(5/6)}{\Gamma(1/6)}\,S^{1/3}\,Q\,P^{1/6}.$$

7.14 Consider the effect of plasma viscosity on the Sweet-Parker reconnection scenario. The viscosity is conveniently parameterized in terms of the magnetic Prandtl number

$$P = \frac{\mu_0\,\mu}{\eta\,\rho},$$

where μ is the dynamic viscosity. Demonstrate that if $P \ll 1$ then the conventional Sweet-Parker reconnection scenario remains valid, but that if $P \gg 1$ then the scenario is modified such that

$$\frac{v_*}{V_A} \sim \frac{1}{P^{1/2}},$$

$$\frac{\delta}{L} \sim \left(\frac{P}{S^2}\right)^{1/4},$$

$$M_0 \sim \frac{1}{(S^2\,P)^{1/4}}.$$

7.15 Derive Equations (7.239)–(7.246) from the MHD equations, (7.1)–(7.4), and Maxwell's equations.

7.16 Derive Equations (7.258)–(7.261) from the MHD Rankine-Hugoniot relations.

7.17 Demonstrate that for a parallel MHD shock the downstream Mach number has the following relation to the upstream Mach number:

$$M_2 = \left[\frac{2 + (\Gamma - 1)\, M_1^2}{2\,\Gamma\, M_1^2 - (\Gamma - 1)}\right]^{1/2}.$$

Hence, deduce that if $M_1 > 1$ then $M_2 < 1$.

7.18 Derive Equations (7.271)–(7.274) from the MHD Rankine-Hugoniot relations.

7.19 Demonstrate that Equation (7.274) is equivalent to

$$-\Gamma\,(\Gamma + 1)\beta_2\, M_2^2\, r^2 + \Gamma\left[2\,(1 + \beta_2) + (\Gamma - 1)\beta_2\, M_2^2\right] r + 2\,(2 - \Gamma)\} = 0.$$

Hence, deduce that if the second law of thermodynamics requires the positive root of this equation to be such that $r > 1$ then

$$M_2^2 < 1 + \frac{2}{\Gamma\,\beta_2}:$$

that is,

$$V_2 < V_{+\,2},$$

where $V_{+\,2} = (V_{S\,2}^2 + V_{A\,2}^2)^{1/2}$ is the downstream fast wave velocity.

7.20 Derive Equations (7.280)–(7.286) from the MHD Rankine-Hugoniot relations combined with Equations (7.278) and (7.279).

8

Waves in Warm Plasmas

8.1 Introduction

In this chapter, we shall investigate electromagnetic wave propagation through a warm collisionless plasma, extending the discussion presented in Chapter 5 to take thermal effects into account. It turns out that the thermal modifications to wave propagation are not particularly well described by fluid equations. We shall, therefore, adopt a kinetic description of the plasma. The appropriate governing kinetic equation is, of course, the Vlasov equation introduced in Section 4.1.

8.2 Landau Damping

Let us begin our study of the Vlasov equation by examining what appears, at first sight, to be a fairly simple and straightforward problem: namely, the propagation of small amplitude plasma waves through a uniform plasma possessing no equilibrium magnetic field. For the sake of simplicity, we shall only consider electron motion, assuming that the ions form an immobile, neutralizing background. The ions are also assumed to be singly charged. We shall search for electrostatic plasma waves of the type discussed in Section 5.7. Such waves are longitudinal in nature (i.e., $\mathbf{E} \propto \mathbf{k}$), and possess a perturbed electric field, but no perturbed magnetic field.

Our starting point is the Vlasov equation for an unmagnetized, collisionless plasma:

$$\frac{\partial f_e}{\partial t} + \mathbf{v}\cdot\nabla f_e - \frac{e}{m_e}\,\mathbf{E}\cdot\nabla_v f_e = 0, \tag{8.1}$$

where $f_e(\mathbf{r}, \mathbf{v}, t)$ is the ensemble-averaged electron distribution function. The electric field satisfies

$$\mathbf{E} = -\nabla\phi. \tag{8.2}$$

where

$$\nabla^2\phi = -\frac{e}{\epsilon_0}\left(n - \int f_e\, d^3\mathbf{v}\right). \tag{8.3}$$

Here, n is the number density of ions (which is the same as the equilibrium number density of electrons).

Because we are dealing with small amplitude waves, it is appropriate to linearize the Vlasov equation. Suppose that the electron distribution function is written

$$f_e(\mathbf{r}, \mathbf{v}, t) = f_0(\mathbf{v}) + f_1(\mathbf{r}, \mathbf{v}, t). \tag{8.4}$$

Here, f_0 represents the equilibrium electron distribution, whereas f_1 represents the small perturbation due to the wave. Of course, $\int f_0\, d^3\mathbf{v} = n$, otherwise the equilibrium state would not be quasi-neutral. The electric field is assumed to be zero in the unperturbed state, so that $\mathbf{E}(\mathbf{r}, t)$ can be regarded as a small quantity. Thus, linearization of Equations (8.1) and (8.3) yields

$$\frac{\partial f_1}{\partial t} + \mathbf{v}\cdot\nabla f_1 - \frac{e}{m_e}\,\mathbf{E}\cdot\nabla_v f_0 = 0, \tag{8.5}$$

and

$$\nabla^2\phi = \frac{e}{\epsilon_0}\int f_1\, d^3\mathbf{v}, \tag{8.6}$$

respectively.

Let us now follow the standard procedure for analyzing small amplitude waves, by assuming that all perturbed quantities vary with $\mathbf{r}$ and t like $\exp[\,{\rm i}\,(\mathbf{k}\cdot\mathbf{r} - \omega\,t)]$. Equations (8.5) and (8.6) reduce to

$$-{\rm i}\,(\omega - \mathbf{k}\cdot\mathbf{v})\,f_1 + {\rm i}\,\frac{e}{m_e}\,\phi\,\mathbf{k}\cdot\nabla_v f_0 = 0, \tag{8.7}$$

and

$$-k^2\,\phi = \frac{e}{\epsilon_0}\int f_1\, d^3\mathbf{v}, \tag{8.8}$$

respectively. Solving the first of these equations for f_1, and substituting into the integral in the second, we conclude that if ϕ is non-zero then we must have

$$1 + \frac{e^2}{\epsilon_0\, m_e\, k^2}\int \frac{\mathbf{k}\cdot\nabla_v f_0}{\omega - \mathbf{k}\cdot\mathbf{v}}\, d^3\mathbf{v} = 0. \tag{8.9}$$

We can interpret Equation (8.9) as the dispersion relation for electrostatic plasma waves, relating the wavevector, $\mathbf{k}$, to the frequency, ω. However, in doing so, we run up against a serious problem, because the integral has a singularity in velocity space, where $\omega = \mathbf{k}\cdot\mathbf{v}$, and is, therefore, not properly defined.

The way to resolve this problem was first explained by Landau in a very influential paper that was the foundation of much subsequent work on plasma oscillations and instabilities (Landau 1946). Landau showed that, instead of simply assuming that f_1 varies in time as $\exp(-{\rm i}\,\omega\,t)$, the problem must be regarded as an "initial value problem" in which f_1 is specified at $t = 0$, and calculated at later times. We may still Fourier analyze with respect to $\mathbf{r}$, so we write

$$f_1(\mathbf{r}, \mathbf{v}, t) = f_1(\mathbf{v}, t)\,\exp(\,{\rm i}\,\mathbf{k}\cdot\mathbf{r}\,). \tag{8.10}$$

It is helpful to define u as the velocity component along $\mathbf{k}$ (i.e., $u = \mathbf{k}\cdot\mathbf{v}/k$), and to also define $F_0(u)$ and $F_1(u,t)$ as the integrals of $f_0(\mathbf{v})$ and $f_1(\mathbf{v},t)$, respectively, over the velocity components perpendicular to $\mathbf{k}$. Thus, Equations (8.5) and (8.6) yield

$$\frac{\partial F_1}{\partial t} + \mathrm{i}\,k\,u\,F_1 - \frac{e}{m_e}\,E\,\frac{\partial F_0}{\partial u} = 0, \tag{8.11}$$

and

$$\mathrm{i}\,k\,E = -\frac{e}{\epsilon_0}\int_{-\infty}^{\infty} F_1(u)\,du, \tag{8.12}$$

respectively, where $\mathbf{E} = E\,\mathbf{k}/k$.

In order to solve Equations (8.11) and (8.12) as an initial value problem, we introduce the Laplace transform of F_1 with respect to t (Riley 1974):

$$\bar{F}_1(u,p) = \int_0^{\infty} F_1(u,t)\,\mathrm{e}^{-p\,t}\,dt. \tag{8.13}$$

If the rate of increase of F_1 with increasing t is no faster than exponential, then the integral on the right-hand side of the previous equation converges, and defines $\bar{F}_1$ as an analytic function of p, provided that the real part of p is sufficiently large.

Noting that the Laplace transform of $\partial F_1/\partial t$ is $p\,\bar{F}_1 - F_1(u, t=0)$ (as is easily shown by integration by parts), we can Laplace transform Equations (8.11) and (8.12) to obtain

$$p\,\bar{F}_1 + \mathrm{i}\,k\,u\,\bar{F}_1 = \frac{e}{m_e}\,\bar{E}\,\frac{\partial F_0}{\partial u} + F_1(u,t=0), \tag{8.14}$$

and

$$\mathrm{i}\,k\,\bar{E} = -\frac{e}{\epsilon_0}\int_{-\infty}^{\infty}\bar{F}_1(u)\,du, \tag{8.15}$$

respectively. The previous two equations can be combined to give

$$\mathrm{i}\,k\,\bar{E} = -\frac{e}{\epsilon_0}\int_{-\infty}^{\infty}\left[\frac{e}{m_e}\,\bar{E}\,\frac{\partial F_0/\partial u}{p+\mathrm{i}\,k\,u} + \frac{F_1(u,t=0)}{p+\mathrm{i}\,k\,u}\right]du, \tag{8.16}$$

yielding

$$\bar{E}(p) = -\frac{(e/\epsilon_0)}{\mathrm{i}\,k\,\epsilon(k,p)}\int_{-\infty}^{\infty}\frac{F_1(u,t=0)}{p+\mathrm{i}\,k\,u}\,du, \tag{8.17}$$

where

$$\epsilon(k,p) = 1 + \frac{e^2}{\epsilon_0\,m_e\,k}\int_{-\infty}^{\infty}\frac{\partial F_0/\partial u}{\mathrm{i}\,p - k\,u}\,du. \tag{8.18}$$

The function $\epsilon(k,p)$ is known as the *plasma dielectric function*. Of course, if p is replaced by $-\mathrm{i}\,\omega$ then the dielectric function becomes equivalent to the left-hand side of Equation (8.9). However, because p possesses a positive real part, the integral on the right-hand side of the previous equation is well defined.

According to Equations (8.14) and (8.17), the *Laplace transform* of the distribution function is written

$$\bar{F}_1 = \frac{e}{m_e}\,\bar{E}\,\frac{\partial F_0/\partial u}{p+\mathrm{i}\,k\,u} + \frac{F_1(u,t=0)}{p+\mathrm{i}\,k\,u}, \tag{8.19}$$

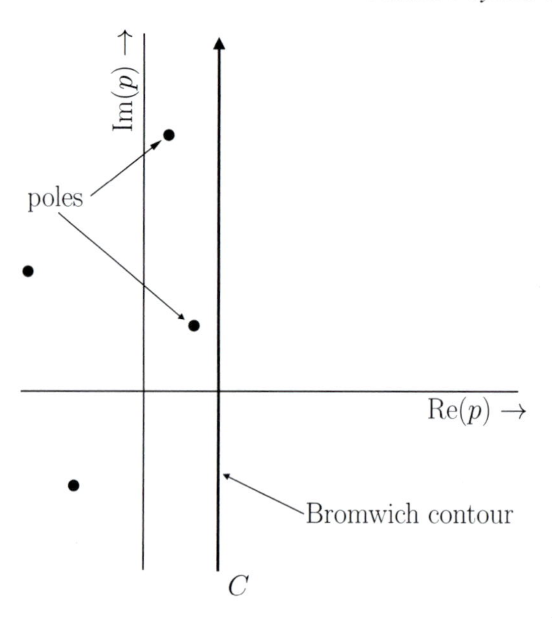

Figure 8.1
The Bromwich contour.

or

$$\bar{F}_1(u,p) = -\frac{e^2}{\epsilon_0\, m_e\, \mathrm{i}\, k}\,\frac{\partial F_0/\partial u}{\epsilon(k,p)\,(p+\mathrm{i}\,k\,u)}\int_{-\infty}^{\infty}\frac{F_1(u',t=0)}{p+\mathrm{i}\,k\,u'}\,du' + \frac{F_1(u,t=0)}{p+\mathrm{i}\,k\,u}. \quad (8.20)$$

Having found the Laplace transforms of the electric field and the perturbed distribution function, we must now invert them to obtain E and F_1 as functions of time. The *inverse Laplace transform* of the distribution function is given by (Riley 1974)

$$F_1(u,t) = \frac{1}{2\pi\,\mathrm{i}}\int_C \bar{F}_1(u,p)\,\mathrm{e}^{\,p\,t}\,dp, \quad (8.21)$$

where C—the so-called *Bromwich contour*—is a contour running parallel to the imaginary axis, and lying to the right of all singularities (otherwise known as poles) of $\bar{F}_1$ in the complex-p plane. (See Figure 8.1.) There is an analogous expression for the parallel electric field, $E(t)$.

Rather than trying to obtain a general expression for $F_1(u,t)$, from Equations (8.20) and (8.21), we shall concentrate on the behavior of the perturbed distribution function at large times. Looking at Figure 8.1, we note that if $\bar{F}_1(u,p)$ has only a finite number of simple poles in the region $\mathrm{Re}(p) > -\sigma$ (where σ is real and positive) then we may deform the contour as shown in Figure 8.2, with a loop around each of the singularities. A pole at p_0 gives a contribution that varies in time as $\mathrm{e}^{\,p_0\,t}$, whereas the vertical part of the contour gives a contribution that varies as $\mathrm{e}^{-\sigma\,t}$. For sufficiently large times, the latter contribution is negligible, and the behavior is dominated by contributions from the poles furthest to the right.

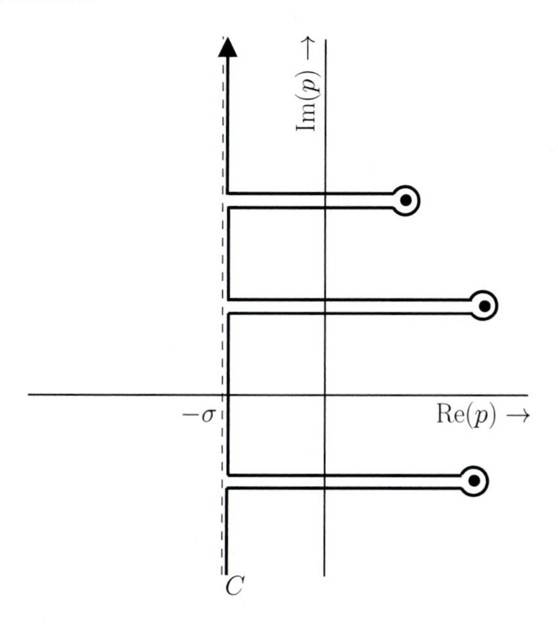

Figure 8.2
The distorted Bromwich contour.

Equations (8.17), (8.18), and (8.20) all involve integrals of the form

$$\int_{-\infty}^{\infty} \frac{G(u)}{u - \mathrm{i}\,p/k}\,du. \tag{8.22}$$

Such integrals become singular as p approaches the imaginary axis. In order to distort the contour C, in the manner shown in Figure 8.2, we need to continue these integrals smoothly across the imaginary p-axis. As a consequence of the way in which the Laplace transform was originally defined—that is, for Re(p) sufficiently large—the appropriate way to do this is to take the values of these integrals when p lies in the right-hand half-plane, and to then find the analytic continuation into the left-hand half-plane (Flanigan 2010).

If $G(u)$ is sufficiently well-behaved that it can be continued off the real axis as an analytic function of a complex variable u, then the continuation of (8.22) as the singularity crosses the real axis in the complex u-plane, from the upper to the lower half-plane, is obtained by letting the singularity take the contour with it, as shown in Figure 8.3 (Cairns 1985).

Note that the ability to deform the Bromwich contour into that of Figure 8.3, and so to find a dominant contribution to $E(t)$ and $F_1(u,t)$ from a few poles, depends on $F_0(u)$ and $F_1(u,t=0)$ having smooth enough velocity dependences that the integrals appearing in Equations (8.17), (8.18), and (8.20) can be analytically continued sufficiently far into the lower half of the complex u-plane (Cairns 1985).

If we consider the electric field given by the inversion of Equation (8.17), then

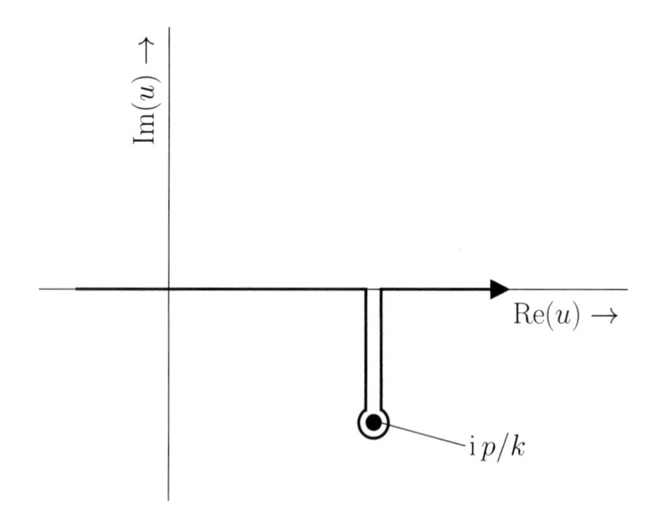

Figure 8.3
The Bromwich contour for Landau damping.

we see that its behavior at large times is dominated by the zero of $\epsilon(k, p)$ that lies furthest to the right in the complex p-plane. According to Equations (8.20) and (8.21), $F_1(u, t)$ has a similar contribution, as well as a contribution that varies in time as $e^{-i k u t}$. Thus, for sufficiently long times after the initial excitation of the wave, the electric field depends only on the positions of the roots of $\epsilon(k, p) = 0$ in the complex p-plane. The distribution function, on the other hand, has corresponding components from these roots, as well as a component that varies in time as $e^{-i k u t}$. At large times, the latter component of the distribution function is a rapidly oscillating function of velocity, and its contribution to the charge density, obtained by integrating over u, is negligible.

As we have already noted, the function $\epsilon(k, p)$ is equivalent to the left-hand side of Equation (8.9), provided that p is replaced by $-i\,\omega$. Thus, the dispersion relation, (8.9), obtained via Fourier transformation of the Vlasov equation, gives the correct behavior at large times, as long as the singular integral is treated correctly. Adapting the procedure that we discovered using the complex variable p, we see that the integral is defined as it is written for $\text{Im}(\omega) > 0$, and analytically continued, by deforming the contour of integration in the u-plane (as shown in Figure 8.3), into the region $\text{Im}(\omega) < 0$. The simplest way to remember how to do the analytic continuation is to observe that the integral is continued from the part of the ω-plane corresponding to growing perturbations to that corresponding to damped perturbations. Once we know this rule, we can obtain kinetic dispersion relations in a fairly direct manner, via Fourier transformation of the Vlasov equation, and there is no need to attempt the more complicated Laplace transform solution.

In Chapter 5, where we investigated the cold-plasma dispersion relation, we found that for any given k there were a finite number of values of ω, say ω_1, ω_2,

$\cdots$, and a general solution was a linear superposition of functions varying in time as $e^{-i\omega_1 t}$, $e^{-i\omega_2 t}$, et cetera. The set of values of ω corresponding to a given value of k is called the *spectrum* of the wave. It is clear that the cold-plasma equations yield a discrete wave spectrum. On the other hand, in the kinetic problem, we obtain contributions to the distribution function that vary in time as $e^{-i k u t}$, with u taking any real value. In other words, the kinetic equation yields a continuous wave spectrum. All of the mathematical difficulties of the kinetic problem arise from the existence of this continuous spectrum (Cairns 1985). At short times, the behavior is very complicated, and depends on the details of the initial perturbation. It is only asymptotically that a mode varying in time as $e^{-i\omega t}$ is obtained, with ω determined by a dispersion relation that is solely a function of the unperturbed state. As we have seen, the emergence of such a mode depends on the initial velocity disturbance being sufficiently smooth.

Suppose, for the sake of simplicity, that the background plasma state is a Maxwellian distribution. Working in terms of ω, rather than p, the kinetic dispersion relation for electrostatic waves takes the form

$$\epsilon(k,\omega) = 1 + \frac{e^2}{\epsilon_0\, m_e\, k}\int_{-\infty}^{\infty}\frac{\partial F_0/\partial u}{\omega - k\,u}\,du = 0, \tag{8.23}$$

where

$$F_0(u) = \frac{n}{(2\pi\,T_e/m_e)^{1/2}}\,\exp\left(-\frac{m_e\,u^2}{2\,T_e}\right). \tag{8.24}$$

Suppose that, to a first approximation, ω is real. Letting ω tend to the real axis from the domain $\mathrm{Im}(\omega) > 0$, we obtain

$$\int_{-\infty}^{\infty}\frac{\partial F_0/\partial u}{\omega - k\,u}\,du = P\int_{-\infty}^{\infty}\frac{\partial F_0/\partial u}{\omega - k\,u}\,du - \frac{i\,\pi}{k}\left(\frac{\partial F_0}{\partial u}\right)_{u=\omega/k}, \tag{8.25}$$

where P denotes the *Cauchy principal part* of the integral (Flanigan 2010). The origin of the two terms on the right-hand side of the previous equation is illustrated in Figure 8.4. The first term—the principal part—is obtained by removing an interval of length $2\,\epsilon$, symmetrical about the pole, $u = \omega/k$, from the range of integration, and then letting $\epsilon \rightarrow 0$. The second term comes from the small semi-circle linking the two halves of the principal part integral. Note that the semi-circle deviates below the real u-axis, rather than above, because the integral is calculated by letting the pole approach the axis from the upper half-plane in u-space.

Incidentally, because Equation (8.25) holds for any well-behaved distribution function, it follows that

$$\frac{1}{\omega - k\,u} = P\,\frac{1}{\omega - k\,u} - i\,\pi\,\delta(\omega - k\,u). \tag{8.26}$$

This famous expression is known as the *Plemelj formula* (Plemelj 1908).

Suppose that k is sufficiently small that $\omega \gg k\,u$ over the range of u where $\partial F_0/\partial u$ is non-negligible. It follows that we can expand the denominator of the principal part integral in a Taylor series:

$$\frac{1}{\omega - k\,u} \simeq \frac{1}{\omega}\left(1 + \frac{k\,u}{\omega} + \frac{k^2\,u^2}{\omega^2} + \frac{k^3\,u^3}{\omega^3} + \cdots\right). \tag{8.27}$$

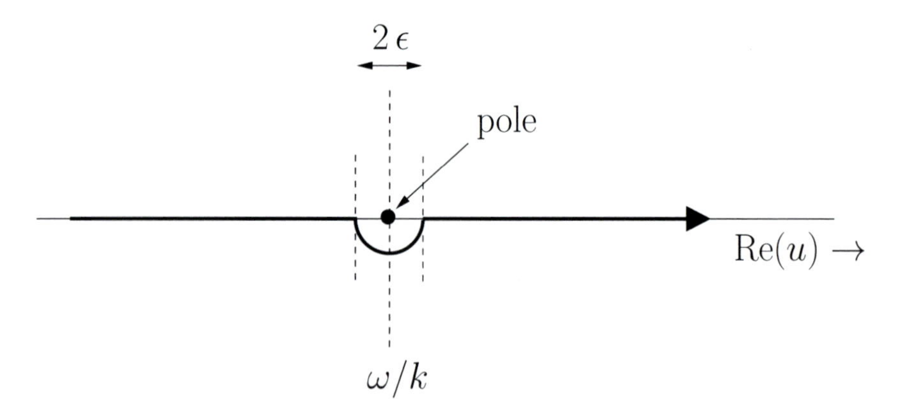

Figure 8.4
Integration path about a pole.

Integrating the result term by term, and remembering that $\partial F_0/\partial u$ is an odd function, Equation (8.23) reduces to

$$1 - \frac{\Pi_e^2}{\omega^2} - 3\,k^2\,\frac{T_e\,\Pi_e^2}{m_e\,\omega^4} - \frac{e^2}{\epsilon_0\,m_e}\,\frac{\mathrm{i}\,\pi}{k^2}\left(\frac{\partial F_0}{\partial u}\right)_{u=\omega/k} \simeq 0, \tag{8.28}$$

where $\Pi_e = (n\,e^2/\epsilon_0\,m_e)^{1/2}$ is the electron plasma frequency. Equating the real part of the previous expression to zero yields

$$\omega^2 \simeq \Pi_e^2\left[1 + 3\,(k\,\lambda_D)^2\right], \tag{8.29}$$

where $\lambda_D = (T_e/m_e\,\Pi_e^2)^{1/2}$ is the Debye length, and it is assumed that $k\,\lambda_D \ll 1$. We can regard the imaginary part of ω as a small perturbation, and write $\omega = \omega_0 + \delta\omega$, where ω_0 is the root of Equation (8.29). It follows that

$$2\,\omega_0\,\delta\omega \simeq \omega_0^2\,\frac{e^2}{\epsilon_0\,m_e}\,\frac{\mathrm{i}\,\pi}{k^2}\left(\frac{\partial F_0}{\partial u}\right)_{u=\omega/k}, \tag{8.30}$$

and so

$$\delta\omega \simeq \frac{\mathrm{i}\,\pi}{2}\,\frac{e^2\,\omega_0}{\epsilon_0\,m_e\,k^2}\left(\frac{\partial F_0}{\partial u}\right)_{u=\omega/k}, \tag{8.31}$$

giving

$$\delta\omega \simeq -\mathrm{i}\,\sqrt{\frac{\pi}{8}}\,\frac{\Pi_e}{(k\,\lambda_D)^3}\,\exp\left[-\frac{1}{2\,(k\,\lambda_D)^2}\right]. \tag{8.32}$$

If we compare the previous results with those for a cold plasma, where the dispersion relation for an electrostatic plasma wave was found to be simply $\omega^2 = \Pi_e^2$ (see Section 5.7), we see, first, that ω now depends on k, according to Equation (8.29), so that, in a warm plasma, the electrostatic plasma wave is a propagating mode, with a non-zero group-velocity. Such a mode is known as a *Langmuir wave*. Second, we now have an imaginary part to ω, given by Equation (8.32), corresponding, because it is negative, to the damping of the wave in time. This damping is generally known as *Landau damping*. If $k\,\lambda_D \ll 1$ (i.e., if the wavelength is much larger than the Debye length) then the imaginary part of ω is small compared to the real part, and the wave is only lightly damped. However, as the wavelength becomes comparable to the Debye length, the imaginary part of ω becomes comparable to the real part, and the damping becomes strong. Admittedly, the approximate solution given previously is not very accurate in the short wavelength case, but it is nevertheless sufficient to indicate the existence of very strong damping.

There are no dissipative effects explicitly included in the collisionless Vlasov equation. Thus, it can easily be verified that if the particle velocities are reversed at any time then the solution up to that point is simply reversed in time. At first sight, this reversible behavior does not seem to be consistent with the fact that an initial perturbation dies out. However, we should note that it is only the electric field that decays in time. The distribution function contains an undamped term varying in time as $e^{-i\,k\,u\,t}$. Furthermore, the decay of the electric field depends on there being a sufficiently smooth initial perturbation in velocity space. The presence of the $e^{-i\,k\,u\,t}$ term means that, as time advances, the velocity space dependence of the perturbation becomes more and more convoluted. It follows that if we reverse the velocities after some time then we are not starting with a smooth distribution. Under these circumstances, there is no contradiction in the fact that, under time reversal, the electric field grows initially, until the smooth initial state is recreated, and subsequently decays away (Cairns 1985).

Landau damping was first observed experimentally in the 1960s (Malmberg and Wharton 1964; Malmberg and Wharton 1966; Derfler and Simonen 1966).

8.3 Physics of Landau Damping

We have explained Landau damping in terms of mathematics. Let us now consider the physical explanation for this effect (Cairns 1985). The motion of a charged particle situated in a one-dimensional electric field varying as $E_0\,\exp[\,i\,(k\,x-\omega\,t)]$ is determined by

$$\frac{d^2x}{dt^2}=\frac{e}{m}\,E_0\,e^{\,i\,(k\,x-\omega\,t)}. \tag{8.33}$$

Because we are dealing with a linearized theory in which the perturbation due to the wave is small, it follows that if the particle starts with velocity u_0 at position x_0 then we may substitute $x_0+u_0\,t$ for x in the electric field term. This is actually the position of the particle on its unperturbed trajectory, starting at $x=x_0$ at $t=0$. Thus, we obtain

$$\frac{du}{dt}=\frac{e}{m}\,E_0\,e^{\,i\,(k\,x_0+k\,u_0\,t-\omega\,t)}, \tag{8.34}$$

which yields

$$u - u_0 = \frac{e}{m}\,E_0\left[\frac{\mathrm{e}^{\,\mathrm{i}\,(k\,x_0 + k\,u_0\,t - \omega\,t)} - \mathrm{e}^{\,\mathrm{i}\,k\,x_0}}{\mathrm{i}\,(k\,u_0 - \omega)}\right]. \tag{8.35}$$

As $k\,u_0 - \omega \rightarrow 0$, the previous expression reduces to

$$u - u_0 = \frac{e}{m}\,E_0\,t\,\mathrm{e}^{\,\mathrm{i}\,k\,x_0}, \tag{8.36}$$

showing that particles with u_0 close to ω/k—that is, with velocity components along the x-axis close to the phase-velocity of the wave—have velocity perturbations that grow in time. These so-called *resonant particles* gain energy from, or lose energy to, the wave, and are responsible for the damping. This explains why the damping rate, given by Equation (8.31), depends on the slope of the distribution function calculated at $u = \omega/k$. The remainder of the particles are non-resonant, and have an oscillatory response to the wave field.

To understand why energy should be transferred from the electric field to the resonant particles requires more detailed consideration (Cairns 1985). Whether the speed of a resonant particle increases or decreases depends on the phase of the wave at its initial position, and it is not the case that all particles moving slightly faster than the wave lose energy, while all particles moving slightly slower than the wave gain energy. Furthermore, the density perturbation oscillates out of phase with the wave electric field, so there is no initial wave-generated excess of particles gaining or losing energy. However, if we consider those particles that start off with velocities slightly above the phase-velocity of the wave, then if they gain energy they move away from the resonant velocity whereas if they lose energy they approach the resonant velocity. The result is that the particles which lose energy interact more effectively with the wave, and, on average, there is a transfer of energy from these particles to the electric field. Exactly the opposite is true for particles with initial velocities lying just below the phase-velocity of the wave. In the case of a Maxwellian distribution, there are more particles in the latter class than in the former, so there is a net transfer of energy from the electric field to the particles: that is, the electric field is damped. In the limit that the wave amplitude tends to zero, it is clear that the damping rate is determined by velocity gradient of the distribution function at the wave speed.

The previous argument fails if the magnitude of the initial electric field becomes too large, because nonlinear effects become important (Cairns 1985). The basic requirement for the validity of the linear result is that a resonant particle should maintain its position relative to the phase of the electric field over a sufficiently long time period for the damping to take place. To determine the condition that this be the case, let us consider the problem in the frame of reference in which the wave is at rest, and the potential $-e\,\phi$ seen by an electron is as sketched in Figure 8.5.

If the electron starts at rest (i.e., in resonance with the wave) at x_0, then it begins to move towards the potential minimum, as shown. The time for the electron to shift its position relative to the wave may be estimated as the period with which it bounces back and forth in the potential well. Near the bottom of the well, the equation of

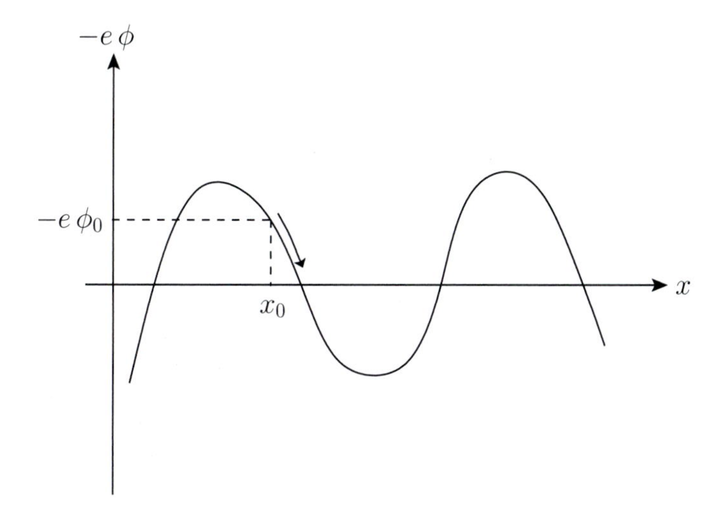

Figure 8.5
Wave-particle interaction.

motion of the electron is written

$$\frac{d^2 x}{dt^2} = -\frac{e}{m_e}\,k^2\,x\,\phi_0, \tag{8.37}$$

where k is the wavenumber, and so the bounce time is

$$\tau_b \sim 2\pi\sqrt{\frac{m_e}{e\,k^2\,\phi_0}} = 2\pi\sqrt{\frac{m_e}{e\,k\,E_0}}, \tag{8.38}$$

where E_0 is the amplitude of the electric field. We may expect the wave to damp according to linear theory if the bounce time, τ_b, is much greater than the damping time. Because the former time varies inversely with the square root of the electric field amplitude, whereas the latter is amplitude independent, this criterion gives us an estimate of the maximum allowable initial electric field amplitude that is consistent with linear damping (Cairns 1985).

If the initial amplitude is large enough for the resonant electrons to bounce back and forth in the potential well a number of times before the wave is damped, then it can be demonstrated that the result to be expected is a non-monotonic decrease in the amplitude of the electric field, as shown in Figure 8.6 (O'Neil 1965; Armstrong 1967). The period of the amplitude oscillations is similar to the bounce time, τ_b.

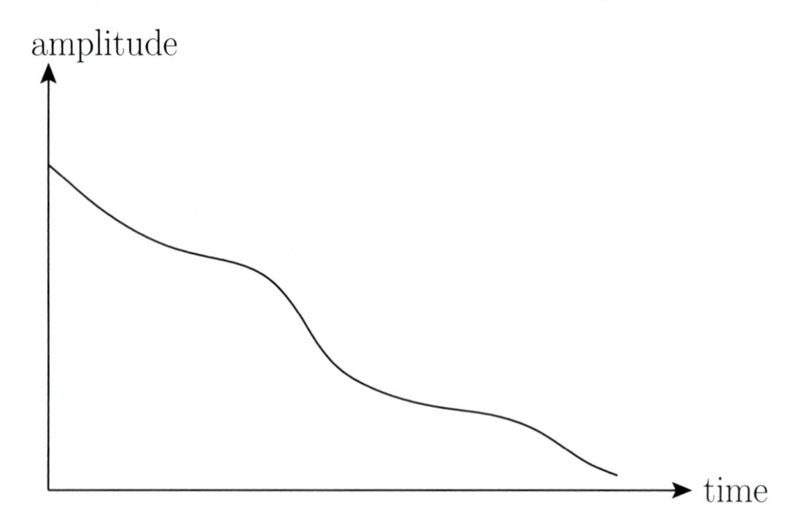

Figure 8.6
Nonlinear Landau damping.

8.4 Plasma Dispersion Function

If the unperturbed distribution function, F_0, appearing in Equation (8.23), is a Maxwellian, then it is readily seen that, with a suitable scaling of the variables, the dispersion relation for electrostatic plasma waves can be expressed in terms of the function

$$Z(\zeta) = \pi^{-1/2} \int_{-\infty}^{\infty} \frac{e^{-t^2}}{t-\zeta}\,dt, \tag{8.39}$$

which is defined as it is written for $\mathrm{Im}(\zeta) > 0$, and is analytically continued for $\mathrm{Im}(\zeta) \leq 0$. This function is known as the *plasma dispersion function*, and very often crops up in problems involving small-amplitude waves propagating through warm plasmas. Incidentally, $Z(\zeta)$ is the Hilbert transform of a Gaussian function.

In view of the importance of the plasma dispersion function, and its regular appearance in the literature of plasma physics, it is convenient to briefly examine its main properties. We, first of all, note that if we differentiate $Z(\zeta)$ with respect to ζ then we obtain

$$Z'(\zeta) = \pi^{-1/2} \int_{-\infty}^{\infty} \frac{e^{-t^2}}{(t-\zeta)^2}\,dt, \tag{8.40}$$

which yields, on integration by parts,

$$Z'(\zeta) = -\pi^{-1/2} \int_{-\infty}^{\infty} \frac{2\,t}{t-\zeta}\,e^{-t^2}\,dt = -2\,(1+\zeta\,Z). \tag{8.41}$$

If we let ζ tend to zero from the upper half of the complex plane, then we get

$$Z(0) = \pi^{-1/2}\,\mathrm{P}\int_{-\infty}^{\infty}\frac{\mathrm{e}^{-t^2}}{t}\,dt + \mathrm{i}\,\pi^{1/2} = \mathrm{i}\,\pi^{1/2}. \tag{8.42}$$

Of course, the principal part integral is zero because its integrand is an odd function of t.

Integrating the linear differential equation (8.41), which possesses an integrating factor $\exp(\zeta^2)$, and using the boundary condition (8.42), we obtain an alternative expression for the plasma dispersion function:

$$Z(\zeta) = \mathrm{e}^{-\zeta^2}\left(\mathrm{i}\,\pi^{1/2} - 2\int_0^{\zeta}\mathrm{e}^{x^2}\,dx\right). \tag{8.43}$$

Making the substitution $t = \mathrm{i}\,x$ in the integral, and noting that

$$\int_{-\infty}^{0}\mathrm{e}^{-t^2}\,dt = \frac{\pi^{1/2}}{2}, \tag{8.44}$$

we finally arrive at the expression

$$Z(\zeta) = 2\,\mathrm{i}\,\mathrm{e}^{-\zeta^2}\int_{-\infty}^{\mathrm{i}\,\zeta}\mathrm{e}^{-t^2}\,dt = \mathrm{i}\,\pi^{1/2}\,\mathrm{e}^{-\zeta^2}\,[1 + \mathrm{erf}(\mathrm{i}\,\zeta)]. \tag{8.45}$$

This formula, which relates the plasma dispersion function to an error function of imaginary argument (Abramowitz and Stegun 1965b), is valid for all values of ζ.

For small ζ, we have the expansion (Huba 2000c)

$$Z(\zeta) = \mathrm{i}\,\pi^{1/2}\,\mathrm{e}^{-\zeta^2} - 2\,\zeta\left[1 - \frac{2\,\zeta^2}{3} + \frac{4\,\zeta^4}{15} - \frac{8\,\zeta^6}{105} + O(\zeta^8)\right]. \tag{8.46}$$

For large ζ, where $\zeta = x + \mathrm{i}\,y$, the asymptotic expansion for $x > 0$ is written (Huba 2000c)

$$Z(\zeta) = \mathrm{i}\,\pi^{1/2}\,\sigma\,\mathrm{e}^{-\zeta^2} - \zeta^{-1}\left[1 + \frac{1}{2\,\zeta^2} + \frac{3}{4\,\zeta^4} + \frac{15}{8\,\zeta^6} + O(\zeta^{-8})\right]. \tag{8.47}$$

Here,

$$\sigma = \begin{cases} 0 & y > 1/|x| \\ 1 & |y| < 1/|x| \\ 2 & y < -1/|x| \end{cases}. \tag{8.48}$$

In deriving our previous expression (8.32) for the Landau damping rate, we, in effect, used the first few terms of the asymptotic expansion (8.47).

The properties of the plasma dispersion function are specified in exhaustive detail in a well-known book by Fried and Conte (Fried and Conte 1961).

8.5 Ion Acoustic Waves

If we now take ion dynamics into account then the dispersion relation (8.23), for electrostatic plasma waves, generalizes to

$$\epsilon(k,\omega) = 1 + \frac{e^2}{\epsilon_0\, m_e\, k}\int_{-\infty}^{\infty}\frac{\partial F_{0\,e}/\partial u}{\omega - k\,u}\,du + \frac{e^2}{\epsilon_0\, m_i\, k}\int_{-\infty}^{\infty}\frac{\partial F_{0\,i}/\partial u}{\omega - k\,u}\,du = 0: \tag{8.49}$$

that is, we simply add an extra term for the ions that has an analogous form to the electron term. Let us search for a wave with a phase-velocity, ω/k, that is much less than the electron thermal velocity, but much greater than the ion thermal velocity. We may assume that $\omega \gg k\,u$ for the ion term, as we did previously for the electron term. It follows that, to lowest order, this term reduces to $-\Pi_i^{\,2}/\omega^2$, where $\Pi_i = (n\,e^2/\epsilon_0\,m_i)^{1/2}$. Conversely, we may assume that $\omega \ll k\,u$ for the electron term. Thus, to lowest order, we may neglect ω in the velocity space integral. Assuming $F_{0\,e}$ to be a Maxwellian with temperature T_e, the electron term reduces to

$$\frac{\Pi_e^{\,2}}{k^2}\frac{m_e}{T_e} = \frac{1}{(k\,\lambda_D)^2}, \tag{8.50}$$

where $\Pi_e = (n\,e^2/\epsilon_0\,m_e)^{1/2}$, and $\lambda_D = (T_e/m_e\,\Pi_e^{\,2})^{1/2}$.

Thus, to a first approximation, the dispersion relation (8.49) can be written

$$1 + \frac{1}{(k\,\lambda_D)^2} - \frac{\Pi_i^{\,2}}{\omega^2} = 0, \tag{8.51}$$

giving

$$\omega^2 = \Pi_i^{\,2}\,\frac{(k\,\lambda_D)^2}{1 + (k\,\lambda_D)^2} = \frac{T_e}{m_i}\,\frac{k^2}{1 + (k\,\lambda_D)^2}. \tag{8.52}$$

For $k\,\lambda_D \ll 1$, we have $\omega = (T_e/m_i)^{1/2}\,k$, a dispersion relation which is like that of an ordinary sound wave, with the pressure provided by the electrons, and the inertia by the ions. However, as the wavelength is reduced towards the Debye length, the frequency levels off and approaches the ion plasma frequency.

Let us check our original assumptions. In the long wavelength limit, we see that the wave phase-velocity, $(T_e/m_i)^{1/2}$, is indeed much less than the electron thermal velocity [by a factor $(m_e/m_i)^{1/2}$], but that it is only much greater than the ion thermal velocity if the ion temperature, T_i, is much less than the electron temperature, T_e. In fact, if $T_i \ll T_e$ then the wave phase-velocity can simultaneously lie on almost flat portions of the ion and electron distribution functions, as shown in Figure 8.7, implying that the wave is subject to very little Landau damping. Indeed, T_e must generally be at least five to ten times greater than T_i before an ion acoustic wave can propagate a distance of a few wavelengths without being strongly damped (Cairns 1985).

Of course, it is possible to obtain the ion acoustic wave dispersion relation, $\omega^2/k^2 = T_e/m_i$, using fluid theory. The kinetic theory used here is an improvement

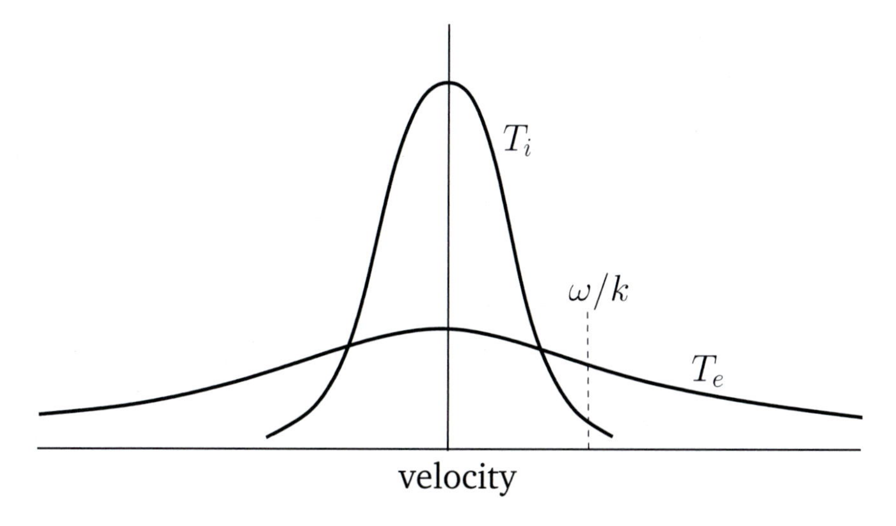

Figure 8.7
Ion and electron distribution functions with $T_i \ll T_e$.

on the fluid theory to the extent that no equation of state is assumed, and also that the former theory makes it clear to us that ion acoustic waves are subject to strong Landau damping (i.e., they cannot be considered normal modes of the plasma) unless $T_e \gg T_i$.

8.6 Waves in Magnetized Plasmas

Consider small amplitude waves propagating through a plasma placed in a uniform magnetic field, $\mathbf{B}_0 \equiv B_0\,\mathbf{e}_z$. Let us take the perturbed magnetic field into account in our calculations, in order to allow for electromagnetic, as well as electrostatic, waves. The linearized Vlasov equation takes the form

$$\frac{\partial f_1}{\partial t} + \mathbf{v}\cdot\nabla f_1 + \frac{e}{m}\,(\mathbf{v}\times\mathbf{B}_0)\cdot\nabla_v f_1 = -\frac{e}{m}\,(\mathbf{E}+\mathbf{v}\times\mathbf{B})\cdot\nabla_v f_0 \tag{8.53}$$

for both ions and electrons, where $\mathbf{E}$ and $\mathbf{B}$ are the perturbed electric and magnetic fields, respectively. Likewise, f_1 is the perturbed distribution function, and f_0 the equilibrium distribution function.

In order to have an equilibrium state at all, we require that

$$(\mathbf{v}\times\mathbf{B}_0)\cdot\nabla_v f_0 = 0. \tag{8.54}$$

Writing the velocity, $\mathbf{v}$, in cylindrical polar coordinates, $(v_\perp,\,\theta,\,v_\parallel)$, aligned with the equilibrium magnetic field, the previous expression can easily be shown to imply that $\partial f_0/\partial\theta = 0$: that is, f_0 is a function only of $v_\perp$ and $v_\parallel$.

Let the trajectory of a particle be $\mathbf{r}(t)$, $\mathbf{v}(t)$. In the unperturbed state,

$$\frac{d\mathbf{r}}{dt} = \mathbf{v}, \tag{8.55}$$

$$\frac{d\mathbf{v}}{dt} = \frac{e}{m}\,(\mathbf{v}\times\mathbf{B}_0). \tag{8.56}$$

It follows that Equation (8.53) can be written

$$\frac{Df_1}{Dt} = -\frac{e}{m}\,(\mathbf{E}+\mathbf{v}\times\mathbf{B})\cdot\nabla_v f_0, \tag{8.57}$$

where Df_1/Dt is the total rate of change of f_1, following the unperturbed trajectories. Under the assumption that f_1 vanishes as $t\rightarrow-\infty$, the solution to Equation (8.57) can be written

$$f_1(\mathbf{r},\mathbf{v},t) = -\frac{e}{m}\int_{-\infty}^{t}[\mathbf{E}(\mathbf{r}',t')+\mathbf{v}'\times\mathbf{B}(\mathbf{r}',t')]\cdot\nabla_v f_0(\mathbf{v}')\,dt', \tag{8.58}$$

where ($\mathbf{r}'$, $\mathbf{v}'$) is the unperturbed trajectory that passes through the point ($\mathbf{r}$, $\mathbf{v}$) when $t'=t$.

It should be noted that the previous method of solution is valid for any set of equilibrium electromagnetic fields, not just a uniform magnetic field. However, in a uniform magnetic field, the unperturbed trajectories are merely helices, whereas in a general field configuration it is difficult to find a closed form for the particle trajectories that is sufficiently simple to allow further progress to be made.

Let us write the velocity in terms of its Cartesian components:

$$\mathbf{v} = (v_\perp\cos\theta,\ v_\perp\sin\theta,\ v_\parallel). \tag{8.59}$$

It follows that

$$\mathbf{v}' = (v_\perp\cos[\Omega(t-t')+\theta],\ v_\perp\sin[\Omega(t-t')+\theta],\ v_\parallel), \tag{8.60}$$

where $\Omega = e\,B_0/m$ is the gyofrequency. The previous expression can be integrated in time to give

$$x'-x = -\frac{v_\perp}{\Omega}\,(\sin[\Omega(t-t')+\theta]-\sin\theta), \tag{8.61}$$

$$y'-y = \frac{v_\perp}{\Omega}\,(\cos[\Omega(t-t')+\theta]-\cos\theta), \tag{8.62}$$

$$z'-z = v_\parallel\,(t'-t). \tag{8.63}$$

Note that both $v_\perp$ and $v_\parallel$ are constants of the motion. This implies that $f_0(\mathbf{v}')=f_0(\mathbf{v})$, because f_0 is only a function of $v_\perp$ and $v_\parallel$. Given that $v_\perp = (v_x'^2+v_y'^2)^{1/2}$, we can write

$$\frac{\partial f_0}{\partial v_x'} = \frac{\partial v_\perp}{\partial v_x'}\frac{\partial f_0}{\partial v_\perp} = \frac{v_x'}{v_\perp}\frac{\partial f_0}{\partial v_\perp} = \cos[\Omega(t'-t)+\theta]\,\frac{\partial f_0}{\partial v_\perp}, \tag{8.64}$$

$$\frac{\partial f_0}{\partial v_y'} = \frac{\partial v_\perp}{\partial v_y'}\frac{\partial f_0}{\partial v_\perp} = \frac{v_y'}{v_\perp}\frac{\partial f_0}{\partial v_\perp} = \sin[\Omega(t'-t)+\theta]\,\frac{\partial f_0}{\partial v_\perp}, \tag{8.65}$$

$$\frac{\partial f_0}{\partial v_z'} = \frac{\partial f_0}{\partial v_\parallel}. \tag{8.66}$$

Let us assume an $\exp[\,{\rm i}\,(\mathbf{k}\cdot\mathbf{r}-\omega\,t)]$ dependence of all perturbed quantities, with $\mathbf{k}$ lying in the x-z plane. Equation (8.58) yields

$$f_1 = -\frac{e}{m}\int_{-\infty}^{t}\left[(E_x + v_y'\,B_z - v_z'\,B_y)\,\frac{\partial f_0}{\partial v_x'} + (E_y + v_z'\,B_x - v_x'\,B_z)\,\frac{\partial f_0}{\partial v_y'}\right.$$
$$\left.+(E_z + v_x'\,B_y - v_y'\,B_x)\,\frac{\partial f_0}{\partial v_z'}\right]\exp\left[\,{\rm i}\,\{\mathbf{k}\cdot(\mathbf{r}'-\mathbf{r}) - \omega\,(t'-t)\}\right]dt'. \tag{8.67}$$

Making use of Equations (8.60)–(8.66), as well as the identity (Abramowitz and Stegun 1965c)

$${\rm e}^{\,{\rm i}\,a\,\sin x} \equiv \sum_{n=-\infty,\infty} J_n(a)\,{\rm e}^{\,{\rm i}\,n\,x}, \tag{8.68}$$

where the J_n are Bessel functions (Abramowitz and Stegun 1965c), Equation (8.67) gives

$$f_1 = -\frac{e}{m}\int_{-\infty}^{t}\left[(E_x - v_\parallel\,B_y)\,\cos\chi\,\frac{\partial f_0}{\partial v_\perp} + (E_y + v_\parallel\,B_x)\,\sin\chi\,\frac{\partial f_0}{\partial v_\perp}\right.$$
$$\left.+(E_z + v_\perp\,B_y\,\cos\chi - v_\perp\,B_x\,\sin\chi)\,\frac{\partial f_0}{\partial v_\parallel}\right]\sum_{n,m=-\infty,\infty} J_n\!\left(\frac{k_\perp\,v_\perp}{\Omega}\right)J_m\!\left(\frac{k_\perp\,v_\perp}{\Omega}\right)$$
$$\exp\left\{\,{\rm i}\left[(n\,\Omega + k_\parallel\,v_\parallel - \omega)\,(t'-t) + (m-n)\,\theta\,\right]\right\}dt', \tag{8.69}$$

where

$$\chi = \Omega\,(t-t') + \theta. \tag{8.70}$$

Maxwell's equations yield

$$\mathbf{k}\times\mathbf{E} = \omega\,\mathbf{B}, \tag{8.71}$$

$$\mathbf{k}\times\mathbf{B} = -{\rm i}\,\mu_0\,\mathbf{j} - \frac{\omega}{c^2}\,\mathbf{E} = -\frac{\omega}{c^2}\,\mathbf{K}\cdot\mathbf{E}, \tag{8.72}$$

where $\mathbf{j}$ is the perturbed current, and $\mathbf{K}$ is the dielectric permittivity tensor introduced in Section 5.2. It follows that

$$\mathbf{K}\cdot\mathbf{E} = \mathbf{E} + \frac{\rm i}{\omega\,\epsilon_0}\,\mathbf{j} = \mathbf{E} + \frac{\rm i}{\omega\,\epsilon_0}\sum_s e_s\int \mathbf{v}\,f_{1\,s}\,d^3\mathbf{v}, \tag{8.73}$$

where $f_{1\,s}$ is the species-s perturbed distribution function.

After a great deal of rather tedious analysis, Equations (8.69) and (8.73) reduce to the following expression for the dielectric permittivity tensor (Harris 1970: Cairns 1985):

$$K_{ij} = \delta_{ij} + \sum_s \frac{e_s^2}{\omega^2\,\epsilon_0\,m_s}\sum_{n=-\infty,\infty}\int \frac{S_{ij}}{\omega - k_\parallel\,v_\parallel - n\,\Omega_s}\,d^3\mathbf{v}, \tag{8.74}$$

where

$$S_{ij} = \begin{pmatrix} v_\perp\,(n\,J_n/a_s)^2\,U, & {\rm i}\,v_\perp\,(n/a_s)\,J_n\,J_n'\,U, & v_\perp\,(n/a_s)\,J_n^2\,W \\ -{\rm i}\,v_\perp\,(n/a_s)\,J_n\,J_n'\,U, & v_\perp\,J_n'^{\,2}\,U, & -{\rm i}\,v_\perp\,J_n\,J_n'\,W \\ v_\parallel\,(n/a_s)\,J_n^2\,U, & {\rm i}\,v_\parallel\,J_n\,J_n'\,U, & v_\parallel\,J_n^2\,W \end{pmatrix}, \tag{8.75}$$

and

$$U = (\omega - k_{\parallel}\, v_{\parallel})\, \frac{\partial f_{0\,s}}{\partial v_{\perp}} + k_{\parallel}\, v_{\perp}\, \frac{\partial f_{0\,s}}{\partial v_{\parallel}}, \tag{8.76}$$

$$W = \frac{n\, \Omega_s\, v_{\parallel}}{v_{\perp}}\, \frac{\partial f_{0\,s}}{\partial v_{\perp}} + (\omega - n\, \Omega_s)\, \frac{\partial f_{0\,s}}{\partial v_{\parallel}}, \tag{8.77}$$

$$a_s = \frac{k_{\perp}\, v_{\perp}}{\Omega_s}. \tag{8.78}$$

The argument of the Bessel functions is a_s. In the previous formulae, $'$ denotes differentiation with respect to argument, and $\Omega_s = e_s\, B_0/m_s$.

The warm-plasma dielectric tensor, (8.74), can be used to investigate the properties of waves in just the same manner as the cold-plasma dielectric tensor, (5.37), was employed in Chapter 5. Note that our expression for the dielectric tensor involves singular integrals of a type similar to those encountered in Section 8.2. In principle, this means that we ought to treat the problem as an initial value problem. Fortunately, we can use the insights gained in our investigation of the simpler unmagnetized electrostatic wave problem to recognize that the appropriate way to treat the singular integrals is to evaluate them as written for $\text{Im}(\omega) > 0$, and by analytic continuation for $\text{Im}(\omega) \leq 0$.

For Maxwellian distribution functions, of the form

$$f_{0\,s} = \frac{n_s}{(2\pi\, T_s/m_s)^{3/2}}\, \exp\left[-\frac{m_s\, (v_{\perp}^2 + v_{\parallel}^2)}{2\, T_s}\right], \tag{8.79}$$

we can explicitly perform the velocity-space integral in Equation (8.74), making use of the identities (Watson 1995)

$$\int_0^{\infty} x\, J_n^2(s\, x)\, \mathrm{e}^{-x^2}\, dx = \frac{1}{2}\, \mathrm{e}^{-s^2/2}\, I_n(s^2/2), \tag{8.80}$$

$$\int_0^{\infty} x^3\, [J_n'(s\, x)]^2\, \mathrm{e}^{-x^2}\, dx = \frac{1}{4}\, \mathrm{e}^{-s^2/2}\, \Big[2\, n^2\, I_n(s^2/2)/s^2 + s^2\, I_n(s^2/2) - s^2\, I_n'(s^2/2)\Big], \tag{8.81}$$

where I_n is a modified Bessel function (Abramowitz and Stegun 1965c). We obtain

$$K_{ij} = \delta_{ij} + \sum_s \frac{\Pi_s^2\, \mathrm{e}^{-\lambda_s}}{\omega\, k_{\parallel}\, v_s} \sum_{n=-\infty,\infty} T_{ij}, \tag{8.82}$$

where $\Pi_s = (n_s\, e_s^2/\epsilon_0\, m_s)^{1/2}$, $v_s = (2\, T_s/m_s)^{1/2}$, and (Harris 1970; Cairns 1985)

$$T_{ij} = \begin{pmatrix} n^2\, I_n\, Z/\lambda_s, & \mathrm{i}\, n\, (I_n' - I_n)\, Z, & -n\, I_n\, Z'/(2\, \lambda_s)^{1/2} \\ -\mathrm{i}\, n\, (I_n' - I_n)\, Z, & (n^2\, I_n/\lambda_s + 2\, \lambda_s\, I_n - 2\, \lambda_s\, I_n')\, Z, & \mathrm{i}\, \lambda_s^{1/2}\, (I_n' - I_n)\, Z'/2^{\,1/2} \\ -n\, I_n\, Z'/(2\, \lambda_s)^{1/2}, & -\mathrm{i}\, \lambda_s^{1/2}\, (I_n' - I_n)\, Z'/2^{\,1/2}, & -I_n\, Z'\, \xi_n \end{pmatrix}. \tag{8.83}$$

Here, λ_s, which is the argument of the modified Bessel functions, is written

$$\lambda_s = \frac{k_\perp^2 v_s^2}{2\,\Omega_s^2}, \tag{8.84}$$

whereas Z and Z' represent the plasma dispersion function and its derivative, both functions being evaluated with the argument

$$\xi_n = \frac{\omega - n\,\Omega_s}{k_\parallel\, v_s}. \tag{8.85}$$

Let us consider the cold-plasma limit, $v_s \to 0$. It follows from Equations (8.84) and (8.85) that this limit corresponds to $\lambda_s \to 0$ and $\xi_n \to \infty$. According to Equation (8.47),

$$Z(\xi_n) \to -\frac{1}{\xi_n}, \tag{8.86}$$

$$Z'(\xi_n) \to \frac{1}{\xi_n^2} \tag{8.87}$$

as $\xi_n \to \infty$. Moreover, (Abramowitz and Stegun 1965c)

$$I_n(\lambda_s) \to \left(\frac{\lambda_s}{2}\right)^{|n|} \tag{8.88}$$

as $\lambda_s \to 0$. It can be demonstrated that the only non-zero contributions to K_{ij}, in this limit, come from $n = 0$ and $n = \pm 1$. In fact,

$$K_{11} = K_{22} = 1 - \frac{1}{2}\sum_s \frac{\Pi_s^2}{\omega^2}\left(\frac{\omega}{\omega - \Omega_s} + \frac{\omega}{\omega + \Omega_s}\right), \tag{8.89}$$

$$K_{12} = -K_{21} = -\frac{\mathrm{i}}{2}\sum_s \frac{\Pi_s^2}{\omega^2}\left(\frac{\omega}{\omega - \Omega_s} - \frac{\omega}{\omega + \Omega_s}\right), \tag{8.90}$$

$$K_{33} = 1 - \sum_s \frac{\Pi_s^2}{\omega^2}, \tag{8.91}$$

and $K_{13} = K_{31} = K_{23} = K_{32} = 0$. It is easily seen, from Section 5.3, that the previous expressions are identical to those found using the cold-plasma fluid equations. Thus, in the zero temperature limit, the kinetic dispersion relation obtained in this section reverts to the fluid dispersion relation derived in Chapter 5.

8.7 Parallel Wave Propagation

Let us consider wave propagation, though a warm plasma, parallel to the equilibrium magnetic field. For parallel propagation, $k_\perp \to 0$, and, hence, from Equation (8.84),

$\lambda_s \to 0$. Making use of Equation (8.88), the matrix T_{ij} simplifies to

$$T_{ij} = \begin{pmatrix} [Z(\xi_1) + Z(\xi_{-1})]/2, & \mathrm{i}\,[Z(\xi_1) - Z(\xi_{-1})]/2, & 0 \\ -\mathrm{i}\,[Z(\xi_1) - Z(\xi_{-1})]/2, & [Z(\xi_1) + Z(\xi_{-1})]/2, & 0 \\ 0, & 0, & -Z'(\xi_0)\,\xi_0 \end{pmatrix}, \tag{8.92}$$

where, again, the only non-zero contributions are from $n = 0$ and $n = \pm 1$. The dispersion relation can be written [see Equations (5.9) and (5.10)]

$$\mathbf{M}\cdot\mathbf{E} \equiv \left[\left(\frac{c}{\omega}\right)^2 \mathbf{k}\mathbf{k} - \left(\frac{c\,k}{\omega}\right)^2 \mathbf{I} + \mathbf{K}\right]\cdot\mathbf{E} = \mathbf{0}, \tag{8.93}$$

where

$$M_{11} = M_{22} = 1 - \frac{k_\parallel^2\,c^2}{\omega^2} + \frac{1}{2}\sum_s \frac{\Pi_s^2}{\omega\,k_\parallel\,v_s}\left[Z\left(\frac{\omega - \Omega_s}{k_\parallel\,v_s}\right) + Z\left(\frac{\omega + \Omega_s}{k_\parallel\,v_s}\right)\right], \tag{8.94}$$

$$M_{12} = -M_{21} = \frac{\mathrm{i}}{2}\sum_s \frac{\Pi_s^2}{\omega\,k_\parallel\,v_s}\left[Z\left(\frac{\omega - \Omega_s}{k_\parallel\,v_s}\right) - Z\left(\frac{\omega + \Omega_s}{k_\parallel\,v_s}\right)\right], \tag{8.95}$$

$$M_{33} = 1 - \sum_s \frac{\Pi_s^2}{(k_\parallel\,v_s)^2}\,Z'\left(\frac{\omega}{k_\parallel\,v_s}\right), \tag{8.96}$$

and $M_{13} = M_{31} = M_{23} = M_{32} = 0$.

The first root of Equation (8.93) is

$$1 + \sum_s \frac{2\,\Pi_s^2}{(k_\parallel\,v_s)^2}\left[1 + \frac{\omega}{k_\parallel\,v_s}\,Z\left(\frac{\omega}{k_\parallel\,v_s}\right)\right] = 0, \tag{8.97}$$

with the eigenvector $(0,\,0,\,E_z)$. Here, use has been made of Equation (8.41). This root evidentially corresponds to a longitudinal, electrostatic plasma wave. In fact, it is easily demonstrated that Equation (8.97) is equivalent to the dispersion relation (8.49) that we found earlier for electrostatic plasma waves, for the special case in which the distribution functions are Maxwellians. The analysis of Section 8.4 implies that the electrostatic wave described by the previous expression is subject to significant damping whenever the argument of the plasma dispersion function becomes less than or comparable with unity: that is, whenever $\omega \lesssim k_\parallel\,v_s$.

The second and third roots of Equation (8.93) are

$$\frac{k_\parallel^2\,c^2}{\omega^2} = 1 + \sum_s \frac{\Pi_s^2}{\omega\,k_\parallel\,v_s}\,Z\left(\frac{\omega + \Omega_s}{k_\parallel\,v_s}\right), \tag{8.98}$$

with the eigenvector $(E_x,\,\mathrm{i}\,E_x,\,0)$, and

$$\frac{k_\parallel^2\,c^2}{\omega^2} = 1 + \sum_s \frac{\Pi_s^2}{\omega\,k_\parallel\,v_s}\,Z\left(\frac{\omega - \Omega_s}{k_\parallel\,v_s}\right), \tag{8.99}$$

with the eigenvector $(E_x, -\mathrm{i}\, E_x, 0)$. The former root evidently corresponds to a right-handed circularly polarized wave, whereas the latter root corresponds to a left-handed circularly polarized wave. The previous two dispersion relations are essentially the same as the corresponding fluid dispersion relations, (5.90) and (5.91), except that they explicitly contain collisionless damping at the cyclotron resonances. Roughly speaking, the damping is significant whenever the arguments of the plasma dispersion functions are less than or of order unity. This corresponds to

$$\omega - |\Omega_e| \lesssim k_\parallel\, v_e \tag{8.100}$$

for the right-handed wave, and

$$\omega - \Omega_i \lesssim k_\parallel\, v_i \tag{8.101}$$

for the left-handed wave.

The collisionless cyclotron damping mechanism is similar to the Landau damping mechanism for longitudinal waves discussed in Section 8.3. In the former case, the resonant particles are those that gyrate about the magnetic field at approximately the same angular frequency as the wave electric field. Note that, in kinetic theory, the cyclotron resonances possess a finite width in frequency space (i.e., the incident wave does not have to oscillate at exactly the cyclotron frequency in order for there to be an absorption of wave energy by the plasma), unlike in the cold plasma model, where the resonances possess zero width.

8.8 Perpendicular Wave Propagation

Let us now consider wave propagation, through a warm plasma, perpendicular to the equilibrium magnetic field. For perpendicular propagation, $k_\parallel \rightarrow 0$, and, hence, from Equation (8.85), $\xi_n \rightarrow \infty$. Making use of the asymptotic expansions (8.86) and (8.87), the matrix T_{ij} simplifies considerably. The dispersion relation can again be written in the form (8.93), where

$$M_{11} = 1 - \sum_s \frac{\Pi_s^2}{\omega}\frac{\mathrm{e}^{-\lambda_s}}{\lambda_s} \sum_{n=-\infty,\infty} \frac{n^2\, I_n(\lambda_s)}{\omega - n\,\Omega_s}, \tag{8.102}$$

$$M_{12} = -M_{21} = -\mathrm{i} \sum_s \frac{\Pi_s^2}{\omega}\, \mathrm{e}^{-\lambda_s} \sum_{n=-\infty,\infty} \frac{n\,[I_n'(\lambda_s) - I_n(\lambda_s)]}{\omega - n\,\Omega_s}, \tag{8.103}$$

$$M_{22} = 1 - \frac{k_\perp^2\, c^2}{\omega^2} - \sum_s \frac{\Pi_s^2}{\omega}\frac{\mathrm{e}^{-\lambda_s}}{\lambda_s} \sum_{n=-\infty,\infty} \frac{n^2\, I_n(\lambda_s) + 2\,\lambda_s^2\, I_n(\lambda_s) - 2\,\lambda_s^2\, I_n'(\lambda_s)}{\omega - n\,\Omega_s}, \tag{8.104}$$

$$M_{33} = 1 - \frac{k_\perp^2\,c^2}{\omega^2} - \sum_s \frac{\Pi_s^2}{\omega}\,\mathrm{e}^{-\lambda_s} \sum_{n=-\infty,\infty} \frac{I_n(\lambda_s)}{\omega - n\,\Omega_s}, \tag{8.105}$$

and $M_{13} = M_{31} = M_{23} = M_{32} = 0$. Here,

$$\lambda_s = \frac{(k_\perp\,\rho_s)^2}{2}, \tag{8.106}$$

where $\rho_s = v_s/|\Omega_s|$ is the species-s gyroradius.

The first root of the dispersion relation (8.93) is

$$n_\perp^2 = \frac{k_\perp^2\,c^2}{\omega^2} = 1 - \sum_s \frac{\Pi_s^2}{\omega}\,\mathrm{e}^{-\lambda_s} \sum_{n=-\infty,\infty} \frac{I_n(\lambda_s)}{\omega - n\,\Omega_s}, \tag{8.107}$$

with the eigenvector $(0,\ 0,\ E_z)$. This dispersion relation obviously corresponds to the electromagnetic plasma wave, or ordinary mode, discussed in Section 5.10. However, in a warm plasma, the dispersion relation for the ordinary mode is strongly modified by the introduction of resonances (where the refractive index, $n_\perp$, becomes infinite) at all the harmonics of the cyclotron frequencies:

$$\omega_{n\,s} = n\,\Omega_s, \tag{8.108}$$

where n is a non-zero integer. These resonances are a finite gyroradius effect. In fact, they originate from the variation of the wave phase across a gyro-orbit (Cairns 1985). Thus, in the cold plasma limit, $\lambda_s \to 0$, in which the gyroradii shrink to zero, all of the resonances disappear from the dispersion relation. In the limit in which the wavelength, λ, of the wave is much larger than a typical gyroradius, ρ_s, the relative amplitude of the nth harmonic cyclotron resonance, as it appears in the dispersion relation (8.107), is approximately $(\rho_s/\lambda)^{|n|}$ [see Equations (8.88) and (8.106)]. It is clear, therefore, that, in this limit, only low-order resonances [i.e., $n \sim O(1)$] couple strongly into the dispersion relation, and high-order resonances (i.e., $|n| \gg 1$) can effectively be neglected. As $\lambda \to \rho_s$, the high-order resonances become increasingly important, until, when $\lambda \lesssim \rho_s$, all of the resonances are of approximately equal strength. Because the ion gyroradius is generally much larger than the electron gyroradius, it follows that the ion cyclotron harmonic resonances are generally more important than the electron cyclotron harmonic resonances.

Observe that the cyclotron harmonic resonances appearing in the dispersion relation (8.107) are of zero width in frequency space: that is, they are just like the resonances that appear in the cold-plasma limit. Actually, this is just an artifact of the fact that the waves we are studying propagate exactly perpendicular to the equilibrium magnetic field. It is clear, from an examination of Equations (8.83) and (8.85), that the cyclotron harmonic resonances originate from the zeros of the plasma dispersion functions. Adopting the usual rule that substantial damping takes place whenever the arguments of the dispersion functions are less than or of order unity, it follows that the cyclotron harmonic resonances lead to significant damping whenever

$$\omega - \omega_{n\,s} \lesssim k_\parallel\,v_s. \tag{8.109}$$

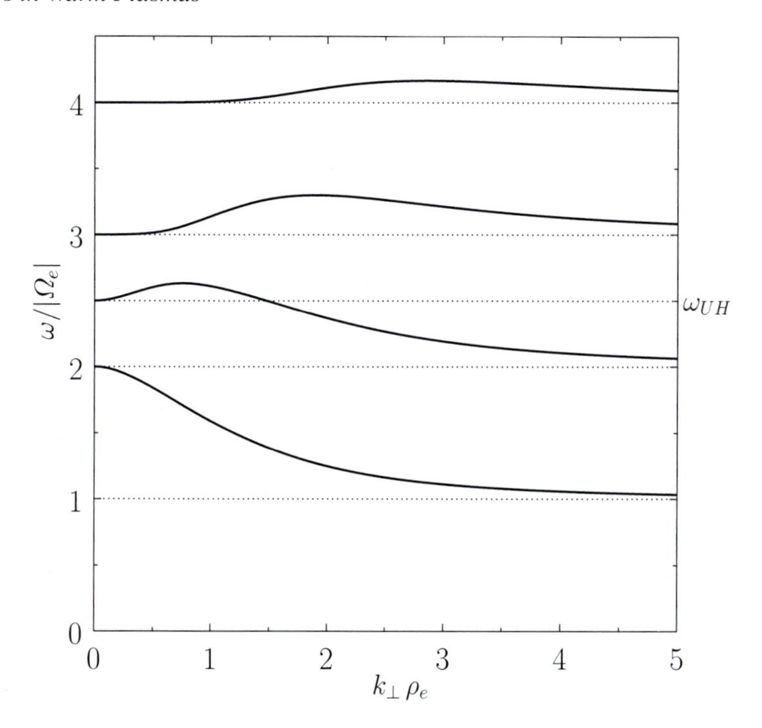

Figure 8.8
Dispersion relation for electron Bernstein waves in a warm plasma for which $\omega_{UH}/|\Omega_e| = 2.5$.

Thus, the cyclotron harmonic resonances possess a finite width in frequency space provided the parallel wavenumber, $k_\parallel$, is non-zero: that is, provided the wave does not propagate exactly perpendicular to the magnetic field.

The appearance of the cyclotron harmonic resonances in a warm plasma is of great practical importance in plasma physics, because it greatly increases the number of resonant frequencies at which waves can transfer energy to the plasma. In magnetic fusion experiments, these resonances are routinely exploited to heat plasmas via externally launched electromagnetic waves (Stix 1992; Swanson 2003).

The other roots of the dispersion relation (8.93) satisfy

$$\left(1-\sum_s \frac{\Pi_s^2}{\omega}\frac{\mathrm{e}^{-\lambda_s}}{\lambda_s}\sum_{n=-\infty,\infty}\frac{n^2\,I_n(\lambda_s)}{\omega-n\,\Omega_s}\right)\left(1-\frac{k_\perp^2\,c^2}{\omega^2}\right.$$
$$\left.-\sum_s \frac{\Pi_s^2}{\omega}\frac{\mathrm{e}^{-\lambda_s}}{\lambda_s}\sum_{n=-\infty,\infty}\frac{n^2\,I_n(\lambda_s)+2\,\lambda_s^2\,I_n(\lambda_s)-2\,\lambda_s^2\,I_n'(\lambda_s)}{\omega-n\,\Omega_s}\right)$$
$$=\left(\sum_s \frac{\Pi_s^2}{\omega}\,\mathrm{e}^{-\lambda_s}\sum_{n=-\infty,\infty}\frac{n\,[I_n'(\lambda_s)-I_n(\lambda_s)]}{\omega-n\,\Omega_s}\right)^2, \tag{8.110}$$

with the eigenvector $(E_x, E_y, 0)$. In the cold plasma limit, $\lambda_s \to 0$, this dispersion relation reduces to that of the extraordinary mode discussed in Section 5.10. This mode, for which $\lambda_s \ll 1$, unless the plasma possesses a thermal velocity approaching the velocity of light, is little affected by thermal effects, except close to the cyclotron harmonic resonances, $\omega = \omega_{n\,s}$, where small thermal corrections are important because of the smallness of the denominators in the previous dispersion relation (Cairns 1985).

However, another mode also exists. In fact, if we look for a mode with a phase-velocity much less than the velocity of light (i.e., $c^2\,k_\perp^2/\omega^2 \gg 1$) then it is clear from (8.102)–(8.105) that the dispersion relation is approximately

$$1 - \sum_s \frac{\Pi_s^2}{\omega}\,\frac{e^{-\lambda_s}}{\lambda_s} \sum_{n=-\infty,\infty} \frac{n^2\,I_n(\lambda_s)}{\omega - n\,\Omega_s} = 0, \tag{8.111}$$

and the associated eigenvector is $(E_x, 0, 0)$. The new waves, which are called *Bernstein waves*—after I.B. Bernstein, who first discovered them (Bernstein 1958)—are a type of slowly propagating, longitudinal, electrostatic wave.

Let us consider electron Bernstein waves, for the sake of definiteness. Neglecting the contribution of the ions, which is reasonable provided that the wave frequencies are sufficiently high, the dispersion relation (8.111) reduces to

$$1 - \frac{\Pi_e^2}{\omega}\,\frac{e^{-\lambda_e}}{\lambda_e} \sum_{n=-\infty,\infty} \frac{n^2\,I_n(\lambda_e)}{\omega - n\,\Omega_e} = 0. \tag{8.112}$$

In the limit $\lambda_e \to 0$ (with $\omega \neq n\,\Omega_e$), only the $n = \pm 1$ terms survive in the previous expression. [See Equation (8.88).] In fact, because $I_{\pm 1}(\lambda_e)/\lambda_e \to 1/2$ as $\lambda_e \to 0$, the dispersion relation yields

$$\omega^2 \to \Pi_e^2 + \Omega_e^2. \tag{8.113}$$

It follows that there is a Bernstein wave whose frequency asymptotes to the upper hybrid frequency (see Section 5.10) in the limit $k_\perp \to 0$. For other non-zero values of n, we have $I_n(\lambda_e)/\lambda_e \to 0$ as $\lambda_e \to 0$. However, a solution to Equation (8.111) can be obtained if $\omega \to n\,\Omega_e$ at the same time. Similarly, as $\lambda_e \to \infty$, we have $e^{-\lambda_e}\,I_n(\lambda_e) \to 0$ (Abramowitz and Stegun 1965c). In this case, a solution can only be obtained if $\omega \to n\,\Omega_e$, for some n, at the same time. The complete solution to Equation (8.111) is plotted in Figure 8.8 for a case where the upper hybrid frequency lies between $2\,|\Omega_e|$ and $3\,|\Omega_e|$. In fact, wherever the upper hybrid frequency lies, the Bernstein modes above and below it behave like those shown in the diagram.

At small values of $k_\perp$, the phase-velocity becomes large, and it is no longer legitimate to neglect the extraordinary mode (Cairns 1985). A more detailed examination of the complete dispersion relation shows that the extraordinary mode and the Bernstein mode cross over near the harmonics of the cyclotron frequency to give the pattern shown in Figure 8.9. Here, the dashed line shows the cold plasma extraordinary mode.

In a lower frequency range, a similar phenomena occurs at the harmonics of the ion cyclotron frequency, producing ion Bernstein waves, with somewhat similar

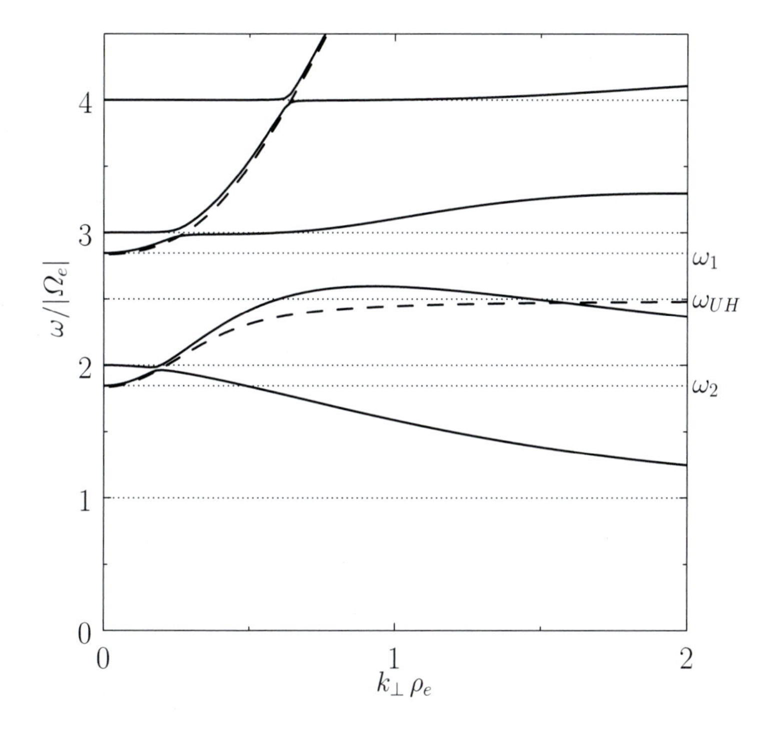

Figure 8.9
Dispersion relation for extraordinary/electron Bernstein waves in a warm plasma for which $\omega_{UH}/|\Omega_e| = 2.5$ and $v_{te}/c = 0.2$. The dashed line indicates the cold plasma extraordinary mode.

properties to electron Bernstein waves. Note, however, that while the ion contribution to the dispersion relation can be neglected for high-frequency waves, the electron contribution cannot be neglected for low-frequency waves, so there is not a complete symmetry between the two types of Bernstein waves.

8.9 Electrostatic Waves

It is instructive to consider the propagation of electrostatic waves through a magnetized plasma. Such waves have purely electrostatic perturbed electric fields of the form

$$\mathbf{E} = -\nabla\phi = -\mathrm{i}\,\phi\,\mathbf{k}. \tag{8.114}$$

Equation (8.8) can be generalized to give

$$k^2\,\phi = \sum_s \frac{e_s}{\epsilon_0}\int f_{1\,s}\,d^3\mathbf{v}. \tag{8.115}$$

Moreover, it follows from Equation (8.71) that

$$\mathbf{B} = \omega^{-1}\,\mathbf{k}\times\mathbf{E} = \mathbf{0}. \tag{8.116}$$

In other words, there is no perturbed magnetic field associated with an electrostatic wave. Equation (8.69) yields

$$f_{1\,s} = \mathrm{i}\,\phi\,\frac{e_s}{m_s}\int_{-\infty}^{t}\left(k_\perp\,\cos\chi\,\frac{\partial f_{0\,s}}{\partial v_\perp} + k_\parallel\,\frac{\partial f_{0\,s}}{\partial v_\parallel}\right)\sum_{n,m=-\infty,\infty} J_n\!\left(\frac{k_\perp\,v_\perp}{\Omega_s}\right) J_m\!\left(\frac{k_\perp\,v_\perp}{\Omega_s}\right)$$
$$\exp\{\mathrm{i}\,[(n\,\Omega_s + k_\parallel\,v_\parallel - \omega)\,(t'-t) + (m-n)\,\theta]\}\,dt'. \tag{8.117}$$

Here, $\Omega_s = e_s\,B_0/m_s$ and $\chi = -\Omega_s\,(t'-t) + \theta$, whereas the Cartesian components of $\mathbf{k}$ and $\mathbf{v}$ are written $(k_\perp,\,0,\,k_\parallel)$ and $(v_\perp\,\cos\theta,\,v_\perp\,\sin\theta,\,v_\parallel)$, respectively. The equilibrium magnetic field takes the form $\mathbf{B}_0 = (0,\,0,\,B_0)$. Equations (8.115) and (8.117) can be combined to give

$$k^2 = \mathrm{i}\sum_s \frac{e_s^2}{\epsilon_0\,m_s}\int_{-\infty}^{t}\int\left(k_\perp\,\cos\chi\,\frac{\partial f_{0\,s}}{\partial v_\perp} + k_\parallel\,\frac{\partial f_{0\,s}}{\partial v_\parallel}\right)\sum_{n,m=-\infty,\infty} J_n\!\left(\frac{k_\perp\,v_\perp}{\Omega_s}\right) J_m\!\left(\frac{k_\perp\,v_\perp}{\Omega_s}\right)$$
$$\exp\{\mathrm{i}\,[(n\,\Omega_s + k_\parallel\,v_\parallel - \omega)\,(t'-t) + (m-n)\,\theta]\}\,d^3\mathbf{v}\,dt'. \tag{8.118}$$

After some tedious analysis, the previous expression reduces to the so-called *Harris dispersion relation* (Harris 1961)

$$1 + \sum_s \frac{e_s^2}{k^2\,\epsilon_0\,m_s}\sum_{n=-\infty,\infty}\int \frac{J_n^2(k_\perp\,v_\perp/\Omega_s)}{\omega - k_\parallel\,v_\parallel - n\,\Omega_s}\left(\frac{n\,\Omega_s}{v_\perp}\,\frac{\partial f_{0\,s}}{\partial v_\perp} + k_\parallel\,\frac{\partial f_{0\,s}}{\partial v_\parallel}\right) d^3\mathbf{v} = 0. \tag{8.119}$$

For Maxwellian distribution functions of the form (8.79), we can explicitly perform the velocity-space integrals in the Harris dispersion relation to give

$$1 + \sum_s \frac{2\,\Pi_s^2}{(k_\parallel^2 + k_\perp^2)\,v_s^2}\left[1 + \xi_0\,\mathrm{e}^{-\lambda_s}\sum_{n=-\infty,\infty} I_n(\lambda_s)\,Z(\xi_n)\right] = 0, \tag{8.120}$$

where $\Pi_s = (n_s\,e_s^2/\epsilon_0\,m_s)^{1/2}$, $v_s = (2\,T_s/m_s)^{1/2}$, $\lambda_s = k_\perp^2\,v_s^2/(2\,\Omega_s^2)$, and $\xi_n = (\omega - n\,\Omega_s)/(k_\parallel\,v_s)$. Here, the I_n are modified Bessel functions (Abramowitz and Stegun 1965c), whereas Z is a plasma dispersion function. (See Section 8.4.) In deriving the previous expression, use has been made of the identity (Watson 1995)

$$\sum_{n=-\infty,\infty} \mathrm{e}^{-\lambda_s}\,I_n(\lambda_s) = 1. \tag{8.121}$$

Consider electrostatic waves propagating parallel to the equilibrium magnetic field. In this case, $k_\perp \to 0$ and $\lambda_s \to 0$, so the dispersion relation (8.120) reduces to

$$1 + \sum_s \frac{2\,\Pi_s^2}{(k_\parallel\,v_s)^2}\,[1 + \xi_0\,Z(\xi_0)] = 0, \tag{8.122}$$

with the eigenvector $(0,\ 0,\ E_z)$. (Recall that $\mathbf{E} \propto \mathbf{k}$ for an electrostatic wave.) It can be seen that this expression is identical to the dispersion relation (8.97) for longitudinal plasma waves. Consider electrostatic waves propagating perpendicular to the equilibrium magnetic field. In this case, $k_\parallel \to 0$ and $\xi_n \to \infty$, so the dispersion relation (8.120) reduces to

$$1 + \sum_s \frac{2\,\Pi_s^2}{(k_\perp\,v_s)^2}\left[1 - \xi_0\,\mathrm{e}^{-\lambda_s} \sum_{n=-\infty,\infty} \frac{I_n(\lambda_s)}{\xi_n}\right] = 0, \tag{8.123}$$

with the eigenvector $(E_x,\ 0,\ 0)$. Making use of the identity (8.121), as well as the fact that $I_{-n}(\lambda_s) = I_n(\lambda_s)$ (Abramowitz and Stegun 1965c), the previous expression can be rearranged to give

$$1 - \sum_s \frac{\Pi_s^2}{\omega}\,\frac{\mathrm{e}^{-\lambda_s}}{\lambda_s} \sum_{n=-\infty,\infty} \frac{n^2\,I_n(\lambda_s)}{\omega - n\,\Omega_s} = 0. \tag{8.124}$$

It can be seen that this expression is identical to the dispersion relation (8.111) for Bernstein waves. Thus, we can now appreciate that plasma waves and Bernstein waves are merely different aspects of a more general type of electrostatic wave. This wave takes the form of a plasma wave when propagating parallel to the equilibrium magnetic field, of a Bernstein wave when propagating perpendicular to the magnetic field, and takes an intermediate form when propagating obliquely to the magnetic field.

8.10 Velocity-Space Instabilities

Up to now, we have mostly concentrated on waves that propagate through warm plasmas possessing Maxwellian velocity distributions. We found that, under certain circumstances, damping occurs because of a transfer of energy from the wave to a group of particles that satisfy some resonance condition. Moreover, the damping rate only depends on the properties of the velocity distribution function in the resonant region of velocity space. It turns out that if the velocity distribution function is not Maxwellian (for instance, if the distribution function possesses multiple maxima) then it is possible for the energy transfer to be reversed, so that the wave grows at the expense of the kinetic energy of the resonant particles. This type of plasma instability, which depends on the exact shape of the velocity distribution function, is generally known as a *velocity-space instability* (Cairns 1985).

Consider the dispersion relation (8.23) for an electrostatic plasma wave in an unmagnetized quasi-neutral plasma with stationary ions. This relation can be written

$$k^2 = \frac{e^2}{\epsilon_0\,m_e} \int_{-\infty}^{\infty} \frac{\partial F_0/\partial u}{u - \omega/k}\,du, \tag{8.125}$$

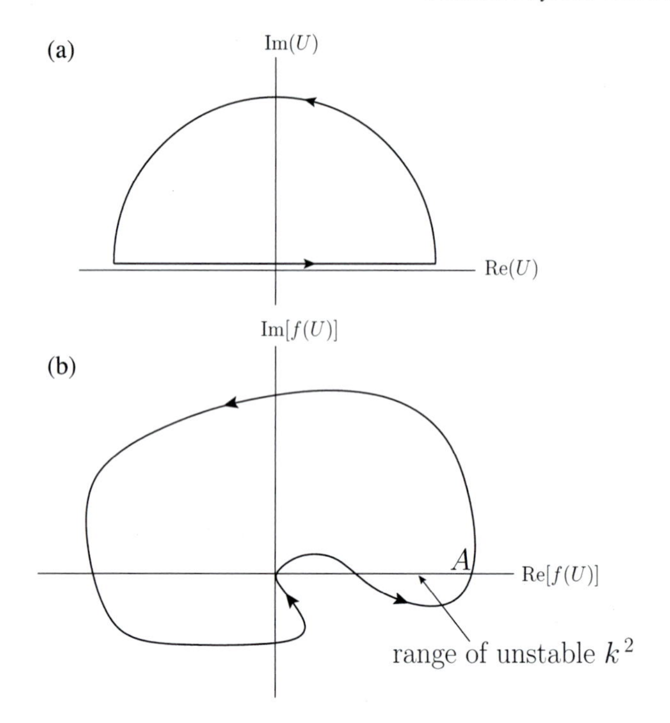

Figure 8.10
A Nyquist diagram.

or

$$k^2 = f(U), \tag{8.126}$$

where

$$f(U) = \frac{e^2}{\epsilon_0\, m_e} \int_{-\infty}^{\infty} \frac{\partial F_0/\partial u}{u - U}\, du, \tag{8.127}$$

and $U = \omega/k$. Taking k to be real and positive, the question of whether the system is stable or not is equivalent to asking whether Equation (8.126) is satisfied for any value of U lying in the upper half of the complex plane.

To answer the previous question, we employ a standard result in complex variable theory which states that the number of zeros minus the number of poles of $f(U) - k^2$ in a given region of the complex U plane is $(2\pi)^{-1}$ times the increase in the argument of $f(U) - k^2$ as U moves once counter-clockwise around the boundary of this region (Flanigan 2010). To determine the latter quantity, we construct what is known as a *Nyquist diagram* (Nyquist 1932). Because the region in which we are interested is the upper-half complex plane, we let U follow the semi-circular path shown in Figure 8.10(a), and plot the corresponding path followed in the complex plane by $f(U)$, as illustrated in Figure 8.10(b). Now, $f(U) \to 0$ as $|U| \to \infty$. Hence, if the radius of the semicircle in Figure 8.10(a) tends to infinity, then only that part of the contour running along the real axis is important, and the $f(U)$ contour starts and finishes at the origin. Because the function $f(U)$ is analytic in the upper-half U plane, by virtue of the way in which it is defined, the number of zeros of $f(U) - k^2$ is equal to the change in argument (divided by 2π) of this quantity as the path shown in Figure 8.10(b) is followed. However, this is just the number of times that the path encircles the point k^2. Hence, the criterion for instability is that the path should encircle part of the positive real axis. Thus, in Figure 8.10(b), the system is unstable for the indicated values of k^2 (Cairns 1985).

In an unstable system, there must exist a point such as A in Figure 8.10(b) where the $f(U)$ contour crosses the real axis going from negative to positive imaginary part. Now, as U moves along the real axis [cf., Equation (8.26)],

$$\frac{\epsilon_0 m_e}{e^2} f(U) = P \int_{-\infty}^{\infty} \frac{F_0'(u)}{u - U} du + \mathrm{i}\,\pi\, F_0'(U). \tag{8.128}$$

Thus, at point A, corresponding to $U = U_0$ (say), it must be the case that $F_0'(U_0) = 0$. Furthermore, $F_0'(U)$ must go from being negative to being positive as U passes through U_0 from below. This implies that $F_0(U)$ attains a minimum at $U = U_0$. In other words, a necessary condition for the distribution function $F_0(u)$ to be unstable is that it should attain a minimum value at some finite value of u. A further condition to be satisfied is that the real part of $f(U)$ be positive at $U = U_0$. In other words,

$$\int_{-\infty}^{\infty} \frac{F_0'(u)}{u - U_0} du > 0. \tag{8.129}$$

Note that the principal part need not be taken in the previous integral, because the numerator vanishes at the same point as the denominator. Integration by parts yields the equivalent condition

$$\int_{-\infty}^{\infty} \frac{F_0(u) - F_0(U_0)}{(u - U_0)^2} du > 0. \tag{8.130}$$

Here, $F_0(U_0)$ has been chosen as the constant of integration in order to again make it unnecessary to take the principal part. The previous relation is called the *Penrose condition*, and is a necessary and sufficient condition for instability, assuming that $f(u)$ attains a minimum value at $u = U_0$ (Penrose 1960).

The previous discussion implies that a single-humped velocity distribution function, such as a Maxwellian, is absolutely stable to velocity-space instabilities (Gardner 1963). This follows because there is no finite value of u at which such a distribution function attains a minimum value. In fact, assuming that the distribution function, $F_0(u)$, is such that $F_0(u) \rightarrow 0$ as $|u| \rightarrow \infty$, we deduce that an unstable distribution function must possess at least one minimum and two maxima for u in the range $-\infty < u < \infty$.

8.11 Counter-Propagating Beam Instability

As an example of a potentially unstable velocity distribution function, consider

$$F_0(u) = n_e \frac{v_e}{2\pi}\left[\frac{1}{v_e^2 + (u-V)^2} + \frac{1}{v_e^2 + (u+V)^2}\right]. \tag{8.131}$$

This function corresponds to two counter-streaming electron beams with so-called *Cauchy velocity distributions* characterized by the mean velocities $\pm V$, and the thermal spreads v_e. Here,

$$n_e = \int_{-\infty}^{\infty} F_0(u)\,du \tag{8.132}$$

is the electron number density. (It is assumed that there is a stationary background ion fluid of charge density $e\,n_e$.) We have seen that a necessary, but not sufficient, criterion for the distribution function (8.131) to be unstable is that it should possess a minimum at finite u. It is easily demonstrated that this is the case provided $v_e < \sqrt{3}\,V$, and, furthermore, that the minimum lies at $u = 0$. Thus, the system is potentially unstable if $v_e < \sqrt{3}\,V$. In order to determine whether the system is actually unstable, we need to evaluate the Penrose condition (8.130) at the minimum. It turns out that the Penrose integral can be evaluated exactly for $U_0 = 0$. In fact,

$$\int_{-\infty}^{\infty} \frac{F_0(u) - F_0(U_0)}{(u-U_0)^2}\,du = n_e\left[\frac{V^2 - v_e^2}{(V^2+v_e^2)^2}\right]. \tag{8.133}$$

The instability criterion is that this integral be positive, which yields $v_e < V$. Assuming that k is real and positive, it can be shown that, in the small-k limit, $k \ll \Pi_e/V$, the growth-rate of the instability is written $\gamma \equiv -\mathrm{i}\,\omega \simeq k\,(V - v_e)$.

8.12 Current-Driven Ion Acoustic Instability

As a second example, consider ion acoustic waves in a plasma with single-charged ions in which the electron velocity distribution function takes the simplified form

$$F_{0\,e}(u_e) = n\,\frac{v_e}{\pi}\,\frac{1}{v_e^2 + (u_e - U_e)^2}, \tag{8.134}$$

and the ion distribution is written

$$F_{0\,i}(u_i) = n\,\frac{v_i}{\pi}\,\frac{1}{v_i^2 + u_i^2}. \tag{8.135}$$

Here, u_e and u_i are the parallel (to $\mathbf{k}$) electron and ion velocities, respectively, n is the particle number density, v_e and v_i are the electron and ion thermal spreads, respectively, and U_e is the electron-ion drift velocity. We saw in Section 8.5 that, in the absence of drift, the ion acoustic wave is damped. We now wish to investigate whether the presence of an electron-ion drift (which is associated with a net current flowing in the plasma) can destabilize the mode. The appropriate dispersion relation is Equation (8.49), which, on integration by parts, can be written

$$k^2 = \frac{e^2}{\epsilon_0\, m_e} \int_{-\infty}^{\infty} \frac{F_{0\,e}}{(u_e - \omega/k)^2}\, du_e + \frac{e^2}{\epsilon_0\, m_i} \int_{-\infty}^{\infty} \frac{F_{0\,i}}{(u_i - \omega/k)^2}\, du_i. \tag{8.136}$$

The previous three equations can be combined together, and the integrals performed as contour integrals in the complex u_e and u_i planes (closed in the lower halves of these planes), making use of the residue theorem (Riley 1974), to give

$$1 = \frac{\Pi_e^2}{(\omega - k\, U_e + \mathrm{i}\, k\, v_e)^2} + \frac{\Pi_i^2}{(\omega + \mathrm{i}\, k\, v_i)^2}, \tag{8.137}$$

where we have assumed that k is real and positive, and that ω/k lies in the upper half of the complex plane. In the limit $k\,\lambda_{D\,e} \ll 1$, where $\lambda_{D\,e} = v_e/\Pi_e$, the left-hand side of the previous expression is negligible compared to the two terms on the right-hand side, and we obtain

$$\omega + \mathrm{i}\, k\, v_i \simeq \pm\mathrm{i} \left(\frac{m_e}{m_i}\right)^{1/2} (\omega - k\, U_e + \mathrm{i}\, k\, v_e). \tag{8.138}$$

Choosing the negative sign, which ensures that the phase-velocity is in the correct direction, we get

$$\omega \simeq k \left(\frac{m_e}{m_i}\right)^{1/2} v_e + \mathrm{i}\, k \left[\left(\frac{m_e}{m_i}\right)^{1/2} U_e - v_i\right]. \tag{8.139}$$

If we write $v_e = (T_e/m_e)^{1/2}$ and $v_i = (T_i/m_i)^{1/2}$, where T_e and T_i are the effective electron and ion temperatures, then the previous expression yields

$$\omega \simeq k \left(\frac{T_e}{m_i}\right)^{1/2} + \mathrm{i}\, k \left(\frac{m_e}{m_i}\right)^{1/2} \left[U_e - \left(\frac{T_i}{m_e}\right)^{1/2}\right]. \tag{8.140}$$

Thus, the phase-velocity of the wave is $(T_e/m_i)^{1/2}$, whereas the growth-rate is

$$\gamma = k \left(\frac{m_e}{m_i}\right)^{1/2} \left[U_e - \left(\frac{T_i}{m_e}\right)^{1/2}\right]. \tag{8.141}$$

It can be seen that the growth-rate becomes positive (i.e., the mode becomes unstable) when the drift velocity exceeds the critical value

$$U_{e\,c} = \left(\frac{T_i}{m_e}\right)^{1/2} = \left(\frac{T_i}{T_e}\right)^{1/2} v_e. \tag{8.142}$$

This calculation indicates that if the electron and ion temperatures are similar then the threshold drift velocity is of order the electron thermal speed, which is usually very large. In other words, a significant current is generally required to drive the ion acoustic wave unstable. The instability threshold (relative to the electron thermal speed) is considerably reduced if the electron temperature greatly exceeds the ion temperature.

If we repeat the previous calculation using the more realistic Maxwellian velocity distributions,

$$F_{0\,e}(u_e) = \frac{n}{(2\pi\,T_e/m_e)^{1/2}}\,\exp\left[-\frac{m_e\,(u_e - U_e)^2}{2\,T_e}\right], \tag{8.143}$$

and

$$F_{0\,i}(u_i) = \frac{n}{(2\pi\,T_i/m_i)^{1/2}}\,\exp\left(-\frac{m_i\,u_i^2}{2\,T_i}\right), \tag{8.144}$$

then the dispersion relation (8.136) yields

$$1 = \frac{Z'(\zeta_e)}{2\,(k\,\lambda_{D\,e})^2} + \frac{Z'(\zeta_i)}{2\,(k\,\lambda_{D\,i})^2}, \tag{8.145}$$

where $\lambda_{D\,s} = (T_s/m_s\,\Pi_s^2)^{1/2}$, $\zeta_e = (m_e/2\,T_e)^{1/2}\,(\omega/k - U_e)$, and $\zeta_i = (m_i/2\,T_i)\,(\omega/k)$. As in Section 8.5, we assume that the phase-velocity of the wave is much less than the electron thermal velocity, but much greater than the ion thermal velocity. This implies that $|\zeta_e| \ll 1$ and $|\zeta_i| \gg 1$. Using the small-argument expansion

$$Z'(\zeta_e) \simeq -\mathrm{i}\,2\sqrt{\pi}\,\zeta_e\,\mathrm{e}^{-\zeta_e^2} - 2, \tag{8.146}$$

and the large-argument expansion

$$Z'(\zeta_i) \simeq -\mathrm{i}\,2\sqrt{\pi}\,\zeta_i\,\mathrm{e}^{-\zeta_i^2} + \frac{1}{\zeta_i^2}, \tag{8.147}$$

we obtain

$$2\,(k\,\lambda_{D\,e})^2 \simeq \frac{T_e}{T_i}\,\frac{1}{\zeta_i^2} - 2 - \mathrm{i}\,2\,\sqrt{\pi}\left(\frac{T_e}{T_i}\,\zeta_i\,\mathrm{e}^{-\zeta_i^2} + \zeta_e\right). \tag{8.148}$$

In the limit $k\,\lambda_{D\,e} \ll 1$, the previous expression yields $\omega = \omega_r + \mathrm{i}\,\gamma$, where

$$\omega_r \simeq k\,c_s, \tag{8.149}$$

and

$$\frac{\gamma}{\omega_r} \simeq -\sqrt{\frac{\pi}{8}}\left[\left(\frac{m_e}{m_i}\right)^{1/2}\left(1 - \frac{U_e}{c_s}\right) + \left(\frac{T_e}{T_i}\right)^{3/2}\,\exp\left(-\frac{T_e}{2\,T_i}\right)\right], \tag{8.150}$$

Here, $c_s = (T_e/m_i)^{1/2}$ is the phase-velocity of the ion acoustic wave, and it is assumed that $|\gamma|/\omega_r \ll 1$. The ion acoustic wave phase-velocity is much less than the electron thermal speed, as previously assumed, but is only much greater than the ion thermal speed if $T_e \gg T_i$. According to Equation (8.150), the threshold electron-ion drift speed above which the ion acoustic wave is destabilized is

$$U_{e\,c} = c_s\left[1 + \left(\frac{m_i}{m_e}\right)^{1/2}\left(\frac{T_e}{T_i}\right)^{3/2}\,\exp\left(-\frac{T_e}{2\,T_i}\right)\right]. \tag{8.151}$$

As before, this formula (which is only accurate when $T_e \gg T_i$) indicates that the threshold is strongly reduced (relative to c_s) as the ratio of the electron to the ion temperature is increased.

8.13 Harris Instability

It is not feasible to give a comprehensive account of velocity-space instabilities in a magnetized plasma, on account of the great number of different instabilities of this type. Rather than trying to analyze the full electromagnetic dispersion relation, we shall concentrate on the stability of electrostatic waves. Instabilities of this type tend to be more important than electromagnetic instabilities, particularly in low-β plasmas (Cairns 1985). Our starting point is the Harris dispersion relation, (8.119):

$$\epsilon(k,\omega) = 1 + \sum_s \frac{e_s^2}{k^2\,\epsilon_0\,m_s} \sum_{n=-\infty,\infty} \int \frac{J_n^2(k_\perp\,v_\perp/\Omega_s)}{\omega - k_\parallel\,v_\parallel - n\,\Omega_s}\left(\frac{n\,\Omega_s}{v_\perp}\,\frac{\partial f_{0\,s}}{\partial v_\perp} + k_\parallel\,\frac{\partial f_{0\,s}}{\partial v_\parallel}\right) d^3\mathbf{v}$$
$$= 0. \tag{8.152}$$

Making use of the Plemelj formula, (8.26), we can write the previous expression in the form

$$\epsilon(k,\omega) = \epsilon_r(k,\omega) + \mathrm{i}\,\epsilon_i(k,\omega) = 0, \tag{8.153}$$

where

$$\epsilon_r(k,\omega) = 1 + \sum_s \frac{e_s^2}{k^2\,\epsilon_0\,m_s} \sum_{n=-\infty,\infty} P\int \frac{J_n^2(k_\perp\,v_\perp/\Omega_s)}{\omega - k_\parallel\,v_\parallel - n\,\Omega_s}\left(\frac{n\,\Omega_s}{v_\perp}\,\frac{\partial f_{0\,s}}{\partial v_\perp} + k_\parallel\,\frac{\partial f_{0\,s}}{\partial v_\parallel}\right) d^3\mathbf{v}, \tag{8.154}$$

and

$$\epsilon_i(k,\omega) = -\pi \sum_s \frac{e_s^2}{k^2\,\epsilon_0\,m_s} \sum_{n=-\infty,\infty} \int J_n^2(k_\perp\,v_\perp/\Omega_s)\,\delta(\omega - k_\parallel\,v_\parallel - n\,\Omega_s)$$
$$\left(\frac{n\,\Omega_s}{v_\perp}\,\frac{\partial f_{0\,s}}{\partial v_\perp} + k_\parallel\,\frac{\partial f_{0\,s}}{\partial v_\parallel}\right) d^3\mathbf{v}. \tag{8.155}$$

Generally speaking, we expect $|\epsilon_i| \ll |\epsilon_r|$. Let us search for an instability whose angular frequency is $\omega = \omega_r + \mathrm{i}\,\gamma$, where ω_r is real and positive, γ is real, and $|\gamma| \ll \omega_r$. Thus, ω_r is the real frequency of the instability, ω_r/k its phase-velocity, and γ its growth-rate. Expanding (8.153) to first-order in γ, we obtain

$$\epsilon(k,\omega_r + \mathrm{i}\,\gamma) \simeq \epsilon_r(k,\omega_r) + \mathrm{i}\,\gamma\,\frac{\partial\epsilon_r(k,\omega_r)}{\partial\omega} + \mathrm{i}\,\epsilon_i(k,\omega_r) = 0. \tag{8.156}$$

Here, the quantities $\epsilon_r(k,\omega_r)$, $\partial\epsilon_r(k,\omega_r)/\partial\omega$, and $\epsilon_i(k,\omega_r)$ are all real. Thus, the real frequency of the instability is determined from

$$\epsilon_r(k,\omega_r) = 0, \tag{8.157}$$

whereas the growth-rate is given by

$$\gamma = -\frac{\epsilon_i(k,\omega_r)}{\partial\epsilon_r(k,\omega_r)/\partial\omega}. \tag{8.158}$$

Consider the so-called *Harris instability*, which occurs for real frequencies close to the ion cyclotron harmonics in a plasma in which the parallel and perpendicular (to the equilibrium magnetic field) temperatures are different (Harris 1970; Cairns 1985). Suppose that the equilibrium velocity distribution functions are two-temperature Maxwellians of the form

$$f_{0\,s} = \frac{n_s}{\left(2\pi\,T_{\perp\,s}^{2/3}\,T_{\parallel\,s}^{1/3}/m_s\right)^{3/2}}\,\exp\left(-\frac{m_s\,v_\perp^2}{2\,T_{\perp\,s}} - \frac{m_s\,v_\parallel^2}{2\,T_{\parallel\,s}}\right). \tag{8.159}$$

Here, $T_{\perp\,s}$ and $T_{\parallel\,s}$ are the species-s perpendicular and parallel temperatures, respectively. It follows that

$$\begin{aligned}\epsilon_r(k,\omega) = 1 + \sum_s \frac{\Pi_s^2}{(k\,v_{\parallel\,s})^2} \sum_{n=-\infty,\infty} \exp\left(-\frac{k_\perp^2\,v_{\perp\,s}^2}{2\,\Omega_s^2}\right) I_n\left(-\frac{k_\perp^2\,v_{\perp\,s}^2}{2\,\Omega_s^2}\right)\\ \left[\frac{2\,n\,\Omega_s}{k_\parallel\,v_{\parallel\,s}}\,\frac{T_{\parallel\,s}}{T_{\perp\,s}}\,Z_r\left(\frac{\omega - n\,\Omega_s}{k_\parallel\,v_{\parallel\,s}}\right) - Z_r'\left(\frac{\omega - n\,\Omega_s}{k_\parallel\,v_{\parallel\,s}}\right)\right],\end{aligned} \tag{8.160}$$

and

$$\begin{aligned}\epsilon_i(k,\omega) = \sum_s \frac{\Pi_s^2}{(k\,v_{\parallel\,s})^2} \sum_{n=-\infty,\infty} \exp\left(-\frac{k_\perp^2\,v_{\perp\,s}^2}{2\,\Omega_s^2}\right) I_n\left(-\frac{k_\perp^2\,v_{\perp\,s}^2}{2\,\Omega_s^2}\right)\\ \left[\frac{2\,n\,\Omega_s}{k_\parallel\,v_{\parallel\,s}}\,\frac{T_{\parallel\,s}}{T_{\perp\,s}}\,Z_i\left(\frac{\omega - n\,\Omega_s}{k_\parallel\,v_{\parallel\,s}}\right) - Z_i'\left(\frac{\omega - n\,\Omega_s}{k_\parallel\,v_{\parallel\,s}}\right)\right],\end{aligned} \tag{8.161}$$

where $\Pi_s = (n_s\,e_s^2/\epsilon_0\,m_s)^{1/2}$, $v_{\perp\,s} = (2\,T_{\perp\,s}/m_s)^{1/2}$, and $v_{\parallel\,s} = (2\,T_{\parallel\,s}/m_s)^{1/2}$. Moreover, Z_r and $\mathrm{i}\,Z_i$ denote the principal part and the remainder of the plasma dispersion function, respectively. However, according to Section 8.4,

$$Z_i(\zeta) = \pi^{1/2}\,\mathrm{e}^{-\zeta^2}, \tag{8.162}$$

$$Z_i'(\zeta) = -2\,\pi^{1/2}\,\zeta\,\mathrm{e}^{-\zeta^2}. \tag{8.163}$$

Hence, we can write

$$\begin{aligned}\epsilon_i(k,\omega) = 2\,\pi^{1/2} \sum_s \frac{\Pi_s^2}{(k\,v_{\parallel\,s})^2} \sum_{n=-\infty,\infty} \exp\left(-\frac{k_\perp^2\,v_{\perp\,s}^2}{2\,\Omega_s^2}\right) I_n\left(-\frac{k_\perp^2\,v_{\perp\,s}^2}{2\,\Omega_s^2}\right)\\ \left[\frac{n\,\Omega_s}{k_\parallel\,v_{\parallel\,s}}\left(\frac{T_{\parallel\,s}}{T_{\perp\,s}} - 1\right) + \frac{\omega}{k_\parallel\,v_{\parallel\,s}}\right]\exp\left[-\left(\frac{\omega - n\,\Omega_s}{k_\parallel\,v_{\parallel\,s}}\right)^2\right].\end{aligned} \tag{8.164}$$

Suppose, for the sake of simplicity, that the electrons are "cold" but the ions are "hot." In other words, $v_{\perp\,s}$, $v_{\parallel\,s} \rightarrow 0$ for the electrons, but not for the ions. In this situation, Equation (8.160) reduces to

$$\epsilon_r(k,\omega) \simeq 1 - \frac{\Pi_e^2}{\omega^2}\,\frac{k_\parallel^2}{k^2}. \tag{8.165}$$

The ion contribution to the previous expression is negligible compared to the electron contribution, because of the m_s^{-1} dependence of Π_s^2. The real frequency of the instability is determined from

$$\epsilon_r(k, \omega_r) \simeq 1 - \frac{\Pi_e^2}{\omega_r^2} \frac{k_\parallel^2}{k^2} = 0, \tag{8.166}$$

which implies that

$$\omega_r \simeq \Pi_e \frac{k_\parallel}{k}. \tag{8.167}$$

Furthermore,

$$\frac{\partial \epsilon_r(k, \omega_r)}{\partial \omega} \simeq \frac{2}{\omega_r}. \tag{8.168}$$

Hence, the growth-rate of the instability is written

$$\gamma = -\frac{\epsilon_i(k, \omega_r)}{\partial \epsilon_r(k, \omega_r)/\partial \omega} = -\frac{\omega_r}{2}\, \epsilon_i(k, \omega_r), \tag{8.169}$$

or

$$\frac{\gamma}{\omega_r} \simeq \pi^{1/2} \sum_s \frac{\Pi_i^2}{(k\, v_{\parallel i})^2} \sum_{n=-\infty,\infty} \exp\left(-\frac{k_\perp^2\, v_{\perp i}^3}{2\,\Omega_i^2}\right) I_n\left(-\frac{k_\perp^2\, v_{\perp i}^2}{2\,\Omega_i^2}\right)$$
$$\left[\frac{n\,\Omega_i}{k_\parallel\, v_{\parallel i}}\left(1 - \frac{T_{\parallel i}}{T_{\perp i}}\right) - \frac{\omega_r}{k_\parallel\, v_{\parallel i}}\right] \exp\left[-\left(\frac{\omega_r - n\,\Omega_i}{k_\parallel\, v_{\parallel i}}\right)^2\right]. \tag{8.170}$$

The electron contribution to the previous expression is negligible compared to the ion contribution, because $v_{\parallel e} \rightarrow 0$.

It can be seen, from the previous formula, that if $T_{\parallel i} = T_{\perp i}$ then $\gamma < 0$ for all values of ω_r (recall that $\omega_r > 0$). In other words, there is no instability if the perpendicular and parallel ion temperatures are equal to one another. On the other hand, if $T_{\parallel i} < T_{\perp i}$ then there is a range of ω_r values for which each term in the sum on the right-hand side of (8.170) is positive. In other words, there is the possibility of an instability. The variation of γ/ω_r with ω_r is shown schematically in Figure 8.11 for a case where $\Omega_i/(k_\parallel\, v_{\parallel i})$ is relatively large. It can be seen that the growth-rate is positive in a narrow range of real frequencies lying on the low frequency side of each harmonic of the ion cyclotron frequency, and negative in a similar range of frequencies on the high frequency side.

According to Equation (8.167), ω_r varies from zero to Π_e. Thus, a necessary condition for obtaining an instability close to the nth ion cyclotron harmonic is $\Pi_e > n\,\Omega_i$. Now, the positive contribution from the nth term in the sum on the right-hand side of Equation (8.170) peaks close to

$$\omega_r = \omega_{r\,n} \equiv n\,\Omega_i\left(1 - \frac{T_{\parallel i}}{T_{\perp i}}\right). \tag{8.171}$$

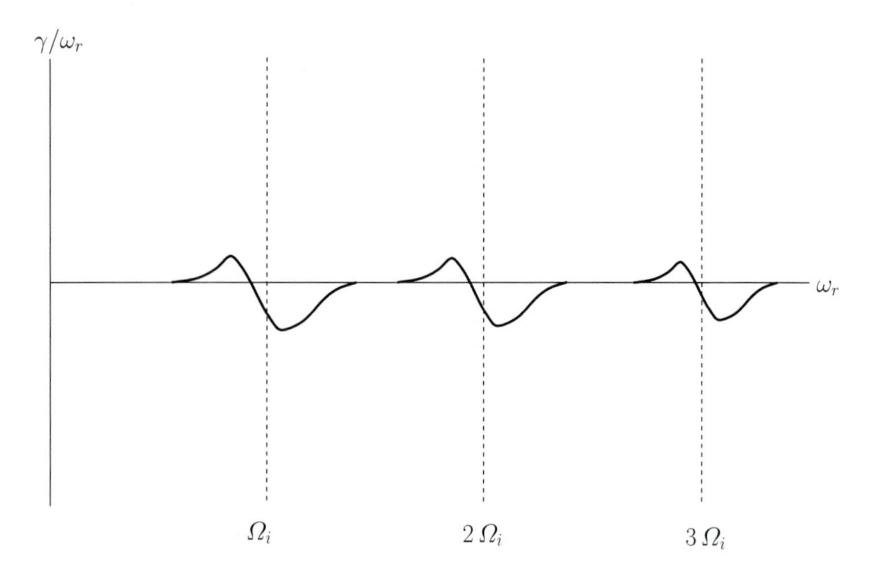

Figure 8.11
Schematic diagram showing the growth-rate of the Harris instability as a function of its real frequency.

In order for the net growth-rate to be positive, we require that

$$\omega_{r\,n} > \left(n - \frac{1}{2}\right)\Omega_i, \tag{8.172}$$

which implies that (Harris 1970)

$$\frac{T_{\parallel\,i}}{T_{\perp\,e}} < \frac{1}{2\,n}. \tag{8.173}$$

If this inequality is not satisfied, then the negative contribution from the $(n-1)$th term in the sum predominates over the positive contribution from the nth term. Observe that the previous inequality becomes harder to satisfy as n increases.

8.14 Exercises

8.1 Derive the dispersion relation (8.28) from Equations (8.23)–(8.27).

8.2 Show that the dispersion relation (8.28) can be written

$$1 - \frac{1}{x} - \frac{3\,y^2}{x^2} + \mathrm{i}\,\epsilon(x, y) = 0,$$

where $x = (\omega/\Pi_e)^2$, $y = k\,\lambda_D$, $\Pi_e = (n\,e^2/\epsilon_0\,m_e)^{1/2}$, $\lambda_D = (T_e/m_e\,\Pi_e^2)^{1/2}$, and

$$\epsilon(x, y) = \left(\frac{\pi}{2}\right)^{1/2} \frac{x^{1/2}}{y^3}\,\exp\left(-\frac{x}{2\,y^2}\right).$$

Demonstrate that, in the limit $y, \epsilon \ll 1$, the approximate solution is

$$x \simeq 1 + 3\,y^2 - \mathrm{i}\,\epsilon(1, y).$$

8.3 Show that, when combined with the Maxwellian velocity distribution (8.24), the dispersion relation (8.23) reduces to

$$1 - \frac{Z'(\zeta)}{2\,(k\,\lambda_D)^2} = 0,$$

where $\zeta = (\omega/\Pi_e)/(k\,\lambda_D)/\sqrt{2}$, $\Pi_e = (n\,e^2/\epsilon_0\,m_e)^{1/2}$, $\lambda_D = (T_e/m_e\,\Pi_e^2)^{1/2}$, and $Z(\zeta)$ is the plasma dispersion function. Hence, deduce from the large argument asymptotic form of the plasma dispersion function that

$$-2\,\mathrm{i}\,\pi^{1/2}\,\zeta\,\mathrm{e}^{-\zeta^2} + \frac{1}{\zeta^2} + \frac{3}{2\,\zeta^4} + O\left(\frac{1}{\zeta^6}\right) = 2\,(k\,\lambda_D)^2$$

in the limit $k\,\lambda_D \ll 1$. Show that the approximate solution of the previous equation is

$$\frac{\omega}{\Pi_e} = \sqrt{2}\,(k\,\lambda_D)\,\zeta \simeq 1 + \frac{3}{2}\,(k\,\lambda_D)^2 - \frac{\mathrm{i}}{2}\left(\frac{\pi}{2}\right)^{1/2} \frac{1}{(k\,\lambda_D)^3}\,\exp\left[-\frac{1}{2\,(k\,\lambda_D)^2}\right].$$

8.4 Show that, when combined with the Maxwellian velocity distribution (8.24), the dispersion relation (8.49) reduces to

$$1 - \frac{Z'(\zeta_e)}{2\,(k\,\lambda_{D\,e})^2} - \frac{Z'(\zeta_i)}{2\,(k\,\lambda_{D\,i})^2} = 0,$$

where $\Pi_s = (n\,e^2/\epsilon_0\,m_s)^{1/2}$, $\lambda_{D\,s} = (T_s/m_s\,\Pi_s^2)^{1/2}$, $\zeta_s = (m_s/2\,T_s)^{1/2}\,\omega/k$, and $Z(\zeta)$ is the plasma dispersion function. Use the large-argument expansion of the plasma dispersion function for the ions,

$$Z'(\zeta_i) \simeq -\mathrm{i}\,2\sqrt{\pi}\,\zeta_i\,\mathrm{e}^{-\zeta_i^2} + \frac{1}{\zeta_i^2},$$

and the small-argument expansion for the electrons,

$$Z'(\zeta_e) \simeq -\mathrm{i}\,2\sqrt{\pi}\,\zeta_e\,\mathrm{e}^{-\zeta_e^2} - 2.$$

Substituting these expansions into the dispersion relation, writing $\omega = \omega_r + \mathrm{i}\,\gamma$, where ω_r and γ are both real, and $|\gamma| \ll \omega_r$, demonstrate that

$$\frac{\omega_r}{k} \simeq \left(\frac{T_e}{m_i}\right)^{1/2} \frac{1}{[1 + (k\,\lambda_{D\,e})^2]^{1/2}},$$

and

$$\frac{\gamma}{\omega_r} \simeq -\frac{(\pi/8)^{1/2}}{[1+(k\,\lambda_{De})^2]^{3/2}}\left[\left(\frac{m_e}{m_i}\right)^{1/2}+\left(\frac{T_e}{T_i}\right)^{3/2}\exp\left(-\frac{T_e}{2\,T_i}\,\frac{1}{[1+(k\,\lambda_{De})^2]}\right)\right].$$

8.5 Derive Equation (8.74) from Equations (8.69) and (8.73).

8.6 Derive Equation (8.82) from Equations (8.74) and (8.79).

8.7 Derive Equations (8.89)–(8.91) from Equation (8.82).

8.8 Derive Equations (8.94)–(8.96) from Equation (8.82).

8.9 Derive Equations (8.102)–(8.105) from Equation (8.82).

8.10 Derive Equation (8.119) from Equation (8.118).

8.11 Derive Equation (8.120) from Equations (8.79) and (8.119).

8.12 Derive Equation (8.124) from Equation (8.123).

8.13 Demonstrate that the distribution function (8.131) possesses a minimum at $u = 0$ when $v_e < \sqrt{3}\,V$, but not otherwise.

8.14 Verify formula (8.133).

8.15 Consider an unmagnetized quasi-neutral plasma with stationary ions in which the electron velocity distribution function takes the form

$$F_0(u) = n_e\,\frac{v_e}{2\pi}\left[\frac{1}{v_e^2+(u-V)^2}+\frac{1}{v_e^2+(u+V)^2}\right].$$

Demonstrate that the dispersion relation for electrostatic plasma waves can be written

$$k^2 = \Pi_e^2\,\frac{v_e}{2\pi}\left[\int_{-\infty}^{\infty}\frac{du}{(u-\omega/k)^2\,[v_e^2+(u-V)^2]}\right.$$
$$\left.+\int_{-\infty}^{\infty}\frac{du}{(u-\omega/k)^2\,[v_e^2+(u+V)^2]}\right],$$

where $\Pi_e = (n_e\,e^2/\epsilon_0\,m_e)^{1/2}$. Assuming that k is real and positive, and that ω/k lies in the upper half of the complex plane, show that when the integrals are evaluated as contour integrals in the complex u-plane (closed in the lower half of the plane), making use of the residue theorem (Riley 1974), the previous dispersion relation reduces to

$$2 = \Pi_e^2\left[\frac{1}{(k\,V-\zeta)^2}+\frac{1}{(k\,V+\zeta)^2}\right],$$

where $\zeta = \omega + \mathrm{i}\,k\,v_e$. Finally, in the small-$k$ limit, $k \ll \Pi_e/V$, demonstrate that the growth-rate of the most unstable mode is

$$\gamma \equiv -\mathrm{i}\,\omega \simeq k\,(V - v_e).$$

8.16 Derive Equation (8.137) from Equations (8.134)–(8.136).

8.17 Derive Equations (8.160) and (8.161) from Equations (8.154), (8.155), and (8.159).

Bibliography

Abramowitz, M., and Stegun, I. (eds.) 1965a. *Handbook of Mathematical Functions: with Formulas, Graphs, and Mathematical Tables*. Dover. Chapter 6.

Abramowitz, M., and Stegun, I. (eds.) 1965b. *Handbook of Mathematical Functions: with Formulas, Graphs, and Mathematical Tables*. Dover. Chapter 7.

Abramowitz, M., and Stegun, I. (eds.) 1965c. *Handbook of Mathematical Functions: with Formulas, Graphs, and Mathematical Tables*. Dover. Chapter 9.

Abramowitz, M., and Stegun, I. (eds.) 1965d. *Handbook of Mathematical Functions: with Formulas, Graphs, and Mathematical Tables*. Dover. Chapter 10.

Abramowitz, M., and Stegun, I. (eds.) 1965e. *Handbook of Mathematical Functions: with Formulas, Graphs, and Mathematical Tables*. Dover. Chapter 19.

Alfvén, H. 1942. *Existence of Electromagnetic-Hydrodynamic Waves.* Nature **150**, 405.

Armstrong, T.P. 1967. *Numerical Studies of the Nonlinear Vlasov Equation.* Physics of Fluids **10**, 1269.

Atzeni, S., and Meyer-ter-Vehn, J. 2009. *The Physics of Inertial Fusion: Beam Plasma Interaction, Hydrodynamics, Hot Dense Matter*. Oxford.

Bateman, G. 1978. *Magnetohydrodynamics Instabilities*. MIT.

Baumjohan, W., and Treumann, R.A. 1996. *Basic Space Plasma Physics*. Imperial College.

Bernstein, I.B. 1958. *Waves in a Plasma in a Magnetic Field.* Physical Review **109**, 10.

Bernstein, I.B. 1974. *Transport in Axisymmetric Plasmas.* Physics of Fluids **17**, 547.

Biskamp, D. 1986. *Magnetic Reconnection via Current Sheets.* Physics of Fluids **29**, 1520.

Boyd, T.J.M., and Sanderson, J.J. 2003. *The Physics of Plasmas*. Cambridge.

Braginskii, S.I. 1965. *Transport Processes in a Plasma.* In *Reviews of Plasma Physics*. Consultants Bureau. Volume 1, 205.

Brillouin, L. 1926. *La Mécanique Ondulatoire de Schrödinger une Method Générale de Resolution par Approximations Successives.* Comptes Rendus des Seances de l'Académie des Sciences Paris **183**, 24.

Brillouin, L. 1960. *Wave Propagation and Group-Velocity*. Academic Press.

Budden, K.G. 1985. *The Propagation of Radio Waves: The Theory of Radio Waves of Low Power in the Ionosphere and Magnetosphere*. Cambridge.

Cairns, R.A. 1985. *Plasma Physics*. Blackie.

Chapman, S. 1957. *Notes on the Solar Corona and the Terrestrial Ionosphere.* Smithsonian Contributions to Astrophysics **2**, 1.

Chapman, S., and Cowling, T.G. 1953. *The Mathematical Theory of Non-Uniform Gases*. Cambridge.

Childress, S., and Gilbert, A.D. 1995. *Stretch, Twist, Fold: The Fast Dynamo*. Springer.

Cowling, T.G. 1934. *The Magnetic Field of Sunspots.* Monthly Notices of the Royal Astronomical Society **94**, 39.

Cowling, T.G. 1957a. *Magnetohydrodynamics*. Interscience.

Cowling, T.G. 1957b. *The Dynamo Maintenance of Steady Magnetic Fields.* Quarterly Journal of Mechanics and Applied Mathematics **10**, 129.

Davidson, R.C. 2001. *Physics of Nonneutral Plasmas*, 2nd Edition. World Scientific.

Derfler, H., and Simonen, T.C. 1966. *Landau Waves: An Experimental Fact.* Physical Review Letters **17**, 172.

Doolittle, J.S. 1959. *Thermodynamics for Engineers*. International Textbook Company.

Dunlop, D.J., and Özdemir, O. 2001. *Rock Magnetism: Fundamentals and Frontiers*. Cambridge.

Erdélyi, A. (ed.) 1954. *Tables of Integral Transforms*. McGraw-Hill. Volume 1.

Fitzpatrick, R. 2008. *Maxwell's Equations and the Principles of Electromagnetism*. Jones & Bartlett.

Fitzpatrick, R. 2013. *Oscillations and Waves: An Introduction*. CRC.

Flanigan, F.J. 2010. *Complex Variables: Harmonic and Analytic Functions*. Dover.

Fortov, V., Iakubov, I., and Khrapak, A. 2007. *Physics of Strongly Coupled Plasma*. Oxford.

Fowler, T.K. 1997. *The Fusion Quest*. Johns Hopkins.

Freidberg, J.P. 2008. *Plasma Physics and Fusion Energy*. Cambridge.

Fried, B.D., and Conte, S.D. 1961. *The Plasma Dispersion Function*. Academic Press.

Furth, H.P., Killeen, J., and Rosenbluth, M.N. 1963. *Finite-Resistivity Instabilities of a Sheet Pinch.* Physics of Fluids **6**, 459.

Gailitis, A., Lielausis, O., Dement'ev, S., Platacis, E., Cifersons, A., Gerbeth, G., Gundrum, T., Stefani, F., Christen, M., Hänel, H., and Will, G. 2000. *Detection of a Flow Induced Magnetic Field Eigenmode in the Riga Dynamo Facility.* Physical Review Letters **84**, 4365.

Gardner, C.S. 1963. *Bound on the Energy Available from a Plasma.* Physics of Fluids **6**, 839.

Goldstein, H., Poole, C., and Safko, J. 2002. *Classical Mechanics*, 3rd Edition. Addison-Wesley.

Green. G. 1837. *On the Motion of Waves in a Variable Canal of Small Depth and Width.* Cambridge Philosophical Transactions **6**, 457.

Gurnett, D.A., and Bhattacharjee, A. 2005. *Introduction to Plasma Physics*. Cambridge.

Haas, F. 2011. *Quantum Plasmas: An Hydrodynamic Approach*. Springer.

Hansen, C.J., Kawaler, S.D., and Trimble, V. 2004. *Stellar Interiors - Physical Principles, Structure, and Evolution*, 2nd Edition. Springer.

Harris, E.G. 1961. *Plasma Instabilities Associated with Anisotropic Velocity Distributions.* Journal of Nuclear Energy C **2**, 138.

Harris, E.G. 1970. *Plasma Instabilities.* In *Physics of Hot Plasmas*, Rye, B.J., and Taylor, J.C. (eds.). Plenum. Chapter 4.

Hazeltine, R.D., and Waelbroeck, F.L. 2004. *The Framework of Plasma Physics.* Westview.

Heading, J. 1962. *An Introduction to Phase-Integral Methods*. Meuthuen.

Hess, W.M. 1968. *The Radiation Belt and the Magnetosphere*. Blaisdell.

Hinton, F.L., and Hazeltine, R.D. 1976. *Theory of Plasma Transport in Toroidal Confinement Systems.* Reviews of Modern Physics **48**, 239.

Huba, J.D. 2000a. *NRL Plasma Formulary*. Naval Research Laboratory. 6–7.

Huba, J.D. 2000b. *NRL Plasma Formulary*. Naval Research Laboratory. 8–9.

Huba, J.D. 2000c. *NRL Plasma Formulary*. Naval Research Laboratory. 30.

Huba, J.D. 2000d. *NRL Plasma Formulary*. Naval Research Laboratory. 34–35.

Jackson, J.D. 1998. *Classical Electrodynamics*, 3rd Edition. Wiley.

Jeffries, H. 1924. *On Certain Approximate Solutions of Linear Differential Equations of the Second Order*. Proceedings of the London Mathematical Society **23**, 428.

Ji, H., Yamada, M., Hsu, S., and Kulsrud, R.M. 1998. *Experimental Test of the Sweet-Parker Model of Magnetic Reconnection.* Physical Review Letters **80**, 3256.

Jones, C.A., Thompson, M.J., and Tobais, S.M. 2010. *The Solar Dynamo.* Space Science Reviews **152**, 591.

Joshi, C. 2006. *Plasma Accelerators.* Scientific American **294**, 40.

Kallenrode, M.-B. 2010. *Space Physics: An Introduction to Plasmas and Particles in the Heliosphere and Magnetospheres*, 3rd Edition. Springer.

Knobloch, E. 1981. *Chaos in the Segmented Disc Dynamo.* Physics Letters **82A**, 439.

Kramers, H.A. 1926. *Wellenmechanik und halbzahlige Quantisierung.* Zeitschrift für Physik **39**, 828.

Krause, F., and Rädler, K.-H. 1980. *Mean-Field Magnetohydrodynamics and Dynamo Theory*. Pergamon.

Kruer, W. 2003. *The Physics Of Laser Plasma Interactions.* Westview.

Kruskal, M.D. 1962. *Asymptotic Theory of Hamiltonian and Other Systems with all Solutions Nearly Periodic.* Journal of Mathematical Physics **3**, 806.

Kruskal, M.D., and Oberman, C.R. 1958. *On the Stability of Plasma in Static Equilibrium.* Physics of Fluids **1**, 275.

Kulsrud, R.M. 2004. *Plasma Physics for Astrophysics.* Princeton.

Landau, L.D. 1936. *Kinetic Equation for the Coulomb Effect.* Physik Z. Sowjetunion **10**, 154.

Landau, L.D. 1946. *On the Vibration of the Electronic Plasma.* Soviet Physics–JETP **10**, 25.

Larmor, J. 1919. *Possible Rotational Origin of Magnetic Fields of Sun and Earth.* Electrical Review **85**, 412.

Lieberman, M.A., and Lichtenberg, A.J. 2005. *Principles of Plasma Discharges and Materials Processing*, 2nd Edition. Wiley-Interscience.

Liouville, J. 1837. *Sur le Développement des Fonctions et Séries.* Journal de Mathématiques Pures et Appliquées **1**, 16.

Longair, M.S. 2008. *Galaxy Formation*, 2nd Edition. Springer.

Lundquist, S. 1949. *Experimental Demonstration of Magneto-Hydrodynamic Waves.* Nature **164**, 145.

Malmberg, J.H., and Wharton, C.B. 1964. *Collisionless Damping of Electrostatic Plasma Waves.* Physical Review Letters **13**, 184.

Malmberg, J.H., and Wharton, C.B. 1966. *Dispersion of Electron Plasma Waves.* Physical Review Letters **17**, 175.

Mestel, L. 2012. *Stellar Magnetism*, 2nd Edition. Oxford.

Moffatt, H.K. 1978. *Magnetic Field Generation in Electrically Conducting Fluids.* Cambridge.

Morozov, A.I., and Solev'ev, L.S. 1966. *Motion of Charged Particles in Electromagnetic Fields.* In *Reviews of Plasma Physics.* Consultants Bureau. Volume 2.

Neugebauer, M., and Snyder, C.W. 1966. *Mariner 2 Observations of the Solar Wind: 1. Average Properties.* Journal of Geophysical Research **71**, 4469.

Northrop, T.G. 1963. *The Adiabatic Motion of Charged Particles.* Interscience.

Northrop, T.G., and Teller, E. 1960. *Stability of the Adiabatic Motion of Charged Particles in the Earth's Field.* Physical Review **117**, 215.

Nyquist, H. 1932. *Regeneration Theory.* Bell System Technical Journal **1**, 126.

Ogg, J.G., 2012. *Geomagnetic Polarity Time Scale.* In *The Geologic Time Scale 2012*, Gradstein, F. et al. (eds.). Elsevier. Chapter 5.

O'Neil, T.M. 1965. *Collisionless Damping of Nonlinear Plasma Oscillations.* Physics of Fluids **8**, 2255.

Parker, E.N. 1957. *Sweet's Mechanism for Merging Magnetic Fields in Conducting Fluids.* Journal of Geophysical Research **62**, 509.

Parker, E.N. 1958. *Dynamics of the Interplanetary Gas and Magnetic Fields.* Astrophysical Journal **128**, 664.

Penrose, O. 1960. *Electrostatic Instabilities of a Uniform Non-Maxwellian Plasma.* Physics of Fluids **3**, 258.

Petschek, H.E. 1964. *Magnetic Field Annihilation.* In *AAS-NASA Symposium on the Physics of Solar Flares.* NASA. 425.

Plemelj, J. 1908. *Riemannsche Funktionenscharen mit gegebener Monodromiegruppe.* Monatshe für Mathematik und Physik **19**, 211.

Ponomarenko, Y.B. 1973. *Theory of the Hydromagnetic Generator.* Journal of Applied Mechanics and Technical Physics **14**, 775.

Priest, E.R. 1984. *Solar Magnetohydrodynamics.* Springer.

Priest, E.R., and Forbes, T.G. 1992. *Does Fast Magnetic Reconnection Exist?* Journal of Geophysical Research **97**, 16757.

Priest, E.R., and Forbes, T.G. 2007. *Magnetic Reconnection: MHD Theory and Applications*. Cambridge.

Ratcliffe, J.A. 1972. *An Introduction to the Ionosphere and Magnetosphere*. Cambridge.

Rayleigh, Lord 1912. *On the Propagation of Waves through a Stratified Medium, with Special Reference to the Question of Reflection.* Proceedings of the Royal Society A **86**, 207.

Reif, F. 1965. *Fundamentals of Statistical and Thermal Physics*. McGraw-Hill.

Riley, K.F. 1974. *Mathematical Methods for the Physical Sciences*. Cambridge.

Roberts, P.H., and King, E.M. 2013. *On the Genesis of the Earth's Magnetism.* Reports on Progress in Physics **76**, 096801.

Rose, D.J., and Clark, M., Jr. 1961. *Plasmas and Controlled Fusion*. MIT.

Rosenbluth, M.N., MacDonald, W.M., and Judd, D.L. 1957. *Fokker-Planck Equation for an Inverse-Square Force.* Physical Review Letters **107**, 1.

Rosenbluth, M.N., and Rostoker, N. 1959. *Theoretical Structure of Plasma Equations.* Physics of Fluids **2**, 23.

Rowe, D.M. (ed.) 2006. *Thermoelectrics Handbook: Macro to Nano*. CRC.

Russell, C.T. 1991. *Planetary Magnetospheres.* Science Progress **75**, 93.

Rutherford, P.H. 1973. *Nonlinear Growth of the Tearing Mode.* Physics of Fluids **16**, 1903.

Spiegel, M.R., Liu, J., and Lipschutz, S. 1999. *Mathematical Handbook of Formulas and Tables*, 2nd Edition. McGraw-Hill. Formula 17.13.1.

Spitzer, L., Jr. 1956. *Physics of Fully Ionized Gases*. Interscience.

Stix, T.H. 1992. *Waves in Plasmas*. American Institute of Physics.

Storey, L.R.O. 1953. *An Investigation of Whistling Atmospherics.* Philosophical Transactions of the Royal Society A **246**, 113.

Suess, S.T. 1990. *The Heliopause.* Reviews of Geophysics **28**, 97.

Swanson, D.G. 2003. *Plasma Waves*, 2nd Edition. Taylor & Francis.

Sweet, P.A. 1958. *The Neutral Point Theory of Solar Flares.* In *Electromagnetic Phenomena in Cosmical Physics*, Lehnert, B. (ed.). Cambridge. 123.

Vainshtein, S., and Zel'dovich, Y.B. 1972. *Origin of Magnetic Fields in Astrophysics (Turbulent 'Dynamo' Mechanisms).* Soviet Physics Uspekhi **15**, 159.

Valet, J.-P., Meynadier, L., and Guyodo, Y. 2005. *Geomagnetic Dipole Strength and Reversal Rate over the Past Two Million Years.* Nature **435**, 802.

Verhille, G., Plihon, N., Bourgoin, M., Odier, P., and Pinton, J.-F. 2009. *Laboratory Dynamo Experiments.* Space Science Reviews **152**, 543.

Watson, G.N. 1995. *A Treatise on the Theory of Bessel Functions*, 2nd Edition. Cambridge.

Webber, W.R., and McDonald, F.B. 2013. *Recent Voyager 1 Data Indicate that on 25 August 2012 at a Distance of 121.7 AU from the Sun, Sudden and Unprecedented Intensity Changes Were Observed in Anomalous and Galactic Cosmic Rays.* Geophysical Research Letters **40**, 1665.

Weber, E.J., and Davis, L., Jr. 1967. *The Angular Momentum of the Solar Wind.* Astrophysical Journal **148**, 217.

Wentzel, G. 1926. *Eine Verallgemeinerung der Quantenbedingungen für die Zwecke der Wellenmechanik.* Zeitschrift für Physick **38**, 518.

White, R.B., Monticello, D.A., Rosenbluth, M.N, and Wadell, B.V. 1977. *Saturation of the Tearing Mode.* Physics of Fluids **20**, 800.

Wilcox, J.M., Boley, F.L., and DeSilva, A.W. 1960. *Experimental Study of Alfvén-Wave Properties.* Physics of Fluids **3**, 15.

Yamada, M., Kulsrud, R.M., and Ji, H. 2010. *Magnetic Reconnection.* Reviews of Modern Physics **82**, 603.

Yoder, C.F., 1995. *Astrometric and Geodetic Properties of Earth and the Solar System.* In *Global Earth Physics: A Handbook of Physical Constants*, Ahrens, T. (ed.). American Geophysical Union.

Index